T0074090

Lecture Notes in Physics

Founding Editors: W. Beiglböck, J. Ehlers, K. Hepp, H. Weidenmüller

The Lecture Notes in Physics

The series Lecture Notes in Physics (LNP), founded in 1969, reports new developments in physics research and teaching – quickly and informally, but with a high quality and the explicit aim to summarize and communicate current knowledge in an accessible way. Books published in this series are conceived as bridging material between advanced graduate textbooks and the forefront of research and to serve three purposes:

- to be a compact and modern up-to-date source of reference on a well-defined topic

- to serve as an accessible introduction to the field to postgraduate students and nonspecialist researchers from related areas

- to be a source of advanced teaching material for specialized seminars, courses and schools

Both monographs and multi-author volumes will be considered for publication. Edited volumes should, however, consist of a very limited number of contributions only. Proceedings will not be considered for LNP.

Volumes published in LNP are disseminated both in print and in electronic formats, the electronic archive being available at springerlink.com. The series content is indexed, abstracted and referenced by many abstracting and information services, bibliographic networks, subscription agencies, library networks, and consortia.

Proposals should be sent to a member of the Editorial Board, or directly to the managing editor at Springer:

Christian Caron
Springer Heidelberg
Physics Editorial Department I
Tiergartenstrasse 17
69121 Heidelberg / Germany
christian.caron@springer.com

J.S. Al-Khalili
E. Roeckl (Eds.)

The Euroschool Lectures on Physics with Exotic Beams, Vol. III

 Springer

J.S. Al-Khalili
University of Surrey
Dept. Physics
Guildford, Surrey
United Kingdom GU2 5XH
j.al-khalili@ surrey.ac.de

E. Roeckl
Gesellschaft for
Schwerionenforschung mbH
(GSI)
Planckstr. 1
64291 Darmstadt
Germany
R.Roeckl@gsi.de

Al-Khalili, J.S., Roeckl, E. (Eds.), *The Euroschool Lectures on Physics with Exotic Beams, Vol. III*, Lect. Notes Phys. 764 (Springer, Berlin Heidelberg 2009), DOI 10.1007/978-3-540-85839-3

ISBN: 978-3-540-85838-6   e-ISBN: 978-3-540-85839-3

DOI 10.1007/978-3-540-85839-3

Lecture Notes in Physics ISSN: 0075-8450   e-ISSN: 1616-6361

Library of Congress Control Number: 2008935358

Cover design: Integra Software Services Pvt Ltd.

Printed on acid-free paper

9 8 7 6 5 4 3 2 1

springer.com

Preface

This is the third and final volume in a series of Lecture Notes based on the highly successful Euro Summer School on Exotic Beams that has been running yearly since 1993 (apart from 1999) and is planned to continue to do so. It is the aim of the series to provide an introduction to Radioactive Ion Beam (RIB) physics at the level of graduate students and young postdocs starting out in the field. Each volume contains lectures covering a range of topics from nuclear theory to experiment to applications.

Our understanding of atomic nuclei has undergone a major re-orientation over the past two decades and seen the emergence of an exciting field of research: the study of 'exotic' nuclei. The availability of energetic beams of short-lived nuclei, referred to as 'radioactive ion beams' (RIBs), has opened the way to the study of the structure and dynamics of thousands of nuclear species never before observed in the laboratory. This field has now become one of the most important and fast-moving in physics worldwide. And it is fair to say that Europe leads the way with a number of large international projects starting up in the next few years, such as the FAIR facility at GSI in Germany. From a broader perspective, one must also highlight just how widely RIB physics impacts on other areas, from energy and the environment to medicine and materials science. There is little doubt that RIB physics has transformed not only nuclear physics itself but many other areas of science and technology too, and will continue to do so in the years to come.

While the field of RIB physics is linked mainly to the study of nuclear structure under extreme conditions of isospin, mass, spin and temperature, it also addresses problems in nuclear astrophysics, solid-state physics and the study of fundamental interactions. Furthermore, important applications and spin-offs also originate from this basic research. The development of new production, acceleration and ion storing techniques and the construction of new detectors adapted to work in the special environment of energetic radioactive beams is also an important part of the science. And, due to the fact that one is not limited anymore to the proton/neutron ratio of stable-isotope beams, virtually the whole chart of the nuclei opens up for research, so theoretical models can be tested and verified all the way up to the limits of nuclear existence: the proton and neutron 'drip lines'.

The beams of rare and 'exotic' nuclei being produced are via two complementary techniques: in-flight separation and post-acceleration of low-energy radioactive beams. Both methods have been developed in a number of European Large Scale Facilities such as ISOLDE (CERN, Switzerland), GANIL (Caen, France), GSI (Darmstadt, Germany), the Accelerator Laboratory of the University of Jyväskylä (Finland), INFN Laboratori Nazionali di Legnaro (Italy) and the Cyclotron Research Centre (Louvain-la-Neuve, Belgium). Indeed, so important is the continued running and success of the School that a number of these European facilities have committed to providing financial support over the coming years.

While the field of RIB physics is linked mainly to the study of nuclear structure under extreme conditions of isospin, mass, spin and temperature, it also addresses problems in nuclear astrophysics, solid-state physics and the study of fundamental interactions. Furthermore, important applications and spin-offs also originate from this basic research. The development of new production, acceleration and ion storing techniques and the construction of new detectors adapted to work in the special environment of energetic radioactive beams is also an important part of the science. And, due to the fact that one is not limited anymore to the proton/neutron ratio of stable beams, virtually the whole chart of the nuclei opens up for research, so theoretical models can be tested and verified all the way up to the limits of nuclear existence: the proton and neutron 'drip lines'.

Volumes I and II of this series have proved to be highly successful and popular with many researchers reaching for it for information or providing it for their PhD students as an introduction to a particular topic. They are now even available to download from the Euro School Website (http://www.euroschoolonexoticbeams.be/eb/pages/lecture_notes). We stress that the contributions in these volumes are not review articles and so are not meant to contain all the latest results or to provide an exhaustive coverage of the field but are written instead in the pedagogical style of graduate lectures and thus have a reasonably long 'shelf life'. As with the first two volumes, the contributions here are by leading scientists in the field who have lectured at the School. They were chosen by the editors to provide a range of topics within the field and will have updated their material delivered at the School (sometimes several years ago) to incorporate recent advances and results.

Finally, we wish to thank the lectures who have contributed to this volume for their hard work and diligence, and indeed for their patience, at a time when everyone finds it difficult to find the time to lay out their subject in such a careful, thorough and readable style. We also wish to thank Dr. Chris Caron and his colleagues at Springer-Verlag for their help, fruitful collaboration and continued support on this project.

Guildford, UK,   *J. Al-Khalili*
Darmstadt, Germany   *E. Roeckl*

Contents

Shell Structure of Exotic Nuclei

T. Otsuka 1

1   Basics of Shell Model 1

2   Construction of an Effective Interaction and an Example in the pf
Shell 16

3   The $N = 20$ Problem: Does the Gap Change? 18

4   Summary 20

References 20

**Testing the Structure of Exotic Nuclei via Coulomb
Excitation of Radioactive Ion Beams at Intermediate Energies**

T. Glasmacher 27

1   Introduction 27

2   Experimental Considerations 31

3   Extraction of Transition Matrix Elements from Cross Sections 41

4   Recent Experimental Results 43

5   Accuracy of the Technique 48

6   Outlook and Summary 50

References 50

Test of Isospin Symmetry Along the $N=Z$ Line

S.M. Lenzi, M.A. Bentley 57

1   Introduction 57

2   Background 59

3   Experimental and Theoretical Tools 67

4   Description of Excitation Energy Differences 73

5   Isobaric Multiplets in the $f_{\frac{7}{2}}$ Shell 84

6   Isobaric Multiplets in the $s\overset{\textstyle}{d}$ and fp Shells 92

7   Conclusions and Outlook 95

References 97

Beta Decay of Exotic Nuclei
B. Rubio and W. Gelletly 99
1   Introduction 99
2   Beta Decay and Nuclear Structure 103
3   Experimental Considerations 114
4   Some Illustrative Examples 122
5   Future Measurements 147
References 148

One- and Two-Proton Radioactivity
B. Blank 153
1   Introduction 153
2   One-Proton Radioactivity 155
3   Two-Proton Radioactivity 174
4   Other Exotic Decay Channels 190
5   Conclusions 192
References 193

Superheavy Elements
S. Hofmann 203
1   Introduction 203
2   Experimental Techniques 205
3   Experimental Results 213
4   Nuclear Structure and Decay Properties 232
5   Nuclear Reactions 238
6   Summary and Outlook 246
References 248

Experimental Tools for Nuclear Astrophysics
C. Angulo 253
1   Understanding the Universe 253
2   Relevant Quantities at Stellar Energies 256
3   Stellar Cycles and Some Key Reactions 260
4   Experimental Techniques in Nuclear Astrophysics 262
5   Future Challenges and Conclusions 279
References 279

Index 283

Color Plate Section 289

Shell Structure of Exotic Nuclei

T. Otsuka[1,2,3,4]

[1] Department of Physics, University of Tokyo, Hongo, Tokyo, 113-0033, Japan
[2] Center for Nuclear Study, University of Tokyo, Hongo, Tokyo, 113-0033, Japan
[3] RIKEN, Hirosawa, Wako-shi, Saitama
[4] National Superconducting Cyclotron Laboratory, Michigan State University, East Lansing, MI, USA

Abstract A basic introduction to the nuclear shell model is presented, without going to any details of many-body theories. First, we explain how magic numbers and shell structures appear from fundamental properties of nuclei such as the short-range attractive interaction and density saturation. Some concepts needed to understand the shell model are explained from scratch. After a general introduction we focus on a topic of particular current interest, the evolution of shell structure, and discuss the importance of the tensor force.

1 Basics of Shell Model

Nuclear theory has been developed in order to construct many-body systems from basic ingredients such as *nucleons* and *nuclear forces* (nucleon–nucleon interactions). The nuclear shell model has been an important part of nuclear theory, and should make crucial input to this end. We begin by asking three basic questions:

(i) What is the shell model?
(ii) Why is it useful?
(iii) How do we perform calculations?

1.1 What is the Shell Model?

We begin with some very basic points about nuclear shell structure and the shell model. Figure 1 shows somewhat schematically the nucleon–nucleon potential as a function of the distance between the two nucleons, for the spin-singlet (two interacting nucleons coupled to total spin $S=0$) and $L=0$ (L, relative orbital angular momentum of the two nucleons) state. This state must have isospin $T=1$ because of the antisymmetric coupling of the nucleons. This is one of the most important states for the nucleon–nucleon potential, as the potential contains a strongly attractive part. Note that the nucleon–nucleon potential depends generally on S, L, and J with $\vec{J} = \vec{L} + \vec{S}$. We

Otsuka, T.: *Shell Structure of Exotic Nuclei.* Lect. Notes Phys. **764**, 1–25 (2009)
DOI 10.1007/978-3-540-85839-3_1   © Springer-Verlag Berlin Heidelberg 2009

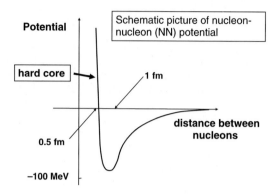

Fig. 1. Schematic illustration of the nucleon–nucleon potential

find, in Fig. 1, a hard-core repulsion inside a distance of 0.5 fm, while it is strongly attractive around 1 fm. These two features are found also in other important states, e.g., of spin-triplet and even L states with $T=0$. The potentials for such states are the origin of the proton–neutron binding in the deuteron.

From these two features, a hard repulsive core and a strong attraction around 1 fm, we can easily expect that the nuclear potential is such that the balance between the attractive part around 1 fm and the inner repulsive part conspires to give a rather constant distance (~ 1 fm) between nucleons, which are strongly bound together. The saturation of the density is thus realized at the same time. Although the actual mechanism contains more sophisticated dynamics – for instance, the density dependence of the potential – we will not go into such details. As the nucleon density should be rather constant, the surface can be defined clearly, despite the fact that the nucleus is such a complex quantum system with complicated interaction.

The nucleon–nucleon interaction is very complicated, but can produce a simple mean potential. Figure 2 depicts this situation. A nucleon (*open circle* in Fig. 2) well inside the nucleus feels the nucleon–nucleon interaction from the surrounding nucleons within reach (or *range*) of the interaction, which is about 1 fm. The sphere within this range is shown in the dashed line in Fig. 2. As the density of the nucleon is constant inside the nucleus, the mean effect from surrounding nucleons should be almost constant, and the mean potential should be almost flat.

Figure 3 indicates the effect from surrounding nucleons for a nucleon at the surface (shown again by an *open circle*). Otherwise the legend of the figure is the same as Fig. 2. The number of the surrounding nucleons becomes smaller, as this nucleon (*open circle*) moves out, resulting in less binding. Thus, the mean potential becomes shallower quickly at the surface.

Figure 4 displays schematically what the mean potential looks like. The single-particle motion inside this potential can be solved. This is just an eigenvalue problem with the eigenstates corresponding to various orbital motions,

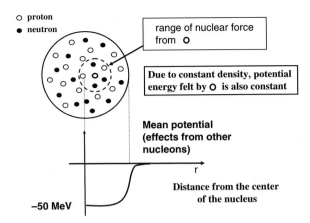

Fig. 2. Schematic illustration of the mean potential for a nucleon *open circle* inside the nuclear surface

similar to electrons in a hydrogen-like atom. The eigenstate is referred to as an *orbit*, having its classical image in mind. Figure 4 shows energy eigenvalues of such orbits. These are usually called single-particle energies (SPEs). At this point, we assume that the nucleus is spherical, and the mean potential appears to be spherical too. The spherical potential gives us quantum numbers of these eigenstates such as the orbital angular momentum denoted by l, the total angular momentum denoted by j, and the number of nodes of the radial wave function denoted by n. Since the potential is spherical, orbits differing only by the z-component of j, called j_z, are degenerate. Thus, there is a degeneracy of $(2j+1)$ magnetic substates for a given j. Because of this degeneracy,

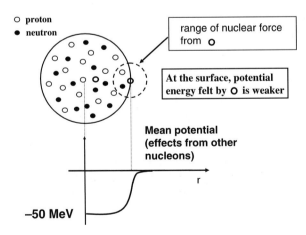

Fig. 3. Schematic illustration of the mean potential for a nucleon *open circle* at the nuclear surface

the SPE is referred to j (with l and n implicitly). If we discuss an "orbit" in a spherical potential, it means j, having $(2j+1)$ degenerate substates.

Each orbit has $(2j+1)$ substates with the same SPE. These substates are called single-particle states. Some orbits (with different j's) are grouped as shown in Fig. 4. Such a group of orbits is called a *shell*. The energy spacing between two shells is called the *shell gap*. If all orbits below a given shell gap are occupied by protons, they form a proton *closed shell*. The number of protons in a closed shell is called the proton magic number. Figure 4 indicates that the magic numbers are actually 2, 8, 20, ... We shall discuss later why these are magic numbers. Likewise, neutron closed shells and magic numbers are defined. The closed shell is also called a *core*, and the pattern of the orbits stated above is known as *shell structure*. We now turn to how to obtain the shell structure from simple arguments, without going to details.

The mean potential can be described, to a good approximation, in terms of the so-called Woods–Saxon (WS) potential. This is specified by three parameters: depth, radius, and diffuseness. The eigenstates of the WS potential can be obtained only numerically. In order to make physics more transparent, we can introduce a harmonic oscillator (HO) potential

$$V_{\mathrm{HO}} = m\omega^2 r^2/2 \,, \tag{1}$$

where m is the mass of the nucleon, ω is the oscillator frequency, and r stands for the distance from the center of the nucleus.

In fact, WS and HO potentials can be set to overlay each other as shown in Fig. 1 if the bottom of the HO potential is adjusted and an appropriate value of ω is chosen. The properties of the eigenstates of a HO potential are much simpler than those of the WS potential and can be described analytically. The eigenvalues are equally separated, as is well known, and are shown in Fig. 2. Mayer and Jensen [1, 2] proposed the shell structure of atomic nuclei by adding a *spin–orbit coupling* to the HO potential, explaining experimental data known at that time without a consistent theoretical description. The spin–orbit coupling is written as

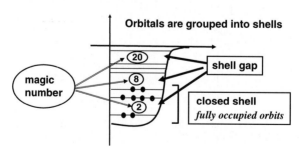

Fig. 4. Single-particle motion in the mean potential. The *horizontal lines* indicate single-particle energies (SPE's), which stand for energy eigenvalues of the orbits. The orbits form shells, and gaps between shells define magic numbers

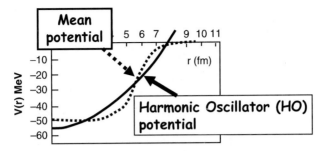

Fig. 5. Comparison between a harmonic oscillator potential and a Woods–Saxon potential. HO is simpler, and can be treated analytically (See also Plate 1 in the Color Plate Section)

$$V_{ls}(r) \;=\; f(r)(\vec{l}\cdot\vec{s}), \tag{2}$$

where f is a function of r, and $\vec{l}$, $\vec{s}$ denote the orbital angular momentum and spin operators of a nucleon, respectively.

The function $f(r)$ is usually given by the derivative of the density divided by r with an appropriate strength. Naturally, $f(r)$ has a peak at the surface because the density changes most rapidly at the surface.

The spin–orbit potential in Eq. (2) can be included as a first-order perturbation and its effect is to lower the energy of the j-*upper* state

$$j_> \;=\; l + 1/2 \tag{3}$$

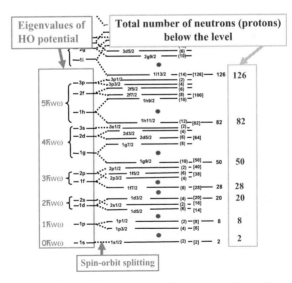

Fig. 6. Energy eigenvalues of harmonic oscillator potential with spin–orbit force, and Mayer–Jensen's magic numbers (See also Plate 2 in the Color Plate Section)

and raise the energy of the *j-lower* state

$$j_< = l - 1/2. \tag{4}$$

These notations will be used later.

After including the spin–orbit coupling, the degenerate orbits in the HO potential are split. The orbit is identified, for example, as $1f_{7/2}$ where the first number (integer) is the number of the node plus one, the second character is the usual notation of l, and the last part represents j.

The energy splitting due to the spin–orbit potential is approximately proportional to the value of l as expected from Eq. (2). So it becomes more and more important for l larger. Mayer and Jensen included this effect, and predicted magic numbers as shown in Fig. 2. The magic numbers 2, 8, and 20 are independent of the spin–orbit coupling. The orbits $1p_{3/2}$ and $1p_{1/2}$ are split, but the splitting is not large enough to break the magic numbers. Likewise, $1d_{5/2}$ and $1d_{3/2}$ are also split to a modest extent. However, the magic number 28 appears because the orbits $1f_{7/2}$ is pushed down significantly, whereas the next orbit $2p_{3/2}$ is lowered only moderately. So a gap is created between them, giving rise to a magic number 28. Similar situations occur for higher magic numbers 50, 82, and 126. These gaps are shown by the filled dots in Fig. 2.

The success of Mayer–Jensen's magic number is tremendous. It really dominates the structure of nuclei at lower energies. There is much experimental evidence for magic numbers, one of which is the separation energy. Figure 3 shows the observed neutron separation energy S_n. Figure 8 shows how magic numbers are related to the separation energy. On the left, neutrons occupy orbits up to a shell gap, as the number of these neutrons is equal to a magic number. On the right, there is another neutron occupying an orbit above the shell gap. In order to take away one of the neutrons on the left, one needs more energy as compared to the right part, where there is one neutron in a less bound orbit above the gap. The neutron separation energy means the minimum energy to take out one neutron from a given nucleus. So, it becomes suddenly smaller, by the amount of the shell gap, as the number of neutron goes beyond a magic number. This phenomenon can be seen in many places in Fig. 3, where vertical lines indicate magic numbers and the separation energy decreases suddenly over these lines, particularly for neutron numbers $N = 50$, 82, 126. This is one of the pieces of evidence for magic numbers. However, if one looks at Fig. 3 carefully, there are cases where the decrease in separation energy is not so large, or even an increase is seen. Thus, it can be expected that the magic numbers may not be perfect. However, such an idea has never been seriously discussed until recently. We will come back to possible changes of magic numbers later in this chapter.

At this point, we summarize this subsection. From its derivation, the magic numbers of Mayer and Jensen are direct and robust consequence of basic properties of the nuclear force and density, and there seems to be no way

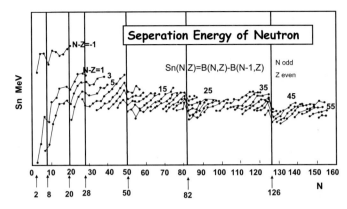

Fig. 7. Observed neutron separation energy, partly taken from Ref. [3] (See also Plate 3 in the Color Plate Section)

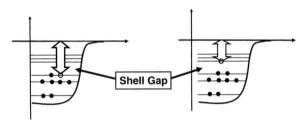

Fig. 8. Relation of neutron magic numbers to neutron separation energy. *Arrow* indicates separation energy. Separation energy of neutron (S_n) or proton (S_p), the minimum energy to take a neutron or proton out, decreases suddenly after the shell gap

out. So, the magic numbers have been believed to remain the same for all (or almost all) nuclei, stable or unstable.

1.2 Why Is the Shell Model Useful?

We now discuss how one can carry out shell model calculations. Through this, we would like to find an answer to the question as to why the shell model can be useful.

The shell model assumes that the orbits are already given as in Fig. 2. The orbits of the proton (neutron) closed shell is completely occupied by protons (neutrons). In general, there can be some protons (or neutrons) occupying the next shell just above the closed shell. This shell is called the *valence shell*, and its nucleons are referred to as *valence nucleons*. The valence shell is, by definition, only partially occupied. If all orbits between two magic numbers are considered, this valence shell is called a *major shell*. For instance, four orbits, $2p_{3/2}$, $2p_{1/2}$, $1f_{5/2}$, and $1g_{9/2}$, form a major shell between 28 and 50. (On the other hand, one might take a part of them as a valence shell. But the shell is not a major shell any more.) In the shell model calculation, the closed shells are treated as a vacuum because the nucleons cannot change their single-particle states as long as they are in the closed shell. If one wants, one can make particle–hole excitations from the closed shell to valence shell. However, just for the sake of simplicity, we do not include such excitations for the time being. Namely, we look at degrees of freedom only in the valence shell. Because of the similarity between the closed-shell and the vacuum, the closed shell is often called the *(inert) core.* When we discuss dynamical properties, the (inert) core sounds more appropriate, but its meaning is the same as the closed shell.

We include two-body interaction between valence nucleons. Three-body interaction, etc. are not included, however. Usually, it is supposed that effects of higher body (> 2 body) interactions are small enough in the energy scale of interest and/or their effects are renormalized into effective two-body interactions somehow. This is an approximation/assumption, but turns out to be reasonable from the viewpoint of comparison to experiment. The Hamiltonian then consists of the following terms,

$$H = \sum_i \epsilon_i n_i + \sum_{i,j,k,l} v_{ij,kl} a_i^\dagger a_j^\dagger a_l a_k, \tag{5}$$

where ϵ_i is the SPE of the orbit i, n_i stands for the number operator of the orbit i, $v_{ij,kl}$ denotes *two-body matrix element* (TBME) of the nucleon–nucleon (effective) interaction for orbits i, j, k, l, and $a^\dagger$ and a mean usual creation and annihilation operators, respectively.

In the single-particle picture of Fig. 2, a nucleon stays in one of the orbits forever. This is true in the closed shell, because all orbits are occupied, and a nucleon cannot move from one orbit to another within the closed shell. The situation differs in the valence shell. Figure 9 indicates how nucleons move via the nucleon–nucleon interaction. The occupancy pattern of nucleons over

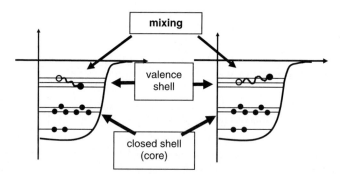

Fig. 9. Mixing of different configurations due to the scattering between valence nucleons. A nucleon does not stay in an orbit forever. The interaction between nucleons changes their occupations as a result of scattering. The pattern of occupation is called a configuration.

different orbits is called *configuration*. In Fig. 9, the configuration on the left-hand side is changed to the one on the right-hand side due to the interaction between two valence nucleons. In other words, the interaction scatters two nucleons into a different pair of orbits compared to the pair of orbits before the scattering. Such scatterings occur an infinite number of times, and all possible configurations are mixed until a kind of equilibrium is achieved. The eigenstate of the shell-model Hamiltonian as in Eq. (5) is thus obtained. It contains, in general, many components corresponding to different configurations. Such a mixing is called *configuration mixing*.

We illustrate how this can be carried out computationally. As step 1, we first calculate matrix elements of the Hamiltonian for the states of various configurations.

$$\langle \phi_1 | H | \phi_1 \rangle, \quad \langle \phi_1 | H | \phi_2 \rangle, \quad \langle \phi_1 | H | \phi_3 \rangle, \cdots$$

These various states can be represented by Slater determinants, ϕ_1, ϕ_2, ϕ_3, ... Each Slater determinant is a product of single-particle states, α, β, γ, ... In the second quantization picture,

$$\phi_1 = a_\alpha^\dagger a_\beta^\dagger a_\gamma^\dagger \cdots |0\rangle, \quad \phi_2 = a_{\alpha'}^\dagger a_{\beta'}^\dagger a_{\gamma'}^\dagger \cdots |0\rangle$$
$$\phi_3 = a_{\alpha''}^\dagger a_{\beta''}^\dagger a_{\gamma''}^\dagger \cdots |0\rangle, \tag{6}$$

where $|0\rangle$ means a closed core, and the Hamiltonian is written as Eq. (5).

In step 2, we construct the matrix of H and diagonalize it.

$$\mathcal{H} = \begin{bmatrix} \langle \phi_1 | H | \phi_1 \rangle & \langle \phi_1 | H | \phi_2 \rangle & * & * & * & \cdot & \cdot \\ \langle \phi_2 | H | \phi_1 \rangle & \langle \phi_2 | H | \phi_2 \rangle & * & * & \cdot & \cdot & \cdot \\ \langle \phi_3 | H | \phi_1 \rangle & * & & * & \cdot & \cdot & \cdot \\ * & & * & & \cdot & \cdot & \cdot \\ * & & & & \cdot & \cdot & \cdot \\ \cdot & & \cdot & & \cdot & \cdot & \cdot \\ \cdot & & \cdot & & \cdot & \cdot & \cdot \end{bmatrix} \xrightarrow{\text{Diagonalize}} \begin{bmatrix} \epsilon_1 & & & & \\ & \epsilon_2 & & & \\ & & \epsilon_3 & & \\ & & & \cdot & \\ & & & & \cdot \\ & & & & & \cdot \end{bmatrix} \tag{7}$$

We thus solve the eigenvalue problem,

$$H\Psi = E\Psi, \tag{8}$$

where E is the energy eigenvalue and its eigenfunction is Ψ. The Ψ wave function is expanded in terms of Slater determinants with probability amplitudes c_1, c_2, c_3, ...,

$$\Psi = c_1 \phi_1 + c_2 \phi_2 + c_3 \phi_3 + \dots. \tag{9}$$

Accordingly, the eigenvector of the matrix eigenvalue problem for Eq. (8) is written as $(c_1, c_2, c_3, \dots)$.

The precise evaluation of the energy eigenvalue and the configuration mixing is one of the major tasks of the shell model calculations. Once we obtain the wave function Ψ, we can calculate a variety of physical observables.

1.3 Some Remarks on Shell Model Calculations

The direct diagonalization of the matrix as in Eq. (7) is what the conventional shell model calculation does. We remark on some features of the shell model calculations so as to give some feeling for the actual calculation. We first briefly discuss the *M-scheme*. Since the Hamiltonian in Eq. (5) is rotationally invariant, the z-component of the total angular momentum, J, is conserved for all eigenstates. The quantum number of J_z of Ψ in Eq. (8) is denoted as M hereafter.

Each single-particle state has a good quantum number j_z, the z-component of j, because of the spherical mean potential. Each Slater determinant in Eq. (9) can have a good J_z, if it is constructed from such single-particle states with good j_z's. Naturally, Slater determinants, ϕ_1, ϕ_2, ϕ_3, ... in Eq. (9) should have the same value of J_z as the full Ψ. As the Hamiltonian conserves J_z, the matrix elements in Eq. (7) are finite only between the same values of J_z. Thus, one can construct the hamiltonian matrix like Eq. (10).

$$
\mathcal{H} =
\begin{array}{cccc}
M=0 & M=1 & M=-1 & M=2
\end{array}
\left(
\begin{array}{cccccc}
\begin{matrix} * & * & * & * \\ * & * & * & * \\ * & * & * & * \\ * & * & * & * \end{matrix} & 0 & 0 & 0 \\
0 & \begin{matrix} * & * & * \\ * & * & * \\ * & * & * \end{matrix} & 0 & 0 \\
0 & 0 & \begin{matrix} * & * & * \\ * & * & * \\ * & * & * \end{matrix} & 0 \\
 & & & \cdots\cdots\cdots
\end{array}
\right)
\tag{10}
$$

The Slater determinant is the product of single-particle states as seen in Eq. (6). The next question is how the *conservation* of J can be achieved. This question can be answered by taking a simple example. Equation (11) shows the case of two neutrons in the $f_{7/2}$ orbit.

$$
\begin{array}{|cc|}
\hline
m_1 & m_2 \\
\hline
7/2 & -7/2 \\
5/2 & -5/2 \\
3/2 & -3/2 \\
1/2 & -1/2 \\
\\
M = 0 \\
\hline
\end{array}
\xrightarrow{J_+}
\begin{array}{|cc|}
\hline
m_1 & m_2 \\
\hline
7/2 & -5/2 \\
5/2 & -3/2 \\
3/2 & -1/2 \\
\\
M = 1 \\
\hline
\end{array}
\xrightarrow{J_+}
\begin{array}{|cc|}
\hline
m_1 & m_2 \\
\hline
7/2 & -3/2 \\
5/2 & -1/2 \\
3/2 & -1/2 \\
\\
M = 2 \\
\hline
\end{array}
\tag{11}
$$

The j_z values of the two neutrons are denoted as m_1 and m_2. There are four states in the $M=0$ space as shown in the left column of the figure. Here, the space is a set of the states belonging to a given M ($M=0$ in this case). The dimension of this space is four. By acting with the angular momentum raising operator, J_+, where $J_+|j, m\rangle \propto |j, m+1\rangle$, we obtain an $M=1$ space comprised of three states. They are shown in the middle column in Eq. (11). The dimension is less by one than that of $M=0$ space. We then know that there is one $J=0$ state in the $M=0$ space, and it was eliminated by the J_+ operation (there is only an $M=0$ state for $J=0$).

By repeating the J_+ operation, we obtain an $M=2$ space. This space is shown in the right column of Eq. (11). The dimension is still three, meaning that there is no $J=1$ state in the spaces we are working on.

By doing a similar analysis, we can see what J states are contained in each M space as shown in Table 1. The diagonalization of the Hamiltonian matrix is done for each M space separately. Since the Hamiltonian conserves J in addition to M, the diagonalization of the Hamiltonian matrix produces eigenstates with *good* J's. This situation is depicted in Eq. (12) for the $M=0$ case of two neutrons in $f_{7/2}$ orbit.

$$M = 0$$

$$\mathcal{H} = \begin{bmatrix} * & * & * & * \\ * & * & * & * \\ * & * & * & * \\ * & * & * & * \end{bmatrix} \xrightarrow{\text{diagonalize}} \begin{bmatrix} e_{J=0} & 0 & 0 & 0 \\ 0 & e_{J=2} & 0 & 0 \\ 0 & 0 & e_{J=4} & 0 \\ 0 & 0 & 0 & e_{J=6} \end{bmatrix} \tag{12}$$

Such a restoration of J quantum number is a general one, and is achieved in general as a result of the diagonalization of the Hamiltonian, as illustrated in Eq. (13).

$$M$$

$$\mathcal{H} = \begin{bmatrix} * & * & * & * \\ * & * & * & * \\ * & * & * & * \\ * & * & * & * \end{bmatrix} \xrightarrow{\text{diagonalize}} \begin{bmatrix} e_J & 0 & 0 & 0 \\ 0 & e_{J'} & 0 & 0 \\ 0 & 0 & e_{J''} & 0 \\ 0 & 0 & 0 & e_{J'''} \end{bmatrix} \tag{13}$$

Table 1. J contents of M spaces of two neutron in $1f_{7/2}$

	Dimension	Components of J value
$M = 0$	4	$J = 0, 2, 4, 6$
$M = 1$	3	$J = 2, 4, 6$
$M = 2$	3	$J = 2, 4, 6$
$M = 3$	2	$J = 4, 6$
$M = 4$	2	$J = 4, 6$
$M = 5$	1	$J = 6$
$M = 6$	1	$J = 6$

In other words, the eigenstate Ψ is given by a linear combination of Slater determinants, and the same linear combination is a simultaneous eigenstate of the J_z operator and the $(\vec{J} \cdot \vec{J})$ operator.

We now see some properties of *TBME* of the nucleon–nucleon interaction. Two nucleons can have total angular momentum J and total isospin T. If nucleons are in single-particle states $|j_1 m_1\rangle$ and $|j_2 m_2\rangle$, they can be coupled to total angular momentum J and its z-projection M by Clebsch–Gordon coefficients as

$$|j_1, j_2, J, M\rangle = \sum_{m_1, m_2} (j_1, m_1, j_2, m_2 | J, M) |j_1, m_1\rangle |j_2, m_2\rangle, \qquad (14)$$

where the parenthesis indicates a Clebsch–Gordon coefficient. The same coupling occurs before and after the interaction (or scattering). We consider matrix elements between such coupled states,

$$\langle j_1, j_2, J, M | V | j_3, j_4, J', M' \rangle = \sum_{m_1, m_2} (j_1, m_1, j_2, m_2 | J, M)$$

$$\times \sum_{m_3, m_4} (j_3, m_3, j_4, m_4 | J', M')$$

$$\times \langle j_1, m_1, j_2, m_2 | V | j_3, m_3, j_4, m_4 \rangle. \qquad (15)$$

Since the interaction is rotationally invariant, it cannot change J or M, and consequently $J = J'$ and $M = M'$ are satisfied. Moreover, the rotational invariance of the interaction makes all coupled matrix elements independent of M therefore

$$\langle j_1, j_2, J, M | V | j_3, j_4, J', M' \rangle = \delta_{JJ'} \delta_{MM'} \langle j_1, j_2, J | V | j_3, j_4, J \rangle. \qquad (16)$$

Note that the quantum number M is eliminated in Eq. (16). Thus, in this J-coupled scheme matrix elements of two-nucleon states are specified only by j_1, j_2, j_3, j_4, and J (see Eq. (16)). There is certainly another dependence on the isospin, T.

Of course, once the potential between two nucleons is given one can calculate all TBME's. However, up to the present time, the potential between nucleons is not completely known. Moreover, there will be renormalization due to core-polarizations (as explained later), as well as many other different mechanisms that contribute to the TBME's. It is true that no theory has succeeded in a perfect prediction of the interaction to be used in shell model calculations, and this situation will not be altered in the near future. Thus, for the practical use of shell model calculations, we need empirical corrections.

An example of successful calculations of TBMEs is by using the USD interaction. The USD interaction was proposed by Wildenthal and Brown [4] for the sd shell comprised of three orbits $1d_{5/2}$, $1d_{3/2}$, and $2s_{1/2}$. It consists of 63 TBMEs and three SPEs. The TBMEs are based on those given by the G-matrix interaction of Kuo [5]. Figure 10 shows a part of the USD TBMEs.

i	j	k	l	J	T	V
d3/2	d3/2	d3/2	d3/2	0	1	-2.1845
d3/2	d3/2	d3/2	d3/2	1	0	-1.4151
d3/2	d3/2	d3/2	d3/2	2	1	-0.0665
d3/2	d3/2	d3/2	d3/2	3	0	-2.8842
d5/2	d3/2	d3/2	d3/2	1	0	0.5647
d5/2	d3/2	d3/2	d3/2	2	1	-0.6149
d5/2	d3/2	d3/2	d3/2	3	0	2.0337
d5/2	d3/2	d5/2	d3/2	1	0	-6.5058
d5/2	d3/2	d5/2	d3/2	1	1	1.0334
d5/2	d3/2	d5/2	d3/2	2	0	-3.8253
d5/2	d3/2	d5/2	d3/2	2	1	-0.3248
d5/2	d3/2	d5/2	d3/2	3	0	-0.5377
d5/2	d3/2	d5/2	d3/2	3	1	0.5894
d5/2	d3/2	d5/2	d3/2	4	0	-4.5062
d5/2	d3/2	d5/2	d3/2	4	1	-1.4497
d5/2	d3/2	s1/2	d3/2	1	0	-1.7080
d5/2	d3/2	s1/2	d3/2	1	1	0.1874
d5/2	d3/2	s1/2	d3/2	2	0	0.2832
d5/2	d3/2	s1/2	d3/2	2	1	-0.5247
d5/2	d3/2	s1/2	s1/2	1	0	2.1042
d5/2	d5/2	d3/2	d3/2	0	1	-3.1856
d5/2	d5/2	d3/2	d3/2	1	0	0.7221
d5/2	d5/2	d3/2	d3/2	2	1	-1.6221
d5/2	d5/2	d3/2	d3/2	3	0	1.8949
d5/2	d5/2	d5/2	d3/2	1	0	2.5435
d5/2	d5/2	d5/2	d3/2	2	1	-0.2828
d5/2	d5/2	d5/2	d3/2	3	0	2.2216
d5/2	d5/2	d5/2	d3/2	4	1	-1.2363
d5/2	d5/2	d5/2	d5/2	0	1	-2.8197
d5/2	d5/2	d5/2	d5/2	1	0	-1.6321
d5/2	d5/2	d5/2	d5/2	2	1	-1.0020
d5/2	d5/2	d5/2	d5/2	3	0	-1.5012

USD interaction

Fig. 10. Part of the USD TBMEs.

The indices i, j, k, and l stand, respectively, for j_1, j_2, j_3, and j_4 mentioned earlier. One finds the dependences on J and T. These TBMEs are calculated within the G-matrix formalism starting from meson exchange theory of the free nucleon–nucleon interaction. Such a calculation gives us reasonable numbers. But, once one performs shell model diagonalization with those TBMEs, the resultant energy levels are very different from experimental ones, and we do not learn much about the structure of the nucleus. We need to make empirical corrections. This is what was done by Wildenthal and Brown to obtain the USD interaction.

The nucleon–nucleon interactions used in shell model calculations are effective ones. They include effects of multiple scattering between nucleons with high-lying intermediate states in the sense of a second-order perturbation

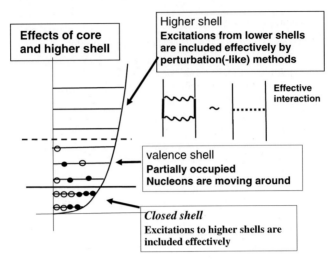

Fig. 11. Corrections to effective interaction used in the shell model calculations. There are two types, one from higher shells, while the other from the closed shell

(Ladder diagram). The upper part of Fig. 11 indicates this contribution schematically as the effects of "higher shell".

Another contribution comes from the excitation of the core (closed shell), as referred to as the core polarization. In a first approximation, the core is completely occupied, but there are excitations in reality. The pairing interaction is enhanced by this mechanism.

Effects of these two types of outer shells are included, ending up with the so-called *effective interaction*.

The core polarization is also important for *effective charge* and the *effective g-factor*. Figure 12 shows how the electric quadrupole moment and magnetic moment are changed due to the excitation of the core. This phenomenon has been proposed by Arima and Horie [6], Blin-Stoyle and Perks [7] in 1954 and by Bohr and Mottelson [3].

We finally note that the single particle in the shell model is an "effective object" (or quasi-particle) with rather complicated correlations behind it like the core polarization. The shell model treats it as if it is a real particle, and includes those correlation effects in terms of the renormalization of effective interaction and operators. The same picture should be taken for mean-field models (or density functional theories), where the renormalization is more severe due to more truncated effective interactions.

The coupling to various correlations reduces the 'purity' of the single particle. Recent experiments show that about 60% of the "single-particle" probability remains in the shell model wave functions after the mixing of more complicated components [8]. However, we emphasize that this is not a catastrophe or anything like that, and is expected.

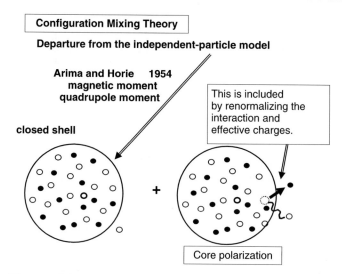

Fig. 12. Schematic illustration of configuration mixing theory

We have seen how the shell model is formulated in simple terms. We skipped many discussions as to how one can obtain good effective interactions from basic theories of the nucleon–nucleon interaction. This is rather complicated and still in progress.

The effective interaction and operators are essential parts of the shell model. These are also crucial in other approaches to nuclear structure. The shell model is constructed very carefully on nuclear forces. In other approaches of nuclear structure, forces are also models. In the shell model, although we use effective interactions, these interactions are built as realistic as possible, particularly in recent large-scale calculations. In this sense, the shell model can reflect various facets of nuclear forces into nuclear structure better than other models.

This is the merit of the shell model. Because of this, the shell model becomes more important in the region of exotica in the nuclear chart, as the predictive power is more needed than in the region of stable nuclei.

If the shell model is so useful, the next question is how to run a shell model calculation; or even whether it is always possible or not?

1.4 How Do We Perform Shell Model Calculations?

Equation (7) indicates the diagonalization of the Hamiltonian matrix. There are several computer programs for performing such shell model calculations. These include OXBASH by Brown, ANTOINE by Caurier et al., MSHELL by Mizusaki, etc. The ANTOINE and MSHELL can be run on parallel computers, and can handle up to 1 billion dimension (as of the year 2008). One can easily imagine that the practical difficulty should increase as the dimension becomes

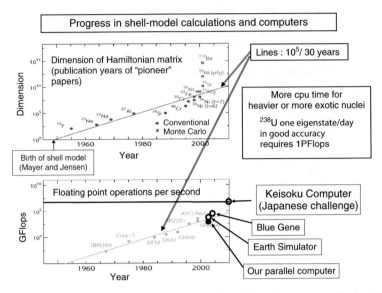

Fig. 13. *Upper panel*: Maximum dimension of the Hamiltonian matrix feasible for the year of publication. *Blue* points are for conventional shell model, while *red* points are for Monte Carlo Shell Model. *Lower panel*: The computer capability (Flops) as a function of the year. The lines in the upper and lower panels indicate an increase of 10^5 times/30 years. (See also Plate 4 in the Color Plate Section)

larger. The difficulties are in the computation and storage of so many matrix elements, and the diagonalization of such a huge matrix.

There have been steady and significant efforts with great success, as reviewed recently in [9]. Figure 4 shows how the maximum feasible dimension has been increased as a function of the year since 1949 (the birth year of the shell model). It is amazing that this growth is almost on a logarithmic scale. However, there exists a limit around 1 billion dimension, at least at present. In order to overcome this limit, the Monte Carlo Shell Model has been proposed [10–12]. Although the computation time becomes longer as the particle number and/or the number of valence orbits increases, this is not a very strong barrier for the Monte Carlo Shell Model. One could keep going to frontier cases with the dimension, for instance, 10^{15} [13]. In exotic nuclei, two conventional shells often merge, and the calculation becomes huge. The Monte Carlo Shell Model can still give us results.

2 Construction of an Effective Interaction and an Example in the *pf* Shell

In this section, we show how one can determine TBMEs.

A simple example of experimental determination is the case of the $1f_{7/2}$ orbit. If this orbit is perfectly isolated, one can extract TBMEs from observed energy levels as follows. First, energies of the states which consists of two valence particles and the closed core are written, using experimental SPE $\epsilon(f7/2)$, as

$$E(J) = 2\epsilon(f7/2) + V_J, \tag{17}$$

where

$$V_J = \langle f7/2, f7/2, J, T = 1|V|f7/2, f7/2, J, T = 1 \rangle. \tag{18}$$

Next, TBMEs are determined so that $E(J)$ in Eq. (17) reproduces the experimental energies of the corresponding states. However, the case of the $1f_{7/2}$ orbit is too simple. Other cases cannot be handled this way. For instance, Arima et al. have carried out a χ^2 fit of TBMEs for $1d_{5/2}$ and $2s_{1/2}$, according to the following procedure.

---
(1) TBMEs are assumed,
(2) Energy eigenvalues are calculated,
(3) χ^2 is calculated between theoretical and experimental energy levels,
(4) TBMEs are modified. Go to Eq. (2), and iterate the process until χ^2 becomes small enough.

They applied the above procedure to $0^+, 2^+$, and 4^+ states in ^{18}O, which was assumed to be a system of two valence orbits, $d_{5/2}$ and $s_{1/2}$, on the top of the closed shell (^{16}O). The TBMEs

$$\langle d5/2, d5/2, J, T = 1|V|d5/2, d5/2, J, T \rangle,$$
$$\langle d5/2, s1/2, J, T = 1|V|d5/2, s1/2, J, T \rangle,$$
$$\langle s1/2, s1/2, J, T = 1|V|s1/2, s1/2, J, T \rangle, \text{etc.}$$

were determined. The idea of the χ^2 fit does not work for the full sd shell, however. In obtaining the USD interaction, Wildenthal and Brown carried out only a partial fit. One can choose some TBMEs whose linear combinations are sensitive to energies of low-lying states. They adjusted 47 linear combinations out of 63 TBMEs and three SPE's. The rest were taken from G-matrix result of Kuo. The USD was thus created [4].

The same idea was taken for determining the GXPF1 interaction, which was created for the description of pf shell nuclei including the middle region [14, 15]. The GXPF1 is based on the G-matrix interaction by H.-Jensen et al. [16]. There is also the KB3 interaction and its family for the pf shell, which are particularly good for the beginning of the pf shell [17, 18].

Figure 14 shows the correlations between TBME of G-matrix of H.-Jensen et al. and the corresponding TBME of GXPF1. If the empirical fit does not change TBMEs, all the points should be lined up on the $y = x$ line in Fig. 14. Indeed all points are near the $y = x$ line, but there are deviations. These

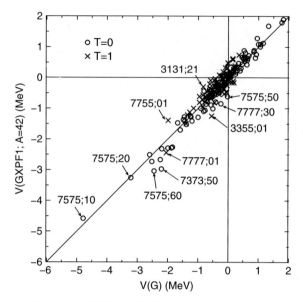

Fig. 14. GXPF1 TBME vs. G-matrix TBME

deviations are results of the fit. The fit was made for 699 levels. One finds some general trends: $T = 0$ TBMEs are shifted to be more attractive as a whole, while $T = 1$ are made more repulsive. In Fig. 14, the orbits, J and T are shown for some points. One sees that the $T = 0$ coupling between $1f_{7/2}$ and $1f_{5/2}$ is strong.

Using this GXPF1 interaction, many interesting results have been obtained [14, 19–56]. In particular, the issues like $N = 32$ and $N = 34$, new magic numbers have been extensively studied as well as deformation of Ti, Cr, and Fe isotopes.

3 The $N = 20$ Problem: Does the Gap Change?

One of the most prominent points around the so-called island of inversion has been the changing shell gap between the sd and pf shells. Figure 15 indicates how the gap depends on the proton number Z. This N=20 gap is about 5–6 MeV for ^{40}Ca. The conventional idea leads us to a constant gap, as shown in Fig. 15. The SDPF-M interaction, which was obtained in [57] and described in [58], reproduces various peculiar phenomena in N=18–22 isotopes of F, Ne, Na, Mg, Al, and Si. This interaction produces a varying gap, as shown in Fig. 15, which turned out to be essential for good agreement with many experimental data obtained in MSU, GANIL, GSI, and RIKEN in recent years [57, 59–72].

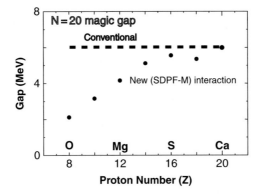

Fig. 15. $N = 20$ gap obtained by the SDPF-M interaction (*red* points). *Blue* line indicates a constant gap set by ^{40}Ca doubly magic nucleus

We discuss briefly below why the gap changes. Before moving there, we would like to point out that, once the gap becomes smaller, two shells tend to merge, and there will be many particle–hole excitations. This means that appropriate shell model calculations can be done only with a huge Hamiltonian matrix, making conventional shell model calculations more difficult. Thus, the Monte Carlo Shell Model plays crucial roles in the studies of exotic nuclei.

The gap change, which is the most prominent consequence of the shell evolution, is now known to be primarily due to the tensor force. For this discussion, the reader is referred to the original papers [73–76]. Figure 16 was taken from [73] with an addition. This figure suggests how SPEs are changed due to a particular part of the nucleon–nucleon interaction. When this paper

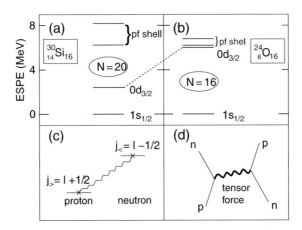

Fig. 16. Mechanism of the shell evolution. Note that the gap shown is the $N = 20$ gap for $N = 16$, whereas the one shown in Fig. 15 is the same gap but for $N = 20$

was published, it was reported that the strong proton–neutron interaction between spin–flip partners ($j_>$ and $j_<$) is the origin of some shell evolution, but its fundamental origin was only speculation. Later, the tensor force has been shown to be the origin. So, we now put a "tensor" force into Fig. 16 by the bold-wavy line. The change of the $N=20$ gap in Fig. 15 is now understood as being due to the tensor force.

4 Summary

The basic points of formulation and the properties of the shell model are presented without going to details or sophisticated theories. These can be found in standard textbooks.

We would like, however, to emphasize one point. Mayer–Jensen's magic numbers are a robust conclusion arising from the short-range attraction of the nuclear force, density saturation and the spin–orbit force. These are sound and cannot be thrown away. Therefore, if there is any deviation, one needs something new. The spin–isospin interactions have been studied in the context of mean potential arguments in the past, but there have been no extensive studies, whereas these interactions have been known and been studied in other contexts. Once their effects were studied, it was found that there are robust and intuitively understood mechanisms that change the shell structure. But, to see this, one needs to change the neutron (or proton) number considerably. It was done for Sb isotopes by Schiffer et al. [77], but generally requires rare isotope beam experiments. So, the shell evolution due to the tensor force, and maybe other as yet unknown physicvs, will open an new era in nuclear structure research.

Acknowledgement

Drs M. Honma, Y. Utsuno, T. Mizusaki, N. Shimizu, T. Suzuki, and N. Akaishi for their help on theoretical parts of works related to this work. There are many experimentalists who helped this work. The author is grateful to Mr. K. Tsukiyama for his assistance in making the manuscript.

This work has been supported in part by the JSPS Core-to-Core project "International Research Network for Exotic Femto Systems".

References

1. M.G. Mayer, Phys. Rev. **75** 1969 (1949).
2. O. Haxel, J.H.D. Jensen and H.E. Suess, Phys. Rev. **75** 1766 (1949).
3. A. Bohr, B.R. Mottelson, *Nuclear Structure*, Vol. 1, (Benjamin, New York 1969).
4. B.A. Brown, B.H. Wildenthal, Annu. Rev. Nucl. Part. Sci. **38**, 29 (1988).

5. T.T.S. Kuo, Nucl. Phys. **A103**, 71 (1967).
6. A. Arima, H. Horie, Prog. Theor. Phys. **11** 509 (1954).
7. R.J. Blin-Stoyle, M.A. Perks, Proc. Phys. Soc. (London) **67A** 885 (1954).
8. A. Gade et al. Phys. Rev. Lett. **93** 042501 (2004).
9. E. Caurier, G. Martinez-Pinedo, F. Nowacki, et al., Rev. Mod. Phys. **77** 427 (2005).
10. M. Honma, T. Mizusaki, T. Otsuka, Phys. Rev. Lett. **75**, 1284 (1995).
11. T. Otsuka, M. Honma, T. Mizusaki, Phys. Rev. Lett. **81**, 1588 (1998).
12. T. Otsuka, M. Honma, T. Mizusaki, N. Shimizu, Y. Utsuno, Prog. Part, Nucl. Phys. **47** 319 (2001).
13. N. Shimizu, T. Otsuka, T. Mizusaki, M. Honma, Phys. Rev. Lett. **86**, 1171 (2001).
14. M. Honma, T. Otsuka, B.A. Brown, T. Mizusaki, Phys. Rev. C **65**, 061301 (2002).
15. M. Honma, T. Otsuka, B.A. Brown, T. Mizusaki, Phys. Rev. C **69**, 034335 (2004).
16. M. Hjorth-Jensen, T.T.S. Kuo, E. Osnes, Phys. Repts. **261**, 125 (1995).
17. A. Poves, A.P. Zuker, Phys. Repts. **70**, 235 (1981).
18. A. Poves, J. Sánchez-Solano, E. Caurier, F. Nowacki, Nucl. Phys. **A694**, 157 (2001).
19. R.V.F. Janssens, B. Fornal, P.F. Mantica, B.A. Brown, R. Broda, P. Bhattacharyya, M.P. Carpenter, M. Cinausero, P.J. Daly, A.D. Davies, T. Glasmacher, Z.W. Grabowski, D.E. Groh, M. Honma, F.G. Kondev, W. Królas, T. Lauritsen, S.N. Liddick, S. Lunardi, N. Marginean, T. Mizusaki, D.J. Morrissey, A.C. Morton, W.F. Mueller, T. Otsuka, T. Pawlat, D. Seweryniak, H. Schatz, A. Stolz, S.L. Tabor, C.A. Ur, G. Viesti, I. Wiedenhöver, J. Wrzesiński, Phys. Lett. B **546**, 55 (2002)
20. P.F. Mantica, A.C. Morton, B.A. Brown, A.D. Davies, T. Glasmacher, D.E. Groh, S.N. Liddick, D.J. Morrissey, W.F. Mueller, H. Schatz, A. Stolz, S.L. Tabor, M. Honma, M. Horoi, T. Otsuka, Phys. Rev. C **67**, 014311 (2003)
21. A.F. Lisetskiy, N. Pietralla, M. Honma, A. Schmidt, I. Schneider, A. Gade, P. von Brentano, T. Otsuka, T. Mizusaki, B.A. Brown, Phys. Rev. C **68**, 034316 (2003)
22. S.N. Liddick, P.F. Mantica, R.V.F. Janssens, R. Broda, B.A. Brown, M.P. Carpenter, B. Fornal, M. Honma, T. Mizusaki, A.C. Morton, W.F. Mueller, T. Otsuka, J. Pavan, A. Stolz, S.L. Tabor, B.E. Tomlin, M. Wiedeking, Phys. Rev. Lett. **92**, 072502 (2004)
23. M. Honma, T. Otsuka, B.A. Brown, T. Mizusaki, Phys. Rev. C **69**, 034335 (2004)
24. S.J. Freeman, R.V.F. Janssens, B.A. Brown, M.P. Carpenter, S.M. Fischer, N.J. Hammond, M. Honma, T. Lauritsen, C.J. Lister, T.L. Khoo, G. Mukherjee, D. Seweryniak, J.F. Smith, B.J. Varley, M. Whitehead, S. Zhu, Phys. Rev. C **69**, 064301 (2004)
25. K.L. Yurkewicz, D. Bazin, B.A. Brown, C.M. Campbell, J.A. Church, D.-C. Dinca, A. Gade, T. Glasmacher, M. Honma, T. Mizusaki, W.F. Mueller, H. Olliver, T. Otsuka, L.A. Riley, J.R. Terry, Phys. Rev. C **70**, 034301 (2004)
26. K.L. Yurkewicz, D. Bazin, B.A. Brown, C.M. Campbell, J.A. Church, D.-C. Dinca, A. Gade, T. Glasmacher, M. Honma, T. Mizusaki, W.F. Mueller, H. Olliver, T. Otsuka, L.A. Riley, J.R. Terry, Phys. Rev. C **70**, 054319 (2004)

27. S.N. Liddick, P.F. Mantica, R. Broda, B.A. Brown, M.P. Carpenter, A.D. Davies, B. Fornal, T. Glasmacher, D.E. Groh, M. Honma, M. Horoi, R.V.F. Janssens, T. Mizusaki, D.J. Morrissey, A.C. Morton, W.F. Mueller, T. Otsuka, J. Pavan, H. Schatz, A. Stoltz, S.L. Tabor, B.E. Tomlin, M. Wiedeking, Phys. Rev. C **70**, 064303 (2004)

28. B. Fornal, S. Zhu, R.V.F. Janssens, M. Honma, R. Broda, P.F.Mantica, B.A. Brown, M.P. Carpenter, P.J. Daly, S.J. Freeman, Z.W.Grabowski, N. Hammond, F.G. Kondev, W. krolas, T. Lauritsen, S.N. Liddick, C.J. Lister, E.F. Moore, T. Otsuka, T. Pawlat, D. Seweryniak, B.E. Tomlin, J. Wrzesinski, Phys. Rev. C **70**,064304 (2004)

29. K.L. Yurkewicz, D. Bazin, B.A. Brown, C.M. Campbell, J.A. Church, D.-C. Dinca, A. Gade, T. Glasmacher, M. Honma, T. Mizusaki, W.F. Mueller, H. Olliver, T. Otsuka, L.A. Riley, J.R. Terry, Phys. Rev. C **70**, 064321 (2004)

30. D.-C. Dinca, R.V.F. Janssens, A. Gade, D. Bazin, R. Broda, B.A. Brown, C.M. Campbell, M.P. Carpenter, P. Chowdhury, J.M. Cook, A.N. Deacon, B. Fornal, S.J. Freeman, T. Glasmacher, M. Honma, F.G. Kondev, J.-L. Lecouey, S.N. Liddick, P.F. Mantica, W.F. Mueller, H. Olliver, T. Otsuka, J.R. Terry, B.A. Tomlin, K. Yoneda, Phys. Rev. C **71**, 041302 (2005)

31. L-L. Andersson, E.K. Johansson, J. Ekman, D. Rudolph, R. du Rietz, C. Fahlander, C.J. Gross, P.A. Hausladen, D.C. Radford, G. Hammond, Phys. Rev. C **71**, 011303 (2005)

32. M. Hagemann, C. Bäumer, A.M. van den Berg, D. De Frenne, D. Frekers, V.M. Hannen, M.N. Harakeh, J. Heyse, M.A. de Huu, E. Jacobs, K. Langanke, G. Martínez-Pinedo, A. Negret, L. Popescu, S. Rakers, R. Schmidt, H.J. Wörtche, Phys. Rev. C **71**, 014606 (2005)

33. A. Bürger, T.R. Saito, H. Grawe, H. Hübel, P. Reiter, J. Gerl, M. Górska, H.J. Wollersheim, A. Al-Khatib, A. Banu, T. Beck, F. Becker, P. Bednarczyk, G. Benzoni, A. Bracco, S. Brambilla, P. Bringel, F. Camera, E. Clément, P. Doornenbal, H. Geissel, A. Görgen, J. Grębosz, G. Hammond, M. Hellström, M. Honma, M. Kavatsyuk, O. Kavatsyuk, M. Kmiecik, I. Kojouharov, W. Korten, N. Kurz, R. Lozeva, A. Maj, S. Mandal, B. Million, S. Muralithar, A. Neußer, F. Nowacki, T. Otsuka, Zs. Podolyák, N. Saito, A.K. Singh, H. Weick, C. Wheldon, O. Wieland, M. Winkler, RISING Collaboration, Phys. Lett. B **622**, 29 (2005)

34. J. Leske, K.-H. Speidel, S. Schielke, O. Kenn, J. Gerber, P. Maier-Komor, S.J. Robinson, A. Escuderos, Y.Y. Sharon, L. Zamick, Phys. Rev. C **71**, 044316 (2005)

35. F. Brandolini, C.A. Ur, Phys. Rev. C **71**, 054316 (2005)

36. M. Honma, T. Otsuka, B.A. Brown, T. Mizusaki, Eur. Phys. J. A **25**, Supp. 1, 499 (2005)

37. M. Honma, T. Otsuka, T. Mizusaki, M. Hjorth-Jensen, B.A. Brown, J. Phys. Conf. **20**, 7 (2005)

38. M.A.G. Silveira, N.H. Medina, J.A. Alcántara-Núñez, E.W. Cybulska, H. Dias, J.R.B. Oliveira, M.N. Rao, R.V. Ribas, W.A. Seale, K.T. Wiedemann, B.A. Brown, M. Honma, T. Mizusaki, T. Otsuka, J. Phys. G **31**, s1577 (2005)

39. R. du Rietz, S.J. Williams, D. Rudolph, J. Ekman, C. Fahlander, C. Andreoiu, M. Axiotis, M.A. Bentley, M.P. Carpenter, C. Chandler, R.J. Charity, R.M. Clark, M. Cromaz, A. Dewald, G. de Angelis, F. Della Vedova, P. Fallon, A. Gadea, G. Hammond, E. Ideguchi, S.M. Lenzi, A.O. Macchiavelli, N. Marginean, M.N. Mineva, O. Möller, D.R. Napoli, M. Nespolo, W. Reviol,

C. Rusu, B. Saha, D.G. Sarantites, D. Seweryniak, D. Tonev, C.A. Ur, Phys. Rev. C **72**, 014307 (2005)

40. L.A. Riley, M.A. Abdelqader, D. Bazin, M.J. Bojazi, B.A. Brown, C.M. Campbell, J.A. Church, P.D. Cottle, D.-C. Dinca, J. Enders, A. Gade, T. Glasmacher, M. Honma, S. Horibe, Z. Hu, K.W. Kemper, W.F. Mueller, H. Olliver, T. Otsuka, B.C. Perry, B.T. Roeder, B.M. Sherrill, T.P. Spencer, J.R. Terry, Phys. Rev. C **72**, 024311 (2005)

41. B. Fornal, S. Zhu, R.V.F. Janssens, M. Honma, R. Broda, B.A. Brown, M.P. Carpenter, S.J. Freeman, N. Hammond, F.G. Kondev, W. Królas, T. Lauritsen, S.N. Liddick, C.J. Lister, S. Lunardi, P.F. Mantica, N. Marginean, T. Mizusaki, E.F. Moore, T. Otsuka, T. Pawlat, D. Seweryniak, B.E. Tomlin, C.A. Ur, I. Wiedenhöver, J. Wrzesinski, Phys. Rev. C **72**, 044315 (2005)

42. A. Poves, F. Nowacki, E. Caurier, Phys. Rev. C **72**, 047302 (2005)

43. A. Costin, N. Pietralla, T. Koike, C. Vaman, T. Ahn, G. Rainovski, Phys. Rev. C **72**, 054305 (2005)

44. S.N. Liddick, P.F. Mantica, R. Broda, B.A. Brown, M.P. Carpenter, A.D. Davies, B. Fornal, M. Horoi, R.V.F. Janssens, A.C. Morton, W.F. Mueller, J. Pavan, H. Schatz, A. Stolz, S.L. Tabor, B.E. Tomlin, M. Wiedeking, Phys. Rev. C **72**, 054321 (2005)

45. T. Adachi, Y. Fujita, P. von Brentano, A.F. Lisetskiy, G.P.A. Berg, C. Fransen, D. De Frenne, H. Fujita, K. Fujita, K. Hatanaka, M. Honma, E. Jacobs, J. Kamiya, K. Kawase, T. Mizusaki, K. Nakanishi, A. Negret, T. Otsuka, N. Pietralla, L. Popescu, Y. Sakemi, Y. Shimbara, Y. Shimizu, Y. Tameshige, A. Tamii, M. Uchida, T. Wakasa, M. Yosoi, K.O. Zell, Phys. Rev. C **73**, 024311 (2006)

46. A. Gade, R.V.F. Janssens, D. Bazin, B.A. Brown, C.M. Campbell, M.P. Carpenter, J.M. Cook, A.N. Deacon, D.-C. Dinca, S.J. Freeman, T. Glasmacher, B.P. Kay, P.F. Mantica, W.F. Mueller, J.R. Terry, S. Zhu, Phys. Rev. C **73**, 037309 (2006)

47. S.N. Liddick, P.F. Mantica, B.A. Brown, M.P. Carpenter, A.D. Davies, M. Horoi, R.V.F. Janssens, A.C. Morton, W.F. Mueller, J. Pavan, H. Schatz, A. Stolz, S.L. Tabor, B.E. Tomlin, M. Wiedeking, Phys. Rev. C **73**, 044322 (2006)

48. M. Horoi, B.A. Brown, T. Otsuka, M. Honma, T. Mizusaki, Phys. Rev. C **73**, 061305 (2006) 061305

49. M. Honma, T. Otsuka, T. Mizusaki, M. Hjorth-Jensen, J. Phys. Conf. **49**, 45 (2006)

50. J. Leske, K.-H. Speidel, S. Schielke, J. Gerber, P. Maier-Komor, S.J.Q. Robinson, A. Escuderos, Y.Y. Sharon, L. Zamick, Phys. Rev. C **74**, 024315 (2006)

51. S. Zhu, A.N. Deacon, S.J. Freeman, R.V.F. Janssens, B. Fornal, M. Honma, F.R. Xu, R. Broda, I.R. Calderin, M.P. Carpenter, P. Chowdhury, F.G. Kondev, W. Królas, T. Lauritsen, S.N. Liddick, C.J. Lister, P.F. Mantica, T. Pawlat, D. Seweryniak, J.F. Smith, S.L. Tabor, B.E. Tomlin, B.J. Varley, J. Wrzesinski, Phys. Rev. C **74**, 064315 (2006)

52. M.A.G. Silveira, N.H. Medina, J.R.B. Oliveira, J.A. Alcántara-Núñez, E.W. Cybulska, H. Dias, M.N. Rao, R.V. Ribas, W.A. Seale, K.T. Wiedemann, B.A. Brown, M. Honma, T. Mizusaki, T. Otsuka, Phys. Rev. C **74**, 064312 (2006)

53. S. Zhu, R.V.F. Janssens, B. Fornal, S.J. Freeman, M. Honma, R. Broda, M.P. Carpenter, A.N. Deacon, B.P. Kay, F.G. Kondev, W. Królas,

J. Kozemczak, A. Larabeee, T. Lauritsen, S.N. Liddick, C.J. Lister, P.F. Mantica, T. Otsuka, T. Pawĺat, A. Robinson, D. Seweryniak, J.F. Smith, D. Steppenbeck, B.E. Tomlin, J. Wrzesinski, X. Wang, Phys. Lett. B **650**, 135 (2007)

54. M. Horoi, S. Stoica, B.A. Brown, Phys. Rev. C **75**, 034303 (2007)

55. C. Dossat, N. Adimi, F. Aksouh, F. Becker, A. Bey, B. Blank, C. Borcea, R. Borcea, A. Boston, M. Caamano, G. Canchel, M. Chartier, D. Cortina, S. Czajkowski, G. de France, F. de Oliveira Santos, A. Fleury, G. Georgiev, J. Giovinazzo, S. Grévy, R. Grzywacz, M. Hellström, M. Honma, Z. Janas, D. Karamanis, J. Kurcewicz, M. Lewitowicz, M.J. López Jiménez, C. Mazzocchi, I. Matea, V. Maslov, P. Mayet, C. Moore, M. Pfützner, M.S. Pravikoff, M. Stanoiu, I. Stefan, J.C. Thomas, Nucl. Phys. A **792**, 18 (2007)

56. A.N. Deacon, S.J. Freeman, R.V. Janssens, M. Honma, M.P. Carpenter, P. Chowdhury, T. Lauritsen, C.J. Lister, D. Seweryniak, J.F. Smith, S.L. Tabor, B.J. Varley, F.R. Xu, S. Zhu, Phys. Rev. C **76**, 054303 (2007)

57. Y. Utsuno, T. Otsuka, T. Mizusaki, M. Honma, Phys. Rev. C **60**, 054315 (1999).

58. Y. Utsuno, T. Otsuka, T. Glasmacher, T. Mizusaki, M. Honma, Phys. Rev. C **70**, 044307 (2004).

59. Y. Utsuno, T. Otsuka, T. Mizusaki, M. Honma, Phys. Rev. C **64**, 011301(R) (2001).

60. Y. Utsuno, T. Otsuka, T. Glasmacher, T. Mizusaki, M. Honma, Phys. Rev. C **70**, 044307 (2004).

61. Z. Elekes, Zs. Dombradi, A. Saito, N. Aoi, H. Baba, K. Demichi, Zs. Fulop, J. Gibelin, T. Gomi, H. Hasegawa, N. Imai, M. Ishihara, H. Iwasaki, S. Kanno, S. Kawai, T. Kishida, T. Kubo, K. Kurita, Y. Matsuyama, S. Michimasa, T. Minemura, T. Motobayashi, M. Notani, T. Ohnishi, H.J. Ong, S. Ota, A. Ozawa, H.K. Sakai, H. Sakurai, S. Shimoura, E. Takeshita, S. Takeuchi, M. Tamaki, Y. Togano, K. Yamada, Y. Yanagisawa, K. Yoneda, Phys. Lett. B **599**, 17 (2004).

62. G. Neyens, M. Kowalska, D. Yordanov, K. Blaum, P. Himpe, P. Lievens, S. Mallion, R. Neugart, Y. Utsuno, T. Otsuka, Phys. Rev. Lett. **94**, ; 022501 (2005).

63. V. Tripathi, S.L. Tabor, P.F. Mantica, C.R. HOffman, M. Wiedeking, A.D. Davies, S.N. Liddick, W.F. Mueller, T. Otsuka, A. Stolz, B.E. Tomlin, Y. Utsuno, A. Volya, Phys. Rev. Lett. **94**, 162501 (2005).

64. P. Mason, N. Marginean, S.M. Lenzi, M. Ionescu-Bujor, F. Della Vedova, D.R. Napoli, T. Otsuka, Y. Utsuno, F. Nowacki, M. Axiotis, D. Bazzacco, P.G. Bizzeti, A. Bizzeti-Sona, F. Brandolini, M. Cardona, G. de Angelis, E. Farnea, A. Gadea, D. Hojman, A. Iordachescu, C. Kalfas, Th.Kroll, S. Lunardi, T. Martinez, C.M. Petrache, B. Quintana, R.V. Ribas, C. Rossi Alvarez, C.A. Ur, R. Vlastou, S. Zilio, Phys. Rev. C **71**, 014316 (2005).

65. M. Belleguic, F. Azaiez, Zs. Dombradi, D. Sohler, M.J. Lopez-Jimenez, T. Otsuka, M.G .Saint-Laurent, O. Sorlin, M. Stanoiu, Y. Utsuno, Yu.-E.Penionzhkevich, N.L. Achouri, J.C. Angelique, C. Borcea, C. Bourgeois, J.M. Daugas, F. De Oliveira-Santos, Z. Dlouhy, C. Donzaud, J. Duprat, Z. Elekes, S. Grevy, D. Guillemaud-Mueller, S. Leenhardt, M. Lewitowicz, S.M. Lukyanov, W. Mittig, M.G. Porquet, F. Pougheon, P. Roussel-Chomaz, H. Savajols, Y. Sobolev, C. Stodel, J. Timar, Phys. Rev. C **72**, 054316 (2005).

66. M. Ionescu-Bujor, A. Iordachescu, D.R. Napoli, S.M. Lenzi, N. Marginean, T. Otsuka, Y. Utsuno, R.V. Ribas, M. Axiotis, D. Bazzacco, A.M. Bizzeti-Sona, P.G. Bizzeti, F. Brandolini, D. Bucurescu, M.A. Cardona, G.de Angelis, M. De Poli, F. Della Vedova, E. Farnea, A. Gadea, D. Hojman, C.A. Kalfas, Th. Kroll, S. Lunardi, T. Martinez, P. Mason, P. Pavan, B. Quintana, C. Rossi Alvarez, C.A. Ur, R. Vlastou, S. Zilio Phys. Rev. C **73**, 024310 (2006).

67. V. Tripathi, S.L. Tabor, C.R. Hoffman, M. Wiedeking, A. Volya, P.F. Mantica, A.D. Davies, S.N. Liddick, W.F. Mueller, A.Stolz, B.E. Tomlin, T. Otsuka, Y. Utsuno, Phys. Rev. C **73**, 054303 (2006).

68. Zs. Dombrádi, Z. Elekes, A. Saito, N. Aoi, H. Baba, K. Demichi, Zs. Fulop, J. Gibelin, T. Gomi, H. Hasegawa, N. Imai, M. Ishihara, H. Iwasaki, S. Kanno, S. Kawai, T. Kishida, T. Kubo, K. Kurita, Y. Matsuyama, S. Michimasa, T. Minemura, T. Motobayashi, M. Notani, T. Ohnishi, H.J. Ong, S. Ota, A. Ozawa, H.K. Sakai, H. Sakurai, S. Shimoura, E. Takeshita, S. Takeuchi, M. Tamaki, Y. Togano, K. Yamada, Y. Yanagisawa, K. Yoneda, Phys. Rev. Lett. **96**, 182501 (2006).

69. J.R. Terry, D. Bazin, B.A. Brown, C.M. Campbell, J.A. Church, J.M. Cook, A.D. Davies, D.-C. Dinca, J. Enders, A. Gade, T. Glasmacher, P.G. Hansen, J.L. Lecouey, T. Otsuka, B. Pritychenko, B.M. Sherrill, J.A. Tostevin, Y. Utsuno, K. Yoneda, H. Zwahlen, Phys. Lett. B **640**, 86 (2006).

70. P. Himpe, G. Neyens, D.L. Balabanski, G. Belier, D. Borremans, J.M. Daugas, F. de Oliveira Santos, M. De Rydt, K. Flanagana, G. Georgievd, M. Kowalskae, S. Mallion, I. Matea, P. Morel, Yu.E. Penionzhkevich, N.A. Smirnovag. Stodel, K. Turzoa, N. Vermeulen, D. Yordanov, Phys. Lett. B **643**, 257 (2006).

71. A. Gade, P. Adrich, D. Bazin, M.D. Bowen, B.A. Brown, C.M. Campbell, J.M. Cook, S. Ettenauer, T. Glasmacher, K.W. Kemper, S. McDaniel, A. Obertelli, T. Otsuka, A. Ratkiewicz, K. Siwek, J.R. Terry, J.A. Tostevin, Y. Utsuno, D. Weisshaar, Phys. Rev. Lett. **99**, 072502 (2007).

72. V. Tripathi, S.L. Tabor, P.F. Mantica, Y.. Utsuno, P. Bender, J. Cook, C.R. Hoffman, S. Lee, T. Otsuka, J. Pereira, M. Perry, K. Pepper, J. Pinter, J. Stoker, A. Volya, D. Weiisshaar, Phys. Rev. C **76**, 021301(R) (2007).

73. T. Otsuka et al., Phys. Rev. Lett. **87**, 082502 (2001).

74. T. Otsuka, T. Suzuki, R. Fujimoto, H. Grawe, Y. Akaishi, Phys. Rev. Lett. **95**, 232502 (2005).

75. T. Otsuka, T. Matsuo, D. Abe, Phys. Rev. Lett. **97**, 162501 (2006).

76. T. Otsuka, M. Honma, D. Abe, Nucl. Phys. A **788**, 3c (2007).

77. J.P. Schiffer et al., Phys. Rev. Lett. **92**, 162501 (2004).

Testing the Structure of Exotic Nuclei via Coulomb Excitation of Radioactive Ion Beams at Intermediate Energies

T. Glasmacher

Department of Physics & Astronomy and National Superconducting Cyclotron Laboratory, Michigan State University, East Lansing, MI 48824, USA

Abstract With the advent of accelerator facilities dedicated to the production of radioactive nuclei, experimenters had to develop new, efficient techniques that can measure observables with the available beam rates. In-flight separated beams offer large luminosity gains through the use of thick secondary targets when combined with the detection of γ-rays to indicate inelastic scattering. Here we review the status of Coulomb excitation at intermediate energies, a technique that allows for the measurement of transition rates in atomic nuclei with beam rates of a few particles per second.

1 Introduction

Both experimental and theoretical nuclear scientists study atomic nuclei in the quest for predictive theoretical descriptions that explain the properties of all nuclei. Progress is made through the unremitting collaboration between theorists and experimentalists – the confrontation of testable hypotheses with precise observables measured under well-controlled conditions [1, 2]. Advances have accelerated in the past decade with the availability of accelerator facilities dedicated to the production of radioactive ions [3].

These facilities make available to experimenters the radioactive atomic nuclei that differ significantly in their properties (e.g., binding energy or proton–to–neutron ratio or radius) from stable nuclei. This in turn enables experiments to test hypotheses with atomic nuclei specifically chosen such that the predicted effect on observables may be most pronounced. Nuclear spectroscopy experiments are typically limited by background, which obscures the signals sought. Being able to work with radioactive atomic nuclei in reactions that yield the largest effect on predicted observables is thus a major advance. This advance, however, comes at a cost and with a major paradigm shift. New experimental techniques need to be developed and their efficacy established to study beams of radioactive nuclei. It is and, impractical indeed, almost always impossible to produce targets made of radioactive nuclei, most of which decay

Glasmacher, T.: *Testing the Structure of Exotic Nuclei via Coulomb Excitation of Radioactive Ion Beams at Intermediate Energies.* Lect. Notes Phys. **764**, 27–55 (2009)
DOI 10.1007/978-3-540-85839-3_2   © Springer-Verlag Berlin Heidelberg 2009

in fractions of a second. The new paradigm in experiments with radioactive beams is that *an experiment's discovery potential is limited by the available beam rate* and nature's cross section, which we desire to measure. With stable beams, beam rate is often not a major concern, rather the cross section to be measured limits the discovery potential of experiments (neglecting at this point practical considerations, such as detectors and other necessities, which apply equally to experiments with radioactive beams). Moles of stable atoms naturally occurring on earth can be ionized during an experiment in efficient ion sources via atomic processes with cross sections which are large compared to those for nuclear processes. In radioactive beam experiments, on the other hand, each single beam particle needs to be made in a nuclear reaction before it can become available for experiments.

For a radioactive ion beam facility with a driver accelerator of given power and production mechanism, the production rate for radioactive ions drops precipitously with each nucleon further away from the valley of stability, often by more than an order of magnitude for each nucleon further away. This observation motivates a corollary to the new paradigm: *Certain observables from reactions with radioactive nuclei cannot be measured, unless an experimental technique exists that can make a meaningful measurement of the observable compatible with the production rate of the radioactive nucleus.* Given today's economics of nuclear science experiments at radioactive beam accelerator facilities (hourly operations costs are of the order of several thousand Euros) prolonging experiments by orders of magnitude is generally not a viable option. Facility upgrades to increase driver power and thus production rate by orders of magnitude can cost tens or hundreds of million Euros. This corollary has thus motivated experimentalists to devise techniques which make the most efficient use of each radioactive atom. For a given observable, the technique which can operate with the lowest beam rate will have furthest scientific reach. In other words, for the most exotic radioactive beams the question of which technique to choose is moot. Instead, it is a question of whether a technique exists at all.

In this chapter I discuss one such technique, namely Coulomb excitation of radioactive ion beams at intermediate energies with γ-ray detection. This technique allows the measurement of Coulomb excitation cross sections between specified initial and final states in atomic nuclei with beam rates of a few particles/s. From the Coulomb excitation cross sections the absolute values of transition matrix elements between the states can be deduced. These latter quantum mechanical observables are calculable in the framework of nuclear theories and can confront measured values.

1.1 Brief History of Coulomb Excitation of Radioactive Beams

Coulomb excitation is one of the oldest [4–6] and best-established experimental probes in nuclear science. The reaction mechanism between a projectile and target interacting electromagnetically is well-known and was used extensively to study electromagnetic transition strengths with stable beams and targets starting in the 1950s [7, 8]. Such experiments were typically performed at beam energies below the Coulomb barrier to allow sufficient physical separation between the projectile and target nuclei to exclude possible nuclear contributions to the excitation mechanism.

The first Coulomb excitation experiment with a radioactive beam was published in 1991 [9]. The excited state at $E_x = 0.98$ MeV in the neutron-rich radioactive nucleus ^{8}Li was populated by scattering a ^{8}Li beam off a 1.1 mg/cm^2 ^{nat}Ni target at a beam energy of 14.6 MeV. The beam was produced in the transfer reaction ^{9}Be(^{7}Li,^{8}Li)^{8}Be at a rate of 10^5–10^7/s and separated in a superconducting solenoid magnet [10, 11] at the University of Notre Dame. Excited ^{8}Li nuclei were detected in a position-sensitive silicon ΔE–E telescope with an energy uncertainty of 400–500 keV, partially due to the beam energy uncertainty.

An alternative approach to detecting scattered particles is the detection of γ-rays to indicate the de-excitation of a bound excited state. This approach yields better energy resolution compared to particle detection, but it can also mean a loss in count rate due to the limited efficiency single germanium detectors. In Chap. 6, we will discuss how this loss in efficiency will be overcome with new detectors towards the end of this decade, almost 20 years after the publication of the first Coulomb excitation experiment with a radioactive beam and γ-ray detection in 1992 [12]. In this first experiment a beam of ^{76}Kr with an energy of 237 MeV and a rate of about 10^6/s was produced in the ^{9}Be(^{70}Ge,3n) reaction at the JAERI tandem accelerator. The ^{76}Kr beam was Coulomb excited through scattering off an enriched ^{208}Pb target of 2.0 mg/cm^2 thickness and deexcitation γ-rays were detected in four germanium detectors. The observed γ-ray yield corresponding to the $2^+ \rightarrow$ g.s. transition in ^{76}Kr agreed with the yield expected from the known $B(E2; 0^+ \rightarrow 2^+)$ value. While this early experiment did not have a high-purity radioactive ion beam available as is now common at dedicated radioactive ion beam facilities, it did demonstrate that Coulomb-excitation cross sections of radioactive beams at below-barrier energies can be measured reliably from γ-ray yields in inverse kinematics. Such studies are now routinely performed at the Holifield Radioactive Ion Beam Facility at Oak Ridge National Laboratory [13–15], at REX-Isolde at CERN [16–19] and are planned in the near future at the ISAC facility at TRIUMF. At these three ISOL facilities radioactive ion beams are produced by the isotope separation on-line (ISOL) technique [20] and reaccelerated to energies below the Coulomb barrier. Radioactive beams produced via the ISOL technique can be very intense and have beam qualities akin to those encountered at stable beam facilities. Beam developments are chemistry-dependent and need thus to be optimized for each element. Refractory elements cannot be produced by the ISOL method. The low-beam energy ensures the absence of nuclear contributions to the excitation process in scattering experiments and requires the use of thin targets with thicknesses of the order of 1 mg/cm^2. This target thickness together with typical Coulomb excitation cross sections necessitates beam rates in excess of 10^3–10^4/s to achieve typical count rates.

1.2 In-Beam γ-ray Spectroscopy Experiments with Fast Beams and Thick Targets

Intermediate-energy Coulomb excitation employs radioactive beams at energies of 30–300 MeV/nucleon ($v \approx 0.25$–$0.65 \, c$) which are separated in-flight by physical means following the fragmentation or fission of a heavy-ion beam on a production target. This approach is complementary to the ISOL technique. In-flight beam developments are fast, chemistry-independent, and applicable to all species. However, with current heavy-ion drivers beam rates are lower for the elements best made via the ISOL technique. ISOL beams also have lower emittance than in-flight separated beams, whose momentum spread is determined by the fragment separator to less than a few percent. If required, the fragment momentum can be determined event-by-event through measurement of each beam particle's position at dispersive images as long as beam rates are compatible with the capabilities of a tracking detector. Cocktails of different isotopes with similar rigidities can be made available in one experiment with each beam particle identified (in charge Z and mass A) on an event-by-event basis. The large beam velocity allows beam tracking and tagging, which can reduce background, and it provides kinematic focusing that allows the efficient detection of scattered beam particles.

Most importantly, the large beam velocity enables the use of thick secondary targets (100–1,000 times thicker than at Coulomb barrier energies) in in-beam γ-ray experiments. In such experiments the number of reactions taking place $N_{\text{reactions}}$ and the number of γ-rays detected, N_γ, are related to the number of atoms per area in the secondary target N_{target}, the number of beam particles impinging onto the target N_{beam}, the detection efficiency ϵ, and the cross section σ to be determined through

$$N_{\text{reactions}} = \frac{N_\gamma}{\epsilon} = \sigma \times N_{\text{target}} \times N_{\text{beam}} \, . \tag{1}$$

In scattering experiments with stable beams, N_{beam} is not a major concern. With the new paradigm, a beam rate that is too low renders an experiment non-feasible. In most radioactive ion beam experiments, experimenters request the maximum beam rate that the accelerator facility can provide. The use of thicker targets (at intermediate energies N_{target} increases by a factor of 100–1,000 relative to low-energy experiments) translates directly into an increase in the number of reactions $N_{\text{reactions}}$ and the number of detected γ-rays. Directly addressing our corollary from above, several experimental techniques have been developed to leverage this luminosity gain with radioactive ion beams at intermediate energies. Notable amongst them are in-beam fragmentation to provide excited state energies [21, 22], single-nucleon knockout reactions [23] to measure configurations in ground state wave functions and spectroscopic factors, two-nucleon knockout reactions [24–26], single-nucleon addition reactions to measure spectroscopic factors [27], and intermediate-energy Coulomb excitation [28, 29]. With the latter technique, in-flight separated beams and thick targets allow us to measure transition matrix elements with beam rates as low as a few particles/s.

The experimenters' task is to determine the cross section σ in Eq. (1) under well-controlled conditions, accurately, and with documented precision. Experimenters communicate their experimental result in a way that enables others to draw conclusions and to reproduce the measurements. The experimental considerations to arrive at cross sections are discussed in Sect. 2. Experimenters or theorists convert the measured cross sections into physics observables that are calculable. This will be discussed in Sect. 3. Considering that a single measurement can cost several hundreds to thousand euros, both steps must be executed with care.

2 Experimental Considerations

In intermediate-energy Coulomb excitation experiments radioactive projectiles are scattered off heavy, stable targets. The scattered projectiles are detected at small scattering angles in coincidence with γ-rays emitted from the target nucleus (which is at rest or slowly recoiling in the laboratory) and the projectile which is moving with close to beam velocity slowed down only by energy loss in the target. This process is schematically illustrated in Fig. 1.

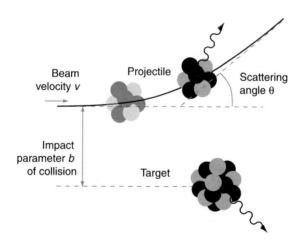

Fig. 1. Schematic illustration of the intermediate-energy Coulomb excitation process. A fast projectile ($v \approx 0.25 - 0.65$ c) impinges on a heavy target at an impact parameter b large enough to avoid nuclear contributions to the excitation process. The projectile and target can excite each other as they pass through each other's electric fields. If the excitation is to a bound excited state with sufficiently short lifetime, a γ-ray is emitted in close proximity to the target and can be detected by γ-ray detectors surrounding the target in coincidence with the scattered beam particle

The success of the measurement (an accurate cross section with defined precision) depends critically on the experimental realization of this concept. Experimenters must implement all assumptions which may be implicit in the schematic and must actively control all external circumstances which can influence the result of the measurement. Students develop these skills by working alongside experienced practitioners in the field and by learning from their peers. These time-honored methods serve experimental nuclear science well. However, it takes about two decades to gain the necessary experience and, with experiments becoming increasingly costly, the old nuclear science model to simply redo an experiment when it has failed may have outlived its timeliness. Novices learn faster and experimental success increases when they work alongside experienced practitioners and if all implicit assumptions and all external circumstances that can affect the experimental outcome are made explicit so that they can be addressed in a considered fashion.[1]

In the following, we closely examine some experimental considerations encountered in intermediate-energy Coulomb excitation experiments. While Eq. (1) does not specify a reaction, the experimenter must implement a specific reaction in the experiment.

2.1 Measuring Coulomb Excitation Cross Sections with Deexcitation γ-rays

The Coulomb excitation cross sections to excite specific states depend for a given projectile and target strongly on the incident beam energy. Figure 2 illustrates for the case of ^{40}S incident on a gold target, that low-lying collective

---

[1] "Considered fashion" means that the effort (or cost) expended to control a possible influence on the experimental outcome be commensurate with the benefit (or worth) derived from controlling the influence. An example may illustrate this: An in-flight separated beam has a momentum spread of 1%. In an intermediate-energy Coulomb excitation experiment, the beam passes through a thick secondary target where it loses 20% of its momentum. A γ-ray can be emitted at any time while the secondary beam traverses the secondary target. The beam velocity assumed for Doppler reconstruction is taken to be that at the mid-point of the secondary target. Should a fast tracking detector be built to measure the beam momentum of the secondary beam on an event-by-event basis to an accuracy of 0.1% to improve the γ-ray resolution? To answer this question, one could study several tracking detector designs and develop cost estimates for them. Alternatively, one can first consider the possible benefit. Since the secondary target introduces a momentum uncertainty of 20%, the initial beam momentum spread is small in comparison and any improvement will yield little benefit in the quality of the data. One concludes that the worth derived by this proposed detector is close to zero. The return on investment of resources (or the value, which is defined as worth/cost) does not warrant the expense. Experienced practitioners perform such value analyses implicitly many, many times in each experiment: Should we interrupt the experiment to repair a bad detector channel? Should we take more data in this configuration or change configurations? ...

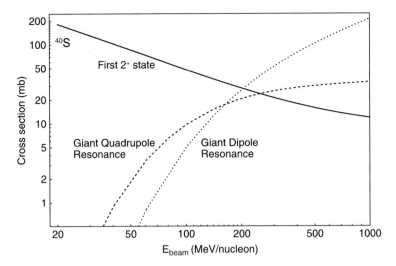

Fig. 2. Calculated cross sections for intermediate-energy Coulomb excitation at different beam energies for a ^{40}S projectile impinging on a gold target. Cross sections are shown for a low-lying collective 2^+ state, the giant quadrupole resonance, and for the giant dipole resonance

states are preferably excited at beam energies below 100–150 MeV/nucleon, while beam energies above 200–300 MeV/nucleon are better suited to excite giant resonances.

The experimenter must ensure that Coulomb excitation dominates the excitation process and that nuclear contributions are either negligible or will be appropriately accounted for. Small nuclear contributions are realized by requiring very forward projectile scattering angles $\theta_{\max}^{\text{lab}}$ in the laboratory and by ensuring that the charge and mass of the reaction product are identical to that of the projectile. This requires that the impact parameter b be larger than a minimum impact parameter $b_{\min}$ which is chosen to ensure a distance between projectile and target that avoids nuclear contributions to the excitation process. The optical model calculation in Fig. 5 illustrates the dominance of the Coulomb excitation cross section over nuclear contributions at small scattering angles.

Commonly used values for $b_{\min}$ are the sum of the projectile and target radii plus 2 fm, which exceeds the interaction radius defined by Wilcke and collaborators [32] by several tens of femtometer for heavy targets.

Since the Coulomb excitation cross section $\sigma_{i \to f}$ from an initial state $|i\rangle$ to a final state $|f\rangle$ in (1) will be determined by measuring the γ-ray yield $I_{f \to i}$ for the deexcitation $|f\rangle \to |i\rangle$ it is important to assess contributions to this yield which are not proportional to the excitation cross section. Some such possibilities are indicated in Fig. 4. The Coulomb excitation process with fast

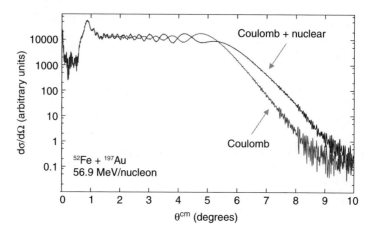

Fig. 3. Calculated excitation cross sections versus center-of-mass scattering angle θ^{cm} for the reaction ^{52}Fe $+$ ^{197}Au at 56.9 MeV/nucleon. Shown are the Coulomb excitation cross section and the Coulomb plus nuclear excitation cross sections. The Coulomb cross section dominates for small scattering angles. Optical model parameters from the ^{40}Ar $+$ ^{208}Pb reaction at 41 MeV/nucleon [30] were used to calculate the cross sections. Figure adapted from [31] (See also Plate 5 in the Color Plate Section)

beams is generally a one-step process and multi-step excitations are highly suppressed. However, multiple low-lying states may be populated (depending on the level density and structure of the nucleus under consideration) and the

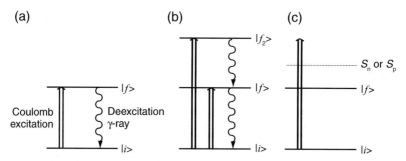

Fig. 4. Schematic illustration of measuring Coulomb excitation cross sections by counting de-excitation γ-rays from bound excited states. Panel **(a)** illustrates the desired process where a nucleus in its ground state $|i\rangle$ is Coulomb excited into a final state $|f\rangle$, which then γ-decays back to the ground state $|i\rangle$. The γ-ray yield $I_{f\rightarrow i}$ is proportional to the Coulomb excitation cross section $\sigma_{i\rightarrow f}$. If other states $|f_2\rangle$ can be Coulomb excited in the experiment they may γ-decay and feed state $|f_2\rangle$ as indicated in panel **(b)**. In this case, the γ-ray yield $I_{f_2\rightarrow f}$ must be subtracted from the yield $I_{f\rightarrow i}$ to deduce the proper Coulomb excitation cross section. Excitations above the particle separation threshold may result in the breakup of the projectile. Such events are excluded from analysis since they will not be identified in the reaction product detector

possibility of feeding must be considered (see Fig. 4(b)). Electron conversion coefficients are most often negligible for the relatively fast transitions encountered with this method.

2.2 Determination of the Number of γ-rays Emitted N_γ Emitted

The γ-rays emitted from the projectile and the target are detected in the laboratory with detectors, which cover a limited solid angle and have an intrinsic γ-ray detection efficiency, which is energy-dependent. The γ-ray spectrum observed is complicated by the Doppler-shift experienced by the γ-rays emitted from the projectile. The γ-ray spectrum is also contaminated by γ-ray background that is either uncorrelated or correlated with the beam – in the latter case the correlated background γ-rays can be emitted both from in-flight sources or at rest. Experimenters determine the γ-ray yield emitted corresponding to a specific transition in the projectile or the target. This yield determination involves several steps.

The γ-ray energy spectra measured in the laboratory in coincidence with a well-identified incoming secondary beam particle and a well-identified scattered beam particle are histogrammed and energy calibrated. Random background is reduced by requiring a tight coincidence between the time at which the γ-ray is emitted and the time at which the projectile impinges on the target. With fast beams, the latter time can be determined to fractions of 1 ns on an event-by-event basis. In most applications the width of this coincidence window is determined by the time resolution of the γ-ray detectors and the discriminator used (typically 10–20 ns for germanium detectors, a few ns for many scintillators, sub-ns for BaF_2 detectors).

Photopeaks (or at high-energy escape peaks) for transitions corresponding to de-excitations in the target are visible in this spectrum. We refer to this spectrum as the laboratory energy spectrum, since γ-ray energies E_γ^{lab} detected in the laboratory are histogrammed. A second Doppler-shifted γ-ray spectrum is prepared. Histogrammed here is each γ-ray observed in the laboratory, but its energy is Doppler-shifted on an event-by-event basis to the energy at emission from the projectile, E_γ^{proj}. The two energies are related through

$$E_\gamma^{proj} = E_\gamma^{lab} \frac{1 - \beta_{emission}^{lab} \cos \theta^{lab}}{\sqrt{1 - (\beta_{emission}^{lab})^2}}, \tag{2}$$

where θ^{lab} is the angle between the γ-ray and the scattered projectile in the laboratory and $\beta_{emission}^{lab}$ is the velocity of the projectile at time of γ-ray emission. Without active targets it is not practical to determine neither the velocity nor the location of γ-ray emission and thus θ^{lab} on an event-by-event basis. An emission source inside the target is generally assumed and an average velocity $\beta_{emission}^{lab}$ is used. This average velocity depends on the lifetime of the excited state. If this lifetime is short compared to the time in which the beam traverses the target, the beam velocity at mid-target is assumed.

If the lifetime of the excited state is long, the beam velocity after traversing the target is used. In practice, experimenters may minimize the width of the photopeak by optimizing $\beta_{\mathrm{emission}}^{\mathrm{lab}}$ between these two limits. The γ-ray yields for each transition in the projectile frame and the laboratory energy spectrum are determined. If the spectrum has few photopeaks and if the background can be well-estimated, the photopeaks can be integrated. If this is not the case, detector response functions for various energies can be simulated to reproduce source spectra measured in the laboratory. The emission function in this simulation can then be modified to simulate in-flight emission from a source with a lifetime corresponding to the state of interest (see bottom row of Fig. 5). In general, the lifetime of the state of interest is not known. If the lifetime is larger than about the time it takes the beam to traverse the target, the width of the photopeak increases since the determination of θ^{lab} assumes γ-ray emission in the target. With increasing lifetime the photopeak disappears and the method becomes no longer viable. Detected, simulated γ-rays are then treated in the same fashion as measured data above to be compared to the γ-ray spectra observed in the laboratory. Starting with the highest energy photopeak, the simulated spectrum for this transition is scaled to the measured spectrum and then subtracted from the measured spectrum. This process is repeated, proceeding towards lower energies until all photopeaks are accounted for and only background remains. The scale factors for each γ-ray are proportional to their individual yields. The yields determined by either method are efficiency corrected, taking into account γ-ray absorption in the target, the intrinsic efficiency of the detector, and the solid angle covered by the detector together with the γ-ray angular distribution [33]. Care must be taken that the energy-dependent efficiency correction is applied at energy E_γ^{lab} and not at E_γ^{proj}. If the γ-ray angular distribution has not been measured, it can be calculated [33] in the projectile frame and converted into the laboratory frame. These steps yield the number of γ-rays N_γ emitted from the projectiles of a specific isotope and detected in coincidence with beam particles scattered into the acceptance of the reaction product identification detector.

2.3 Beam Particles N_{beam} Impinging on Target

To determine the number of beam particles, N_{beam}, in Eq. (1), the experiment must determine the number of particles of a specific species ^{A}Z incident onto the target. In other words, the number of atoms in the radioactive beam must be counted and identified before they interact with the secondary target.

At the National Superconducting Cyclotron Laboratory (NSCL), this can be accomplished by measuring the time-of-flight (and, if desired, energy loss) between two transmission detectors located about 30 m apart as illustrated in Fig. 6. The choice of detectors depends on beam rate, secondary beam purity, and composition. Generally, when beam rates are high, experimenters request secondary beams with one or only a few isotopes in the secondary beam

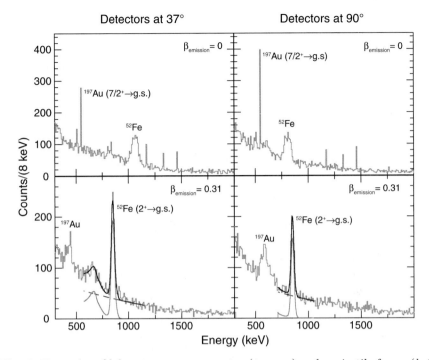

Fig. 5. Examples of laboratory energy spectra (*top row*) and projectile frame (*bottom row*) measured in the intermediate-energy Coulomb excitation of ^{52}Fe on a 257.7 mg/cm^2 (or 0.133 mm) thick ^{197}Au target. The secondary beam energy was 65.2 MeV/nucleon ($\beta = 0.356$) when impinging onto the target. The mid-target beam velocity was $\beta = 0.334$. The emission velocity for reconstruction of the Doppler-shifted γ-ray spectra was $\beta_{\text{emission}} = 0.31$, which is less than the mid-target velocity. The half-life of the 849 keV first excited 2^+ state in ^{52}Fe is $T_{1/2} = 7.8(10)$ ps, which means that the projectile can travel fractions of 1 mm in one half-life. In this experiment detectors were located at two azimuthal angles with respect to the beam axis (*left and right panels*) as indicated in the *inset* of Fig. 6. The laboratory energy spectra (*top row*) show a sharp photopeak for the $7/2^+ \rightarrow$ g.s. transition at 547 keV in the ^{197}Au target, while the photopeaks corresponding to the transition in the ^{52}Fe projectile are very broad. The detectors at $\theta^{\text{lab}} = 37°$ observe the 849 keV transition in ^{52}Fe at energies higher than 849 keV, since this angle corresponds to a forward angle in the center-of-mass (of the projectile–target system). Accordingly, $\theta^{\text{lab}} = 90°$ corresponds to a backward angle in the center-of-mass and the energy observed in the laboratory for the 849 keV transition is less than 849 keV. In the projectile frame spectra (*bottom row*) the photopeaks of the 849 keV transition are sharp and the transitions in the gold target are broad. At 37°, the Compton edge of the photopeak is visible, while it is less pronounced at 90° where the energies detected in the laboratory are lower and the cross section for the Compton effect is thus less. Indicated as *gray solid lines* are scaled simulated response functions, which when added to the background (*dashed gray line*) reproduce the observed γ-ray spectra well. This figure was adapted from [31]

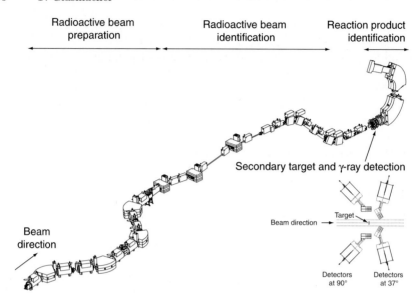

Fig. 6. Design model of the radioactive ion beam facility at NSCL. While details differ, the facilities at GANIL, GSI, and RIKEN have the same functionalities. Radioactive beams are produced by fragmentation or fission of a primary beam on the production target at the entrance to the A1900 fragment separator. The magnets in the A1900 select the isotopes of interest by rigidity ρ ($\rho = p/q$, where p is the beam particle's momentum and q its charge state). During the optimization of beam rate and purity ions are identified and stopped in detectors in the A1900 focal plane. The optimized beam is then transmitted to the secondary target through the beam transport system. A thin transmission scintillator or diamond detector located after the A1900 focal plane records a time signal for each beam particle and a second detector can be located in the intermediate image of the S800 beam analysis line. This detector can either be another timing detector or, if beam rates are low enough, a thin silicon transmission detector to measure energy loss. Located at the focal point of the S800 spectrograph [34] is the secondary target which is surrounded by γ-ray detectors. Indicated here are the 32–fold segmented high-purity germanium detectors from SeGA [35]. The S800 spectrograph identifies each scattered projectile and determines its momentum vector

cocktail in order to enhance the count rate of the primarily desired reaction channel. When only a few isotopes are in the beam cocktail, identification by time-of-flight is often sufficient to resolve the distinct masses of the isotopes. When beam rates for the isotope of interest are low, experimenters often request a number of isotopes in the beam cocktail to leverage the investment of beam time with minimal impact on data acquisition dead-time. In this latter case a thin silicon detector may replace the second timing detector and incoming beam particles may be identified by energy-loss versus time-of-flight measurements.

Experimenters carefully assess what fraction of counted and identified beam particles actually impinges onto the secondary target. While this fraction is ideally unity, transmission through additional beam transport systems after the beam identification and counting must be measured and monitored during the experiment. (At NSCL, there are 13 m of additional beam transport before the secondary target). Locating a counting detector directly adjacent to the secondary target may be advisable in certain experiments, but the possibility of it generating γ-rays must be considered carefully in intermediate-energy Coulomb excitation experiments. Special care should also be taken to ensure that the secondary target is large enough to accommodate the profile of the secondary beam and that the secondary beam does not hit the target holder. As a precaution, one wants to ensure that the energy loss in the target is different enough from all other possible beam paths, so that beam not impinging onto the target either does not get transmitted to or can be identified in the reaction product identification detector. The more general idea is to work hard to avoid possible errors, but to make them as explicit as possible should they occur. This may allow a data set to be saved through more elaborate analysis.

During the experiment certain devices must be monitored more carefully than others, depending on whether or not a failure of the particular device affects the measured cross section or not:

- A change in beam composition, a change in the beam transport system prior to beam identification, or a failure of the incoming beam identification system results in fewer incoming beam particles that are correctly identified. Since correct incoming beam identification will be a condition in the analysis, there will be fewer events, but the cross section will not be affected.
- A change in the transmission of the beam transport system after beam identification, however, can affect the cross section, since the number of beam particles impinging on the target N_{beam} is the product of the particles identified and counted and the transmission after counting.

Most radioactive ion beam facilities have control systems that can capture facility configurations and alert the experimenters of deviations during the experiment. Experimenters should double-check that all relevant optical devices are included in the captured configuration.

If the acceptance of the reaction product detector is large enough to accept the entire elastically and inelastically Coulomb scattered beam, experimenters at times approximate N_{beam} with the number of identified and scattered beam particles. This approximation relies on the Coulomb elastic scattering cross section being dominant over all others.

2.4 Number of Scattering Centers in the Target N_{target}

The number of scattering centers per unit area in the target can be calculated from the thickness d and volume density ρ of the target material when foils are used. Uniform target thickness and density are important, but can be realized relatively easily since most targets used in intermediate-energy Coulomb excitation experiments are self-supporting metal foils. Care must be taken that the target is stably and reproducibly mounted at a known angle with respect to the beam so that the effective thickness remains constant over the experiment. In addition, the position of the target along the beam axis relative to the γ-ray detectors must be known for the Doppler reconstruction of γ-rays discussed earlier.

While isotopic purity of targets is not required for Coulomb excitation of the projectile, knowledge of the isotopic composition is necessary to determine the Coulomb excitation cross section of the isotopes in the target, which provides a valuable cross check. In addition, discrete γ-rays emitted from the target appear very broadened after being Doppler shifted into the projectile frame. For this reason most experiments are performed with monoisotopic (e.g., ^{197}Au and ^{209}Bi) or isotopically enriched targets (e.g., ^{208}Pb). If the energy of the γ-ray to be measured is known or can be anticipated, secondary targets are often chosen so that the energy regions of the target and projectile γ-rays are not close to each other.

2.5 Presentation of experimental results

After the the Coulomb excitation cross section σ in Eq. (1) is determined it must be presented together with sufficient information so that others can deduce a transition rate. We discuss here the information needed and possible sources of error when converting a cross section into a transition rate.

The experimental cross section σ is measured in a particular reaction with an experimental setup, in which scattered beam particles are detected and identified in a reaction product detector, which has a particular acceptance ϵ_{RPD}. Thus

$$\sigma = \int_{\Omega} \frac{d\sigma(\theta')}{d\Omega'} \epsilon_{RPD}(\theta', \phi') d\Omega'. \tag{3}$$

Often the reaction product detector's acceptance is symmetric and uniform of the form

$$\epsilon_{RPD}(\theta, \phi) = \begin{cases} 1 & \text{for } \theta < \theta_{max} \\ 0 & \text{otherwise} \end{cases}. \tag{4}$$

In this case the communication of the maximum scattering angle in the laboratory θ_{max}^{lab} suffices to describe the solid angle over which the cross section was measured. If the reaction product detector acceptance is not uniform in ϕ or θ experimenters must communicate the acceptance as a function of θ and ϕ. Preferably, the acceptance ϵ_{RPD} is expressed as a function of θ alone (see, for example, Fig. 2 in [36]). When deducing transition rates from cross sections, the theoretical cross section must be integrated over the same acceptance ϵ_{RPD} as was realized in the experiment. Care must also be taken to choose the proper frame of reference. Acceptances are generally given in the laboratory frame while calculations are usually performed in the center-of-mass system. The center-of-mass scattering angle θ^{cm} is related to the one in the laboratory θ^{lab} through

$$\tan \theta^{\text{lab}} = \frac{\sin \theta^{\text{cm}}}{\gamma(\cos \theta^{\text{cm}} + \frac{\beta^{\text{cm}}}{\beta^{\text{proj}}})}. \tag{5}$$

Here, while β^{cm} is the center-of-mass velocity of the projectile–target system while the projectile velocity β^{proj} should be taken as the mid-target velocity of the projectile. This approximates the velocity dependence of Eq. (5) appropriately for experiments where the velocity change in the target is small relative to the velocity of the incoming beam. Thus, it is recommended that experimentalists report the mid-target velocity explicitly. The mid-target beam energy should also be reported and used as effective beam energy in theoretical calculations. Evaluating the cross section at mid-target beam energy $E_{\text{beam}}(d_{\text{target}}/2)$ approximates the average of the cross section $\sigma(E_{\text{beam}}(x))$ over target thickness d_{target}.

3 Extraction of Transition Matrix Elements from Cross Sections

The Coulomb excitation process at energies below the Coulomb barrier has been extensively described in the literature [8, 37] and treated fully quantum-mechanically [38]. At low energies the relative motion between projectile and target follows the classical Rutherford trajectories and relativistic effects are negligible. At relativistic energies straight-line trajectories are a very good approximation. At intermediate energies, relativistic effects are still important, but straight-line trajectories can no longer be assumed and one has to consider two relativistic charged particles moving with respect to each other. This problem can be solved analytically only if the mass of one particle in the scattering process is infinite. This is schematically illustrated in Fig. 1 where the target nucleus does not recoil. Winther and Alder and Alder described the relativistic Coulomb excitation process semi-classically in 1979 [28]. To account for the recoil of the target as a first-order deviation from straight-line trajectories the impact parameter b was rescaled to

$$b \rightarrow b + \frac{\pi a}{2\gamma}, \tag{6}$$

where γ is the relativistic Lorentz factor $\gamma = 1/\sqrt{1 - \beta^2}$ and a is the half-distance of closest approach in a non-relativistic head-on collision

$$a = \frac{Z_{\text{proj}} Z_{\text{target}} e^2}{m_0 \beta^2 c^2}. \tag{7}$$

Here, Z_{proj} and Z_{target} are the respective charges of the projectile and target and m_0 is the reduced mass of the projectile–target system. β is the beam velocity relative to the speed of light c.

Winther and Alder decompose the Coulomb excitation cross section into the sum of the allowed multipole matrix elements characteristic of the electromagnetic decay of the nuclear state $|f >$ to state $|i >$ as

$$\sigma_{i \to f} = \sum_{\pi \lambda} \sigma_{\pi \lambda}. \tag{8}$$

The individual contributions of multi-polarity λ and parity π for straight-line trajectories with impact parameters larger than a minimum impact parameter b_{min} are of the form

$$\sigma_{\pi \lambda} \approx \left(\frac{Z_p e^2}{\hbar c} \right)^2 \frac{\pi}{e^2 b_{\text{min}}^{2\lambda - 2}} B(\pi \lambda, 0 \to \lambda) \begin{cases} (\lambda - 1)^{-1} & : \quad \text{for } \lambda \geq 2 \\ 2 \ln (b_a / b_{\text{min}}) & : \quad \text{for } \lambda = 1. \end{cases} \tag{9}$$

Here, b_a denotes the impact parameter at which the adiabatic cutoff of the Coulomb excitation process sets in. This occurs when the time of internal motion in the nucleus $\hbar / E_\gamma$ equals the collision time $b_a / (\gamma c \beta)$, where E_γ is the energy of the excited state $|f >$ relative to the initial state $|i >$. Thus

$$b_a = \frac{\gamma \hbar c \beta}{E_\gamma}. \tag{10}$$

Equation (10) implies that the maximum energy of final states that can be excited in collisions with impact parameter b is of the order of

$$E_\gamma^{\text{max}} \approx \frac{\gamma \hbar c \beta}{b}. \tag{11}$$

Equation (11) illustrates why giant resonance experiments are best-performed at beam energies above 200–300 MeV/nucleon as illustrated in Fig. 2.

b_{min} is the minimum impact parameter realized in the experiment

$$b_{\text{min}} = \frac{a}{\gamma} \cot(\Theta_{\text{max}}^{cm}/2), \tag{12}$$

where theta is the maximum scattering angle Θ_{max}^{cm} of the projectile in the center–of–mass system. The conversion of the maximum scattering angle into the laboratory is given in Eq. (5). The Coulomb excitation cross section $\sigma_{i \to f}$ is directly related to the reduced transition probability $B(\pi \lambda; i \to f)$ as shown in Eq. (9).

The Weizsäcker–Williams method developed in 1934 provides an alternative approach and describes the Coulomb excitation process in terms of equivalent photon numbers [39, 40]. Coulomb excitation is understood as the absorption of virtual photons which are produced by relativistically moving charged particles. The equivalent photon number $n_{\pi\lambda}$, the number of real photons that would have the equivalent net effect on a particular transition, relates to the photo-absorption cross section $\sigma_{i \to f} \propto n_{\pi\lambda}\sigma_{abs}$. The idea underlying the Weizsäcker–Williams method had already been used in 1924 by Fermi to connect the absorption of X-rays by atoms and the energy loss due to ionization [41].

In 1984, Hoffman and Baur [42] showed that the equivalent photon method and the semi-classical approach by Alder and Winther [28] provide the same results for relativistic $E1$ Coulomb excitation cross sections [42]. At the same time Goldberg [43] extended the virtual photon method to all multipolarities [43].

Bertulani and collaborators [44, 45] performed self-contained derivations and showed that a quantum theory leads to minor modifications of the classical results [46]. A coupled channels description of intermediate-energy Coulomb excitation was developed in 2003 [47]. The interplay between relativistic retardation effects, which are included in the relativistic description of Coulomb excitation, and the correct treatment of recoil effects in the classical theory (recoil effects are only approximated through Eq. (6) in the relativistic theory) was investigated in [48].

Extending this work, Bertulani, Stuchberry, and collaborators [49] developed an exact numerical solution for the Coulomb excitation cross section and then reviewed the importance of including relativistic dynamics and the appropriate trajectories over a large range of beam energies. Cross sections to low-lying collective states at intermediate energies are dominated by collisions at large impact parameters and recoil corrections are less important than for high-energy excitation, such as giant resonances, which are dominated by collisions at small-impact parameters. For the first excited state in ^{40}S at 0.89 MeV, the difference between cross sections calculated with the exact numerical solution and semi-classically with the impact parameter rescaled (see Eq. (6)) is less than 5% above 50 MeV/nucleon [49].

The influence of nuclear excitations and the possibility of Coulomb nuclear interference need to be considered in the data analysis, especially for light nuclei and for reactions where experimenters desire to include data at larger scattering angles to increase statistics. These issues are discussed in Sect. 4 in the context of experimental results on the neutron-halo nucleus ^{11}Be.

4 Recent Experimental Results

In the past decade intermediate-energy Coulomb excitation has become a spectroscopic tool in use at all four major facilities that provide in-flight separated radioactive beams: GSI (Germany), GANIL (France), Michigan State University (USA), and RIKEN (Japan). Here we discuss two regions in the nuclear chart, where this method has contributed to discoveries. Unexpected results can be exciting and lead to new insights, or they can be wrong and be disproven at a later time. The beauty of experimental science is that controversies always work themselves out over time.

4.1 The Neutron-Halo Nucleus ^{11}Be

^{11}Be is a loosely bound neutron-halo nucleus. Its neutron-separation energy is $S_n = 504(6)$ keV and only one bound excited state exists at 320 keV ($J^\pi = 1/2^+$). This state decays to the ground state ($J^\pi = 1/2^-$) through the fastest known dipole transition between bound states in atomic nuclei. The strong coupling between the two states was discovered in 1983 by Millener and collaborators in a lifetime measurement at Brookhaven National Laboratory which yielded a transition strength of $B(E1, 1/2^+_{\text{g.s.}} \rightarrow 1/2^-) = 0.116(12)$ e^{2}fm^2 [50]. In 1995 a Coulomb excitation experiment at GANIL (of ^{11}Be on a lead target at 43 MeV/nucleon) reported a cross section that when analyzed in the semi-classical theory of Winther and Alder [28] yielded a transition strength of about 40% of the strength observed in the lifetime measurement [51]. This large discrepancy led to several studies that investigated in detail the influence of certain assumptions made in the semi-classical model.

The neutron-separation energy in ^{11}Be is small (504(6) keV) and coupling to the continuum may affect the deduced transition strength. For example, ^{11}Be after being excited into its bound excited state may be excited into the continuum in a second step. Typel and Baur studied this by extending the single-step theory to multi-step higher-order electromagnetic interactions [55]. These effects could account for a possible reduction of the $B(E1)$ strength observed in [51] to 95.5–89.9% of the value from the lifetime measurement. Similar results were observed by Bertulani and collaborators [56] in a semi-classical coupled channels approach that couples the bound states to the continuum and includes nuclear coupling effects. Only a 5% cross section reduction compared to the cross section anticipated from the lifetime transition strength could be explained and the authors conclude that "first order perturbation theory is appropriate to calculate the cross section" [56].

In the analysis according to Winther and Alder nuclear excitations are excluded in an approximate way through the introduction of a minimum impact parameter b_{min}. The standard prescription to determine b_{min} is the sum of the projectile and target nucleus plus several femtometer or the use of the interaction radius [32]. These definitions of a minimum impact parameter may not be applicable for ^{11}Be with its diffuse neutron halo. This question was investigated by Tarutina, Chamon, and Hussein [57] who multiplied the impact-parameter dependent Coulomb excitation probability by the impact-parameter dependent survival probability during integration, instead of assuming a hard cutoff b_{min}. This more accurate treatment of nuclear absorption yielded an increase in the deduced transition strength of 2% for the result in [51] and increases up to 5% for later experiments [53, 54]. While each of these small corrections improve on the analysis with the Alder and Winther theory, they cannot individually or when combined explain the small cross section observed in [51].

In 1997, intermediate-energy heavy-ion scattering experiments at RIKEN and Michigan State University (MSU) were performed to elucidate the GANIL result. At RIKEN, Nakamura and collaborators scattered ^{11}Be off a lead target ($350\,\mathrm{mg/cm^2}$) at a beam energy of $63.9\,\mathrm{MeV/nucleon}$ and observed a large cross section of $302 \pm 8 \pm 30\,\mathrm{mb}$ corresponding to a transition rate of $B(E1, 1/2^+_{\mathrm{g.s.}} \to 1/2^-) = 0.099(10)\,\mathrm{e^2fm^2}$. At MSU beams of ^{11}Be were scattered of ^{9}Be ($195\,\mathrm{mg/cm^2}$), ^{nat}C ($411\,\mathrm{mg/cm^2}$), ^{197}Au ($533\,\mathrm{mg/cm^2}$), and ^{208}Pb ($80\,\mathrm{mg/cm^2}$) at mid-target beam energies of 58.4, 56.7, 57.6, and $59.4\,\mathrm{MeV/nucleon}$, respectively. Excitation cross sections to the first excited state of 1.7(2)(4), 4.0(2)(5), 244(7)(24), and 304(10)(33) mb were observed, respectively. The thick gold target necessitated a correction of the observed γ-ray yield by 75% due to the strong absorption of the $320\,\mathrm{keV}$ photon in the target. Transition strengths extracted for the gold and lead targets were $B(E1, 1/2^+_{\mathrm{g.s.}} \to 1/2^-) = 0.079(8)$ and $0.094(11)\,\mathrm{e^2fm^2}$, respectively. The measurements at RIKEN and MSU were consistent with each other and the lifetime measurement, while they did not agree with the GANIL result. In addition, the small excitation cross sections on the light targets indicate that nuclear contributions are small. The effect of Coulomb-nuclear interference was investigated [58] for the case of ^{11}Be scattering in a full quantum calculation (with both nuclear and Coulomb potentials) through continuum discretized coupled channels calculations. For the light neutron-halo nucleus ^{11}Be scattering off a heavy target Coulomb nuclear interference can be both constructive or destructive and cannot be neglected even when selecting events with large impact parameters only.

In 2007 a new experiment was performed at GANIL [52] to measure the excitation cross section of ^{11}Be on ^{208}Pb at $38.6\,\mathrm{MeV/nucleon}$. A cross section of 416(66) mb was observed, corresponding to a transition rate of $B(E1, 1/2^+_{\mathrm{g.s.}} \to 1/2^-) = 0.105(12)\mathrm{e^2fm^2}$, consistent with the measurements at RIKEN and at MSU. The deduction of the transition strength from the measured cross section in [52] was performed with the extended discretized coupled channels method, a fully quantum mechanical description of Coulomb excitation with coupling to the continuum. In contrast to the earlier theoretical analyses described earlier and the small cross sections measured on light targets, the authors find that the "excitation process involves significant contributions from nuclear, continuum, and higher-order effects".

Figure 7 summarizes the deduced transition strengths $B(E1, 1/2^+_{\mathrm{g.s.}} \to 1/2^-)$ in ^{11}Be. The measurements on lead targets at RIKEN, MSU and the later measurement at GANIL agree well with each other and the lifetime measurement. The measurement on the gold target is consistent with the other measurements, but may suffer from an underestimation of the systematic error introduced in the 75% correction of the photon yield due to absorption in the target. The low-cross section reported in [51] cannot be reproduced.

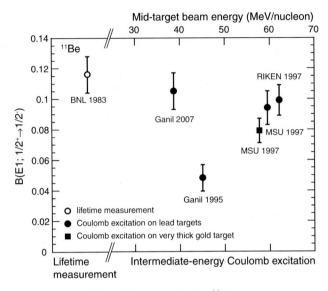

Fig. 7. Transition rates $B(E1, 1/2^+_{\text{g.s.}} \rightarrow 1/2^-)$ in ^{11}Be plotted versus beam energy. The *open circle* denotes the lifetime measurement at Brookhaven National Laboratory [50]; *solid circles* indicate experiments on relatively thin lead targets at GANIL in 2007 [52] and 1995 [51], at RIKEN in 1997 [53] and at Michigan State University (MSU) in 1997 [54]. The *solid square* denotes an experiment at MSU in 1997 [54] on a thick gold target, which required a 75% correction for photon absorption in the target

4.2 ^{30,32,34}Mg and the Island of Inversion

In 1975 Thibault and collaborators [59] found that the neutron-rich sodium isotopes are more tightly bound than expected by shell model calculations in the $\nu(sd)$ model space. Based on Hartree–Fock calculations this observation was attributed by Campi and collaborators [60] to strongly deformed ground states due to the filling of $\nu f_{7/2}$ negative parity orbitals. Shell model calculations [61] suggested that the ground state configurations of $^{30-32}$Ne, $^{31-33}$Na, and $^{32-34}$Mg are dominated by intruder configurations $\nu(sd)^{(N-2)}(f_{7/2})^2$ and form an "island of inversion" in the table of isotopes where such configurations are more energetically favorable rather than the normal $\nu(sd)^N$ configurations.

The large transition rate $B(E2, 0^+_{\text{g.s.}} \rightarrow 2^+_1) = 454(78)\,e^2\text{fm}^4$ in ^{32}Mg measured by Motobayashi and collaborators at RIKEN [36] was successfully explained by shell model calculations with $\nu(sd)^{(N-2)}(f_{7/2})^2$ configurations in the ground state and in the first excited 2^+ state supporting the idea of the island of inversion. Subsequent measurements at MSU [62, 63] and at RIKEN [64] confirmed the original result. The MSU data were also analyzed under the assumption of possible feeding into the 2^+ state via a 1,436 keV γ-ray. Such a γ-ray was observed in β-decay studies of ^{32}Na [65]. It remains an experimental questions as to whether or not this 1,436 keV γ-ray is observed

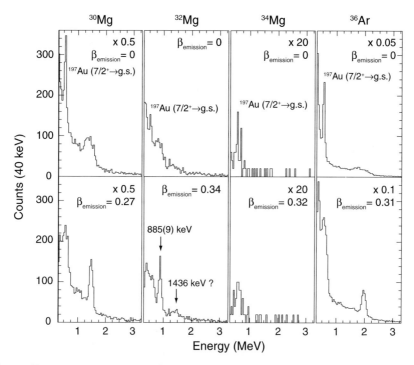

Fig. 8. Energy spectra measured in the intermediate-energy Coulomb excitation of ^{30,32,34}Mg on gold targets (518, 702, and 702 mg/cm^2 thick, respectively) at beam energies of 36.5, 57.8, and 50.6 MeV/nucleon, respectively. The γ-ray spectrum for the ^{36}Ar test case is also shown. The question as to whether or not a 1,436 keV γ-ray is visible in the ^{32}Mg spectrum remains open. Figure adapted from [62]

in intermediate-energy Coulomb excitation. Both the γ-ray spectra at MSU and RIKEN are statistics-limited and inconclusive. The spectra from [62] are shown in Fig. 8 so that the reader may assess the situation. If such a feeding transition is present, it would originate from an excited state with $J^\pi=1^-$, 1^+, or 2^+ and would reduce the cross sections and $B(E2)$ values in both the RIKEN and MSU experiments.

Transition rates measured in ^{34}Mg both at RIKEN [64] and MSU [63] agree with each other and can be understood in calculations where the ground state and the excited state are dominated by intruder configurations (Fig. 9). An intermediate-energy Coulomb excitation measurement at GANIL [66] yielded a transition rate in ^{32}Mg which is about 35% larger than the RIKEN and MSU values. The origin of this difference is currently not understood. If the GANIL value is correct and interpreted in a rotational model, it would indicate a very large charge deformation of $\beta_C = 0.61(4)$ for ^{32}Mg[66]. In the same experiment a transition rate of 435(58) e^{2}fm^4 for ^{30}Mg was deduced which is 47% larger than the MSU value of 295(26) e^{2}fm^4 [63]. The latter value is in agreement with the recent low-energy Coulomb excitation experiment performed at REX-isolde, which found $B(E2, 0^+_{g.s.} \to 2^+_1) = 241(31)\,e^2\text{fm}^4$ in ^{30}Mg.

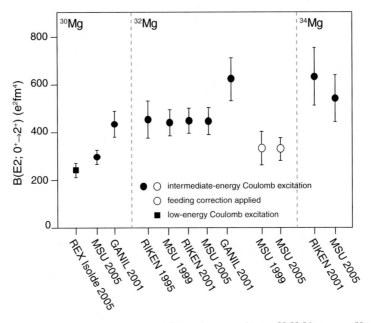

Fig. 9. Measured transition rates $B(E2, 0^+_{g.s.} \to 2^+_1)$ in ^{30,32,34}Mg. In ^{30}Mg the low-energy measurement at REX-isolde [16] agrees with the intermediate-energy measurement at MSU [63], but not with the measurement at GANIL [66]. This latter experiment also yields a larger $B(E2)$ value for ^{32}Mg, compared to the measurements at RIKEN [36, 64] and MSU [62, 63], provided no feeding correction is applied to the photon yield for the 885 keV transition. The transition rates for ^{34}Mg measured at RIKEN [64] and MSU [63] are in agreement with each other

5 Accuracy of the Technique

Whenever a new experimental technique is developed, its efficacy needs to be carefully established to avoid confusion and the unnecessary expense of effort that arises when results with questionable accuracy are published. A good way to establish the credibility of a new technique is to measure well-established observables that have been measured previously with different techniques at various laboratories. Since intermediate-energy Coulomb excitation measurements can easily measure transition rates in isotopic chains, well-known transition rates in stable isotopes have been measured in many experiments that also established new transition rates on radioactive isotopes.

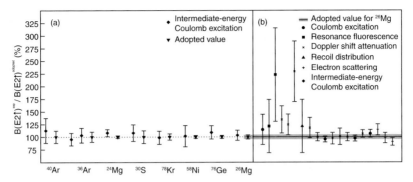

Fig. 10. Ratio of measured to adopted $B(E2; 0_{g.s.}^+ \rightarrow 2_1^+)$ values for eight different stable isotopes (panel (**a**)). The transition rates plotted were measured as stable-beam test cases in intermediate-energy Coulomb excitation experiments (^{40}Ar [67],^{36}Ar [68, 69], ^{24}Mg [70],^{30}S [71],^{78}Kr [72],^{58}Ni [73], ^{76}Ge [74],^{26}Mg [63]) between 1999 and 2005. For each isotope, the transition rates are compared to the adopted values [75] for the same transition. The average difference between the measured and adopted transition rate is 6%. To put this difference into perspective, panel (**b**) compares the same ratio of measured to adopted $B(E2; 0_{g.s.}^+ \rightarrow 2_1^+)$ values for ^{26}Mg. Here, the experimental values have been measured with a variety of "established" experimental probes (data taken from [75]) between 1961 and 1982. The absolute difference between the measured values and the adopted value for ^{26}Mg is 23%, while it is 3% for the intermediate-energy Coulomb excitation result [63]. This figure has been adapted from [76]

Through use of the same experimental setup and an identical analysis procedure, the measurement of these stable-beam "test cases" and comparison of the derived observables to adopted values, lends credence to newly measured observables on radioactive nuclei. Results from eight such measurements on stable isotopes are shown in Fig. 10. Of particular importance is the fact that these transition rates were measured in nuclei moving at about 30–40% of the speed of light in identical conditions to the newly measured transition rates. The comparison between the adopted and measured values then tests the complete analysis procedure. A lesser degree of certainty is provided by a comparison between the transition rate in the target (where the γ-ray was emitted at rest) and an adopted value, since the γ-ray yield from target nuclei does not undergo the kinematic reconstruction needed for γ-rays emitted from the projectile. Agreement between measured and adopted transition rates for target excitations is necessary to demonstrate the efficacy of an experiment, but it is not sufficient. Agreement between measured and adopted transition rates for excitations of the projectile is very close to sufficient.

Figure 10 shows the average difference between adopted transition rates for the stable isotopes and values measured at NSCL. The values presented here set an empirical scale for the overall accuracy of the technique. Two components contribute to the precision and accuracy of the technique. The first arises from the experimental measurement of the cross section and was discussed in Sect. 2. The second component arises from the extraction of a transition rate from the experimental cross section and was discussed in Sect. 3. These two components are largely independent and the overall accuracy of the technique compares very favorably with other established techniques. Proper quantum calculations must be performed to account for Coulomb-nuclear interference when scattering light halo nuclei, such as ^{11}Be as discussed.

6 Outlook and Summary

In the past 10 years intermediate-energy Coulomb excitation of radioactive ions has become an established technique employed at all major radioactive beam facilities that provide in-flight separated beams worldwide. Transition rates have been measured with beam rates as low as 3 atoms/s [77]. The γ-ray detectors used in these experiments have either been scintillation detectors [78, 79] with good efficiency and moderate energy resolution or segmented high-purity germanium detectors [35, 80, 81] with good energy resolution and small photopeak efficiency (2–7% at 1,332 keV). A new concept for the efficient detection of γ-ray radiation with high-photopeak efficiency, large peak-to-background ratio, and very good position resolution is being developed. The γ-ray energy tracking array (GRETA) [82] in the United States and the advanced gamma tracking array (AGATA) [83, 84] in Europe will have more than 40% photopeak efficiency (for a single γ-ray at 1,332 keV) and will be able to determine the first interaction point of a γ-ray in the detector with an accuracy of about 2 mm (rms). The availability of such detectors will increase the sensitivity of current-day intermediate-energy Coulomb excitation experiments by more than an order of magnitude. The precision of the analysis of intermediate energy Coulomb excitation cross sections was long limited to about 5–10% secondary to the simplifying assumptions made in the semi-classical theory. With the advent of a theory that contains relativistic kinematics and dynamics and a correct treatment of the Coulomb trajectories [49], the precision of the analysis has been taken to the next level.

References

1. K. Popper, *Logik der Forschung* (Wien, 1935)
2. K.R. Popper, *The logic of scientific discovery* (Hutchinson, London, 1959)
3. R.I.S.A. Committee, (National Academies Press, Washington, 2007)
4. E. Fermi, Z. Phys. **29**, 315 (1924)
5. C.F. Weizsäcker, Z. Phys. **88**, 612 (1934)
6. E.J. Williams, Phys. Rev. **45**, 729 (1934)
7. K. Alder, A. Bohr, T. Huus, B. Mottelson, A. Winther, Rev. Mod. Phys. **28**, 432 (1956)

8. K. Alder, A. Winther, *Coulomb Excitation* (Academic Press, New York, 1966)
9. J.A. Brown, F.D. Becchetti, J.W. Jänecke, K. Ashktorab, D.A. Roberts, J.J. Kolata, R.J. Smith, K. Lamkin, R.E. Warner, Phys. Rev. Lett. **66**, 2452 (1991)
10. J.J. Kolata, A. Morsad, X.J. Kong, R.E. Warner, F.D. Becchetti, W.Z. Liu, D.A. Roberts, J.W. Jänecke, Nucl. Instrum. Methods B **40**, 503 (1989)
11. F.D. Becchetti, W.Z. Liu, D.A. Roberts, , J.W. Jänecke, J.J. Kolata, A. Morsad, X.J. Kong, R.E. Warner, Phys. Rev. C **40**, R1104 (1989)
12. M. Oshima, Y. Gono, T. Murakami, H. Kusakari, M. Sugawara, S. Ichikawa, Y. Hatsukawa, T. Morikawa, B.J. Min, Nucl. Instrum. Methods A **312**, 425 (1992)
13. D.C. Radford, C. Baktash, J.R. Beene, B. Fuentes, A. Galindo-Uribarri, C.J. Gross, P.A. Hausladen, T.A. Lewis, P.E. Mueller, E. Padilla, D. Shapira, D.W. Stracener, C.H. Yu, C.J. Barton, M.A. Caprio, L. Coraggio, A. Covello, A. Gargano, D.J. Hartley, N.V. Zamfir, Phys. Rev. Lett. **88**(22), 222501 (2002)
14. J.R. Beene, R.L. Varner, C. Baktash, A. Galindo-Uribarri, C.J. Gross, J.G. del Campo, M.L. Halbert, P.A. Hausladen, Y. Larochelle, J.F. Liang, J. Mas, P.E. Mueller, E. Padilla-Rodal, D.C. Radford, D. Shapira, D.W. Stracener, J.P. Urrego-Blanco, C.H. Yu, Nuclear Phys. A **746**, 471C (2004)
15. E. Padilla-Rodal, A. Galindo-Uribarri, C. Baktash, J.C. Batchelder, J.R. Beene, R. Bijker, B.A. Brown, O. Castanos, B. Fuentes, J.G. del Campo, P.A. Hausladen, Y. Larochelle, A.F. Lisetskiy, P.E. Mueller, D.C. Radford, D.W. Stracener, J.P. Urrego, R.L. Varner, C.H. Yu, Phys. Rev. Lett. **94**, 122501 (2005)
16. O. Niedermaier, H. Scheit, V. Bildstein, H. Boie, J. Fitting, R. von Hahn, F. Kock, M. Lauer, U.K. Pal, H. Podlech, R. Repnow, D. Schwalm, C. Alvarez, F. Ames, G. Bollen, S. Emhofer, D. Habs, O. Kester, R. Lutter, K. Rudolph, M. Pasini, P.G. Thirolf, B.H. Wolf, J. Eberth, G. Gersch, H. Hess, P. Reiter, O. Thelen, N. Warr, D. Weisshaar, F. Aksouh, P.V. den Bergh, P.V. Duppen, M. Huyse, O. Ivanov, P. Mayet, J.V. de Walle, J. Aysto, P.A. Butler, J. Cederkall, P. Delahaye, H.O.U. Fynbo, L.M. Fraile, O. Forstner, S. Franchoo, U. Koster, T. Nilsson, M. Oinonen, T. Sieber, F. Wenander, M. Pantea, A. Richter, G. Schrieder, H. Simon, T. Behrens, R. Gernhauser, T. Kroll, R. Krucken, M. Munch, T. Davinson, J. Gerl, G. Huber, A. Hurst, J. Iwanicki, B. Jonson, P. Lieb, L. Liljeby, A. Schempp, A. Scherillo, P. Schmidt, G. Walter, Phys. Rev. Lett. **94**, 172501 (2005)
17. J. Cederkall, A. Ekstrom, C. Fahlander, A.M. Hurst, M. Hjorth-Jensen, F. Ames, A. Banu, P.A. Butler, T. Davinson, U.D. Pramanik, J. Eberth, S. Franchoo, G. Georgiev, M. Gorska, D. Habs, M. Huyse, O. Ivanov, J. Iwanicki, O. Kester, U. Koster, B.A. Marsh, O. Niedermaier, T. Nilsson, P. Reiter, H. Scheit, D. Schwalm, T. Sieber, G. Sletten, I. Stefanescu, J.V. de Walle, P.V. Duppen, N. Warr, D. Weisshaar, F. Wenander, Phys. Rev. Lett. **98**, 172501 (2007)
18. I. Stefanescu, G. Georgiev, F. Ames, J. Aysto, D.L. Balabanski, G. Bollen, P.A. Butler, J. Cederkall, N. Champault, T. Davinson, A.D. Maesschalck, P. Delahaye, J. Eberth, D. Fedorov, V.N. Fedosseev, L.M. Fraile, S. Franchoo, K. Gladnishki, D. Habs, K. Heyde, M. Huyse, O. Ivanov, J. Iwanicki, J. Jolie, B. Jonson, T. Kroll, R. Krucken, O. Kester, U. Koster, A. Lagoyannis, L. Liljeby, G.L. Bianco, B.A. Marsh, O. Niedermaier, T. Nilsson, M. Oinonen, G. Pascovici, P. Reiter, A. Saltarelli, H. Scheit, D. Schwalm, T. Sieber, N. Smirnova,

J.V.D. Walle, P.V. Duppen, S. Zemlyanoi, N. Warr, D. Weisshaar, F. Wenander, Phys. Rev. Lett. **98**, 122701 (2007)

19. A.M. Hurst, P.A. Butler, D.G. Jenkins, P. Delahaye, F. Wenander, F. Ames, C.J. Barton, T. Behrens, A. Burger, J. Cederkall, E. Clement, T. Czosnyka, T. Davinson, G. de Angelis, J. Eberth, A. Ekstrom, S. Franchoo, G. Georgiev, A. Gorgen, R.D. Herzberg, M. Huyse, O. Ivanov, J. Iwanicki, G.D. Jones, P. Kent, U. Koster, T. Kroll, R. Krucken, A.C. Larsen, M. Nespolo, M. Pantea, E.S. Paul, M. Petri, H. Scheit, T. Sieber, S. Siem, J.F. Smith, A. Steer, I. Stefanescu, N.U.H. Syed, J.V. de Walle, P.V. Duppen, R. Wadsworth, N. Warr, D. Weisshaar, M. Zielinska, Phys. Rev. Lett. **98**, 072501 (2007)

20. P.V. Duppen, The Euroschool Lectures on Physics with Exotic Beams, p. 37, (Springer 2006)

21. S. Wan, P. Reiter, J. Cub, H. Emling, J. Gerl, R. Schubart, D. Schwalm, Z. Phys. A **356**, 231 (1997)

22. M. Belleguic, M.J. Lopez-Jimenez, M. Stanoiu, F. Azaiez, M.G. Saint-Laurent, O. Sorlin, N.L. Achouri, J.C. Angelique, C. Bourgeois, C. Borcea, J.M. Daugas, C. Donzaud, F.D. Oliveira-Santos, J. Duprat, S. Grevy, D. Guillemaud-Mueller, S. Leenhardt, M. Lewitowicz, U.E. Penionzhkevich, Y. Sobolev, Nucl. Phys. A **682**, 136C (2001)

23. P.G. Hansen, J.A. Tostevin, Annu. Rev. Nucl. Part. Sci. **53**, 221 (2003)

24. D. Bazin, B.A. Brown, C.M. Campbell, J.A. Church, D.C. Dinca, J. Enders, A. Gade, T. Glasmacher, P.G. Hansen, W.F. Mueller, H. Olliver, B.C. Perry, B.M. Sherrill, J.R. Terry, J.A. Tostevin, Phys. Rev. Lett. **91**, 012501 (2003)

25. J.A. Tostevin, G. Podolyak, B.A. Brown, P.G. Hansen, Phys. Rev. C **70**, 064602 (2004)

26. J.A. Tostevin, B.A. Brown, Phys. Rev. C **74**, 064604 (2006)

27. S. Michimasa, S. Shimoura, H. Iwasaki, A. Tamaki, S. Ota, N. Aoi, H. Baba, N. Iwasa, S. Kanno, S. Kubono, K. Kurita, A. Kurokawa, T. Minemura, T. Motobayashi, M. Notani, H.J. Ong, A. Saito, H. Sakurai, E. Takeshita, S. Takeuchi, Y. Yanagisawa, A. Yoshida, Phys. Lett. B **638**, 146 (2006)

28. A. Winther, K. Alder, Nucl. Phys. A **319**, 518 (1979)

29. T. Glasmacher, Annu. Rev. Nucl. Part. Sci. **48**, 1 (1998)

30. T. Suomijärvi, D. Beaumel, Y. Blumenfeld, P. Chomaz, N. Frascaria, J.P. Garron, J.C. Roynette, J.A. Scarpaci, J. Barrette, B. Fernandez, J. Gastebois, Nucl. Phys. A **509**, 369 (1990)

31. K.L. Yurkewicz, D. Bazin, B.A. Brown, C.M. Campbell, J.A. Church, D.C. Dinca, A. Gade, T. Glasmacher, M. Honma, T. Mizusaki, W.F. Mueller, H. Olliver, T. Otsuka, L.A. Riley, J.R. Terry, Phys. Rev. C **70**, 034301 (2004)

32. W.W. Wilcke, J.R. Birkelund, H.J. Wollersheim, A.D. Hoover, J.R. Huizenga, W.U. Schroeder, L.E. Tubbs, At. Data Nucl. Data Tables **25**, 391 (1980)

33. H. Olliver, T. Glasmacher, A.E. Stuchbery, Phys. Rev. C **68**, 044312 (2003)

34. D. Bazin, J.A. Caggiano, B.M. Sherrill, J. Yurkon, A. Zeller, Nucl. Instrum. Methods Phys. Res. B **204**, 629 (2003)

35. W.F. Mueller, J.A. Church, T. Glasmacher, D. Gutknecht, G. Hackman, P.G. Hansen, Z. Hu, K.L. Miller, P. Quirin, Nucl. Instrum. Methods Phys. Res. A **466**, 492 (2001)

36. T. Motobayashi, Y. Ikeda, Y. Ando, K. Ieki, M. Inoue, N. Iwasa, T. Kikuchi, M. Kurokawa, S. Moriya, S. Ogawa, H. Murakami, S. Shimoura, Y. Yanagisawa,

T. Nakamura, Y. Watanabe, M. Ishihara, T. Teranishi, H. Okuno, R.F. Casten, Phys. Lett. B **346**, 9 (1995)

37. L. Biederharn, P. Brussard, Coulomb Excitation, (Clarendon Press, Oxford, 1965)
38. P. Morse, H. Feshbach, Methods of theoretical physics, (McGraw-Hill, New York, 1953)
39. C.F. Weizsäcker, Z. Physics **88**, 612 (1934)
40. E.J. Williams, Phys. Rev. **45**, 729 (1934)
41. E. Fermi, Z. Phys. **29**, 315 (1924)
42. B. Hoffman, G. Baur, Phys. Rev. C **30**, 247 (1984)
43. A. Goldberg, Nucl. Phys. A **240**, 636 (1984)
44. C.A. Bertulani, G. Baur, Nucl. Phys. A **442**, 739 (1985)
45. C.A. Bertulani, G. Baur, Phys. Rep. **163**, 299 (1988)
46. C. Bertulani, A. Nathan, Nucl. Phys. A **554**, 158 (1993)
47. C. Bertulani, C. Campbell, T. Glasmacher, Comp. Phys. Comm. **152**, 317 (2003)
48. A.N.F. Aleixo, C.A. Bertulani, Nucl. Phys. A **505**, 448 (1989)
49. C.A. Bertulani, A. Stuchberry, T. Mertzimekis, A. Davies, Phys. Rev. C **68**, 044609 (2003)
50. D.J. Millener, J.W. Olness, E.K. Warburton, S. Hanna, Phys. Rev. C **28**, 497 (1983)
51. R. Anne, D. Bazin, R. Bimbot, M.J.G. Borge, J.M. Corre, S. Dogny, H. Emling, D. Guillemaud-Mueller, P.G. Hansen, P. Hornsøj, P. Jensen, B. Jonson, M. Lewitowicz, A.C. Mueller, R. Neugart, T. Nilsson, G. Nyman, F. Pougheon, M.G. Saint-Laurent, G. Schrieder, O. Sorlin, O. Tengblad, K. Wilhelmsen-Rolander, Z. Phys. A **352**, 397 (1995)
52. N. Summers, S. Pain, N. Orr, W. Catford, J. Angélique, N. Ashwood, N.C. V. Bouchat, N.M. Clarke, M. Freer, B. Fulton, F. Hanappe, M. Labiche, J. Lecouey, R. Lemmon, D. Mahboub, A. Ninane, G. Normand, F. Nunes, N. Soic, L. Stuttge, C. Timis, I. Thompson, J. Winfield, V. Ziman, Phys. Lett. B **650**, 124 (2007)
53. T. Nakamura, T. Motobayashi, Y. Ando, A. Mengoni, T. Nishio, H. Sakurai, S. Shimoura, T. Teranishi, Y. Yanagisawa, M. Ishihara, Phys. Lett. B **394**, 11 (1997)
54. M.C.M. Fauerbach, T. Glasmacher, P. Hansen, R. Ibbotson, D. Morrissey, H. Scheit, P. Thirolf, M. Thoennessen, Phys. Rev. C **56**, 1(R) (1997)
55. S. Typel, G. Baur, Phys. Lett. B **356**, 186 (1995)
56. C.A. Bertulani, L.F. Canto, M.S. Hussein, Phys. Lett. B **353**, 413 (1995)
57. T. Tarutina, L.C. Chamon, M.S. Hussein, Phys. Rev. C **67**, 044605 (2003)
58. M. Husseein, R.L. ad F.M. Nunes, I. Thompson, Phys. Lett. B **640**, 91 (2006)
59. C. Thibault, R. Klapisch, C. Rigaud, A.M. Poskanzer, R. Prieels, L. Lessard, W. Reisdorf, Phys. Rev. C **12**, 644 (1975)
60. X. Campi, H. Flocard, A.K. Kerman, S. Koonin, Nucl. Phys. A **251**, 193 (1975)
61. E.K. Warburton, J.A. Becker, B.A. Brown, Phys. Rev. C **41**, 1147 (1990)
62. B.V. Pritychenko, T. Glasmacher, P.D. Cottle, M. Fauerbach, R.W. Ibbotson, K.W. Kemper, V. Maddalena, A. Navin, R. Ronningen, A. Sakharuk, H. Scheit, V.G. Zelevinsky, Phys. Lett. B **467**, 309 (1999)
63. J.A. Church, C.M. Campbell, D.C. Dinca, J. Enders, A. Gade, T. Glasmacher, Z. Hu, R.V.F. Janssens, W.F. Mueller, H. Olliver, B.C. Perry, L.A. Riley, K.L. Yurkewicz, Phys. Rev. C **72**, 054320 (2005)

54   T. Glasmacher

64. H. Iwasaki, T. Motobayashi, H. Sakurai, K. Yoneda, T. Gomi, N. Aoi, N. Fukuda, Z. Fulop, U. Futakami, Z. Gacsi, Y. Higurashi, N. Imai, N. Iwasa, T. Kubo, M. Kunibu, M. Kurokawa, Z. Liu, T. Minemura, A. Saito, M. Serata, S. Shimoura, S. Takeuchi, Y. Watanabe, K. Yamada, Y. Yanagisawa, M. Ishihara, Phys. Lett. B **522**, 227 (2001)
65. G. Klotz, Phys. Rev. C **47**, 2502 (1993)
66. V. Chiste, A. Gillibert, A. Lepine-Szily, N. Alamanos, F. Auger, J. Barrette, F. Braga, M.D. Cortina-Gil, Z. Dlouhy, V. Lapoux, M. Lewitowicz, R. Lichtenthaler, R.L. Neto, S.M. Lukyanov, M. MacCormick, F. Marie, W. Mittig, N.A. Orr, F.D. Santos, A.N. Ostrowski, S. Ottini, A. Pakou, Y.E. Penionzhkevich, P. Roussel-Chomaz, J.L. Sida, Phys. Lett. B **514**, 233 (2001)
67. R.W. Ibbotson, T. Glasmacher, B.A. Brown, L. Chen, M.J. Chromik, P.D. Cottle, M. Fauerbach, K.W. Kemper, D.J. Morrissey, H. Scheit, M. Thoennessen, Phys. Rev. Lett. **80**, 2081 (1998)
68. B.V. Pritychenko, T. Glasmacher, P.D. Cottle, M. Fauerbach, R.W. Ibbotson, K.W. Kemper, V. Maddalena, A. Navin, R. Ronningen, A. Sakharuk, H. Scheit, V.G. Zelevinsky, Phys. Lett. B **461**, 322 (1999)
69. P.D. Cottle, M. Fauerbach, T. Glasmacher, R.W. Ibbotson, K.W. Kemper, B. Pritychenko, H. Scheit, M. Steiner, Phys. Rev. C **60**, 031301 (1999)
70. P.D. Cottle, V.B. Pritychenko, J.A. Church, M. Fauerbach, T. Glasmacher, R.W. Ibbotson, K.W. Kemper, H. Scheit, M. Steiner, Phys. Rev. C **64**, 057304 (2001)
71. P.D. Cottle, Z. Hu, B.V. Pritychenko, J.A. Church, M. Fauerbach, T. Glasmacher, R.W. Ibbotson, K.W. Kemper, L.A. Riley, H. Scheit, M. Steiner, Phys. Rev. Lett. **88**, 172502 (2002)
72. A. Gade, D. Bazin, A. Becerril, C.M. Campbell, J.M. Cook, D.J. Dean, D.C. Dinca, T. Glasmacher, G.W. Hitt, M.E. Howard, W.F. Mueller, H. Olliver, J.R. Terry, K. Yoneda, Phys. Rev. Lett. **95**, 022502 (2005)
73. K.L. Yurkewicz, D. Bazin, B.A. Brown, C.M. Campbell, J.A. Church, D.C. Dinca, A. Gade, T. Glasmacher, A. Honma, T. Mizusaki, W.F. Mueller, H. Olliver, T. Otsuka, L.A. Riley, J.R. Terry, Phys. Rev. C **70**, 054319 (2004)
74. D.C. Dinca, R.V.F. Janssens, A. Gade, D. Bazin, R. Broda, B.A. Brown, C.M. Campbell, M.P. Carpenter, P. Chowdhury, J.M. Cook, A.N. Deacon, B. Fornal, S.J. Freeman, T. Glasmacher, M. Honma, F.G. Kondev, J.L. Lecouey, S.N. Liddick, P.F. Mantica, W.F. Mueller, H. Olliver, T. Otsuka, J.R. Terry, B.A. Tomlin, K. Yoneda, Phys. Rev. C **71**, 041302 (2005)
75. S. Raman, At. Data Nucl. Data Tables **78**, 1 (2001)
76. J.M. Cook, T. Glasmacher, A. Gade, Phys. Rev. C **73**, 024315 (2006)
77. B.V. Pritychenko et al, Phys. Rev. C **63**, 011305(R) (2001)
78. B. Perry, C. Campbell, J. Church, D. Dinca, J. Enders, T. Glasmacher, Z. Hu, K. Miller, W. Mueller, H. Olliver, Nucl. Instr. Meth. A **505**(1-2), 85 (2003)
79. N. Kaloskamis, K. Chan, A. Chishti, J. Greenberg, C. Lister, S. Freedman, M. Wolanski, J. Last, B. Utts, Nucl. Instr. Meth. A **330**, 447 (1993)
80. H. Wollersheim, D. Appelbe, A. Banu, R. Bassini, T. Beck, F. Becker, P. Bednarczyk, K.H. Behr, M. Bentley, G. Benzoni, C. Boiano, U. Bonnes, A. Bracco, S. Brambilla, A. Brunle, A. Burger, K. Burkard, P. Butler, F. Camera, D. Curien, J. Devin, P. Doornenbal, C. Fahlander, K. Fayz, H. Geissel, J. Gerl, M. Gorska, H. Grawe, J. Grebosz, R. Griffiths, G. Hammond, M. Hellstrom, J. Hoffmann, H. Hubel, J. Jolie, J. Kalben, M. Kmiecik, I. Kojouharov,

R. Kulessa, N. Kurz, I. Lazarus, J. Li, J. Leske, R. Lozeva, A. Maj, S. Mandal, W. Meczynski, B. Million, G. Munzenberg, S. Muralithar, M. Mutterer, P. Nolan, G. Neyens, J. Nyberg, W. Prokopowicz, V. Pucknell, P. Reiter, D. Rudolph, N. Saito, T. Saito, D. Seddon, H. Schaffner, J. Simpson, K.H. Speidel, J. Styczen, K. Sümmerer, N. Warr, H. Weick, C. Wheldon, O. Wieland, M. Winkler, M. Zieblinski, Nucl. Instrum. Methods Phys. Res. A **537**, 637 (2005)

81. S. Shimoura, Nucl. Instr. Methods A **525**, 188 (2004)
82. M.A. Deleplanque, I. Lee, K. Vetter, G.J. Schmid, F.S. Stephens, R.M. Clark, R.M. Diamond, P. Fallon, A.O. Macchiavelli, Nucl. Instrum. Methods A **430** (1999)
83. J. Gerl, Acta Phys. Polonica B **34**, 2481 (2003)
84. J. Simpson, J. Phys. G. **31**, S1801 (2005)

Test of Isospin Symmetry Along the $N = Z$ Line

S.M. Lenzi[1] and M.A. Bentley[2]

[1] Dipartimento di Fisica dell'Università and INFN, Sezione di Padova, I-35131 Padova, Italy
silvia.lenzi@pd.infn.it
[2] Department of Physics, University of York, Heslington, York. YO10 5DD, UK
mab503@york.ac.uk

Abstract The study of isospin symmetry in nuclei as a function of angular momentum has now become established as a very powerful tool to understand nuclear properties in rotating nuclei. These studies have become feasible in the last decade due to recent experimental developments in the identification of proton-rich nuclei produced with very low cross sections. Contemporaneously, state-of-the-art shell-model codes have been produced for the description of these data. The synergy between theory and experiment for the study of energy differences of mirror and isobaric analogue nuclei in the mass region between $A \sim 30$ and ~ 60 has allowed the investigation of the evolution of the nuclear wave functions with increasing spin. The alignment process, changes of the nuclear shape and the intrinsic configuration, together with the evidence of isospinnon-conserving terms of the nuclear interaction are examples of the type of phenomena that can be studied from the analysis of Coulomb energy differences.

1 Introduction

The concept of symmetry in physics is a very powerful tool for the understanding of the behaviour of Nature. Symmetries are intimately related to conservation laws and to conserved quantities which, in quantum mechanics, translate into good quantum numbers. In nuclear physics, several symmetries have been identified. In particular, *isospin* symmetry is related to the identical behaviour of protons and neutrons in the nuclear field. At the very beginning of nuclear physics, only charged particles were known. To explain the nuclear mass, Rutherford suggested in 1920 the existence of a neutral particle with mass very similar to that of the proton. When Chadwick discovered the neutron in 1932, it was clear that the nuclear force acts similarly on protons and neutrons. This induced Heisenberg to propose treating them as the two-quantum states of a particle called the *nucleon*. These two states are characterised by the projection of the *isospin* quantum number t, with

Lenzi, S.M., Bentley, M.A.: *Test of Isospin Symmetry Along the* N = Z *Line*. Lect. Notes
Phys. **764**, 57–98 (2009)
DOI 10.1007/978-3-540-85839-3_3

$t_z = -1/2$ for the proton and $t_z = 1/2$ in the case of a neutron. This is in analogy with the spin quantum number, but whereas the spin of an elementary particle is determined by the projection of its spin in real space along some axis, the isospin state of the nucleon is determined by the projection in an abstract space: the isospin space. In practice, the angular momentum algebra we know for the spin can be easily applied to isospin.

In a nucleus formed by N neutrons and Z protons, the total isospin T is given by the vector sum of the single-nucleon isospins. The isospin projection $T_z = (N - Z)/2$ is well defined and therefore $|N - Z|/2 \leq T \leq (N + Z)/2$. Putting the Coulomb interaction to one side, the concepts of charge-symmetry and independence can result in identical behaviour of two nuclei with the same total number of nucleons (*isobaric* nuclei), but with different numbers of neutrons and protons. Of course, the Pauli principle puts obvious constraints on the available configurations and hence on the range of the symmetries observed. The isospin quantum number, T, directly couples together the two concepts of charge-symmetry/independence and the Pauli principle. Isospin thus becomes a good quantum number to characterise analogue states in isobaric multiplets (Wigner, 1937). These states are termed *isobaric analogue states* (IAS), and the near-identicality of such states demonstrates the power of the isospin concept. In particular, nuclei with the same mass but with the numbers of protons and neutrons interchanged, *mirror nuclei*, would have identical structure, with all analogue states at the same excitation energy. Energy differences between IAS are due to isospin non-conserving forces, such as the Coulomb interaction.

The study of these energy differences, for many decades confined to low excitation energy and angular momentum [1], has been extended in the last decade to high-spin yrast states. This has been possible due to the advance in gamma-ray detection efficiency and resolving power achieved with large Ge multi-detector arrays in combination with other ancillary devices. The most studied isobaric multiplets are those of the $f_{\frac{7}{2}}$ shell where collective structures have been observed up to the band-terminating states. Interestingly, these nuclei can be described with very good accuracy by the shell model. The extension of these calculations to the description of excitation energy differences between IAS allows the origin of isospin-symmetry-breaking (ISB) effects to be investigated. It turns out that such energy differences – usually called Coulomb energy differences (CED) – yield detailed information on changes in *nuclear* structure with increasing energy and angular momentum. The study of CED in the region of nuclei between $A \sim 30$ and ~ 60 will be discussed in these lectures.

We will start with an introduction to the concepts of charge-symmetry and charge-independence and an introduction to the application of the isospin concept to energy differences between analogue states in isobaric multiplets. Technical developments in both experimental technique and the nuclear shell model that have enabled the rapid progress in this field will then be discussed. The origin of energy differences between analogue states in terms of Coulomb, and other isospin-breaking effects as well as a detailed description of how such effects are calculated within the framework of the shell model will be covered. A discussion of some of the experimentally measured energy differences, with a consistent set of calculations from the shell model, for $f_{7/2}$, sd and upper fp shell nuclei is then presented.

2 Background

2.1 Charge Invariance of the Nuclear Force

The general concept of the charge invariance of the nuclear force can be sub-divided into two separate ideas – charge-symmetry and charge-independence. Charge-symmetry requires that the nuclear proton–proton interaction (V_{pp}) is equal to that between neutrons, V_{nn}. Recently, more accurate experimental data have become available from nucleon–nucleon scattering experiments, revealing better evidence for a slight charge asymmetry in the measured scattering lengths of -18.9 ± 0.4 fm (nn) and -17.3 ± 0.4 fm (pp) [2]. Of course, these data refer to free-nucleon interactions, and not the effective nucleon–nucleon interaction in the nuclear medium. The origin of the observed charge-symmetry-breaking (CSB) is not yet fully settled, and models based on the effects of nucleon mass splitting and meson mixing have been applied to this problem ([2, 3], and references therein). The more stringent condition of charge-independence also requires that $(V_{pp} + V_{nn})/2 = V_{np}$, which is also known to be broken slightly [4]. Nevertheless, the concepts of charge-symmetry and charge-independence will be expected to result in clear symmetries in nuclear behaviour.

Electromagnetic transitions and the weak interaction can be used to test isospin invariance, and certain decays are forbidden if isospin is a good quantum number for hadronic forces. Examples are as follows. Firstly, beta-decay: Fermi matrix elements are zero unless along isobaric multiplets. Secondly, electromagnetic transitions: $E\lambda$ and $M\lambda$ transitions only connect states with $\Delta T = 0, \pm 1$, $E1$ transitions between $\Delta T = 0$ states are forbidden in $N = Z$ nuclei, and have equal strengths in mirror nuclei; $E2$ transitions have a linear dependence on T_z in an isobaric multiplet; $M\lambda$ transitions between states of the same T in an $N = Z$ nucleus are hindered.

The value of T is not an observable quantity, though for nuclei near $N = Z$ (where the concept of isospin is most relevant) it can usually be "assigned" using logical arguments. This can be done easily if it is remembered that states of a given T can only occur in a set of nuclei with $T_z = T, T - 1, ..., -T$ (since $|T_z| > T$ is forbidden). We start our consideration of isospin with the simplest system – that of two nucleons: nn, pp and np. Here, T_z is $0, \pm 1$ and so the value of T is restricted to 0 and 1. Charge-independence dictates that any state that can be constructed in the pp (or nn) system must also exist in the np system. However, the inverse statement cannot be made. That is, there are some states in the np system that are forbidden by the Pauli principle in the pp (or nn) system. The ground state of the deuteron (np) with $J^\pi = 1^+$ is such a state and, therefore, must have $T = 0$ as the isospin projection is limited only to $T_z = 0$. Similarly, any state in the nn system (and its equivalent state in pp) must have $T = 1$ as the projection is $T_z = 1$ (-1 for pp). However, as a $T = 1$ state can have a $T_z = 0$ projection, this state must also exist in the np system. Thus, there are three identically constructed $T = 1$ states which can be found in the nn, np and pp systems (i.e., with $T_z = 1, 0, -1$, respectively). These states form an *isospin triplet*, and the lowest $J^\pi = 0^+$ states in these three two-nucleon systems form such a triplet. In fact, as we know, all three of these states are unbound.

These classification arguments can easily be extended to states in many-particle systems, where it is also useful to remember that, in general, the lowest energy states (e.g. the ground state) of a nucleus will have the lowest available value of isospin (i.e. $T = |T_z|$) (exceptions to this "rule" are $N = Z$ odd–odd nuclei in the $f_{\frac{7}{2}}$ shell and ^{34}Cl, where $T_z = 0$ and the ground state has $T = 1$). Thus, for example, one finds four nuclei with $T_z = \pm\frac{1}{2}, \frac{3}{2}$ all four of which contain an identically constructed $T = \frac{3}{2}$ state – an isospin quadruplet. The two "outer" nuclei with $T_z = \pm\frac{3}{2}$ have $T = \frac{3}{2}$ ground states (which are mirror states – see below), and for the other two nuclei with $T_z = \pm\frac{1}{2}$ the $T = \frac{3}{2}$ analogue states are excited states, as the ground states of these two nuclei will be expected to have $T = \frac{1}{2}$.

The simplest example of an isobaric multiplet is a pair of mirror nuclei. In a mirror pair, the total number of pp interactions in one member of the pair is the same as the number of nn interactions in the other. Hence only the charge-symmetry of the nucleon–nucleon interaction is required to provide isospin symmetry in a mirror pair. This is generally not the case for any set of IAS, where the isospin symmetry across a multiplet relies on both charge-symmetry *and* charge-independence.

2.2 The Isobaric Multiplet Mass Equation

The isobaric multiplet mass equation (IMME) is one of the most basic predictions to follow from the concept of isospin in nuclear physics. It was first investigated in 1957 by E.P. Wigner [5] and describes the dependence of the mass (or binding energy) of a set of IAS as a function of Z (or T_z). The largest effect on the splitting is always due to the Coulomb interaction, which lowers the total binding energy of a state in one member of the multiplet relative to the IAS in the neighbouring lower-Z isobar. A full description of this can be found in several papers and reviews (e.g. [1, 6–8]). We present an outline derivation of the IMME here, as it serves as an excellent example of the power of the isospin formalism, as well as having real consequences regarding the interpretation of energy differences between excited states of isobaric multiplets. We start with the eigenstates $|\alpha T T_z\rangle$ of the charge-independent Hamiltonian H_{CI}, where α contains all the additional quantum numbers that define the state. Since H_{CI}, by definition, conserves T, the eigenvalues are independent of T_z – i.e. the isobaric analogue states are completely degenerate. A charge-violating interaction will lift this degeneracy and can be treated as a perturbation if the total energy splitting induced is small compared with the binding due to the nuclear force – as is the case with the Coulomb interaction. The total binding energy can be determined by:

$$\text{BE}(\alpha T T_z) = \langle \alpha T T_z | H_{CI} + H'_{CV} | \alpha T T_z \rangle, \tag{1}$$

where H'_{CV}, represents the charge-violating interaction(s). If two-body forces alone are responsible for the nature of H'_{CV}, then it can be written as,

$$H'_{CV} = \sum_{k=0}^{2} H_{CV}^{(k)}, \tag{2}$$

where $k = 0, 1, 2$ correspond to the isoscalar, isovector and isotensor components of this interaction, respectively. The total energy splitting of the isobaric multiplet is given by

$$\Delta\text{BE}(\alpha T T_z) = \langle \alpha T T_z | \sum_{k=0}^{2} H_{CV}^{(k)} | \alpha T T_z \rangle. \tag{3}$$

The application of the Wigner–Eckart theorem can then extract explicitly the T_z-dependence of the energy splitting of the multiplet:

$$\Delta\text{BE}(\alpha T T_z) = \sum_{k=0}^{2} (-)^{T-T_z} \begin{pmatrix} T & k & T \\ -T_z & 0 & T_z \end{pmatrix} \langle \alpha T \| H_{CV}^{(k)} \| \alpha T \rangle, \tag{4}$$

where the double-bars in the final term denote matrix elements reduced in isospin. The above Wigner 3–j coefficient has a simple analytical form for the three values of k, and we obtain

$$\begin{aligned}
\Delta BE(\alpha T T_z) &= \frac{1}{\sqrt{2T+1}} \Big[M^{(0)} \\
&+ \frac{T_z}{\sqrt{T(T+1)}} M^{(1)} \\
&+ \frac{3T_z^2 - T(T+1)}{\sqrt{T(T+1)(2T+3)(2T-1)}} M^{(2)} \Big] \\
&= a + bT_z + cT_z^2
\end{aligned} \tag{5}$$

where $M^{(k)}$ are the three sets of reduced matrix elements $\langle \alpha T \| H_{CV}^{(k)} \| \alpha T \rangle$, the values of which are independent of T_z but otherwise dependent on T and α.

Here, in the last step, we have re-ordered the terms in the penultimate step, and the IMME acquires a quadratic form. The conclusion then is that the binding energy splitting (and hence mass) of an isobaric multiplet is quadratic in T_z where the coefficients a, b and c depend only on T and sets of reduced matrix elements. In addition, the derivation of Eq. (5) shows that the coefficients are directly related to the three tensor components of the interaction. The a coefficient depends mostly on the isoscalar component, with a small contribution from the isotensor one. The b and c coefficients are related only to the isovector and isostensor components, respectively. Put another way, the values of b and c (which can be determined experimentally) can potentially yield separate information on the charge-symmetry and charge-independence, respectively, of the attractive nucleon–nucleon interaction.

The IMME is valid in the presence of any charge-violating (i.e. isospin non-conserving) interaction or set of interactions. Of course, the Coulomb interaction is expected to be the dominant contributor. However, the quadratic nature of the IMME would be valid even in the presence of charge-asymmetric and charge-dependent components of the attractive nucleon–nucleon potential. Only the values of the coefficients would be affected by the presence of such effects. Deviations from IMME would be expected, of course, if higher-order perturbations and/or the inclusion of three-body terms are important. In addition, as pointed out by Auerbach [9], a significant component of isospin mixing could also result in deviations from the IMME quadratic behaviour.

As the IMME is such a basic prediction leading from the isospin concept, testing the validity of the equation is clearly of fundamental importance. The most effective way is examine isobaric multiplets with at least four members (i.e. $T \geq \frac{3}{2}$), and fitting a cubic expression by inclusion of a $d\,T_z^3$ term. Of course, the value of the d coefficient should be consistent with zero if the quadratic nature of the IMME is valid. New experimental data are available that have now allowed the validity of the IMME to be tested this way – see Britz, Pape and Anthony [10] for a comprehensive compilation. Figure 1 contains data taken from [10] for all the known $T = \frac{3}{2}$ isobaric quadruplets. Firstly, Fig. 1 clearly shows that the d coefficients, extracted from the data following the cubic fit described above, have values consistent with zero over all the measured masses so far. Thus, the agreement with the prediction of the quadratic nature of the IMME is quite remarkable. Only one exception appears, once the magnitude of the error bars is taken into account, that of the $T = \frac{3}{2}$, $J^\pi = \frac{3}{2}^-$ isobaric quadruplet for $A = 9$.

Even if the quadratic form of the IMME is established experimentally, this does not in itself yield any direct information on the nature of the two-body interaction. Put another way, a detailed understanding of the charge-violating components of the interaction is required to reproduce theoretically the values of the coefficients of the IMME. Here, in fact, there is a long-standing historical issue – the famous Nolen–Schiffer anomaly (see Sect. 2.3). However, for now we can make a rough estimate of the coefficients.

If we assume that the nucleus can be treated as a uniformly charged sphere, we can derive a simplistic estimate of the Coulomb coefficients of the IMME as a function of mass. The Coulomb energy of such a uniformly charged sphere

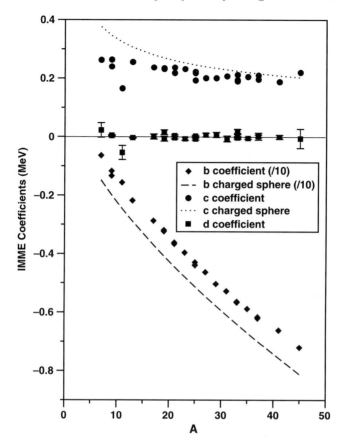

Fig. 1. Coefficients of the IMME of the form $BE(T, T_z) = a + bT_z^2 + cT_z^2 + dT_z^3$ obtained from fits to experimental data for $T = \frac{3}{2}$ quadruplets (data taken from [10]). The experimentally determined b, c and d coefficients along the predictions of the simple charged-sphere model – see text and Eq. (7) – where r_0 is taken to be 1.2 fm

is given by

$$E_C = \frac{3e^2 Z(Z-1)}{5R_C} = \frac{3e^2}{5r_0 A^{\frac{1}{3}}} \left[\frac{A}{4}(A-2) + (1-A)T_z + T_z^2 \right]. \quad (6)$$

Hence we arrive at the expressions

$$a = \frac{3e^2 A(A-2)}{20r_0 A^{\frac{1}{3}}}, \quad b = -\frac{3e^2(A-1)}{5r_0 A^{\frac{1}{3}}}, \quad c = \frac{3e^2}{5r_0 A^{\frac{1}{3}}}. \quad (7)$$

The predictions of this crude estimate are shown in Fig. 1 along with the b and c coefficients extracted from the $T = \frac{3}{2}$ isobaric quadruplets [10]. We note that the proton–neutron atomic mass difference Δ_{nH} also contributes to the b coefficient, and the prediction has been modified to account for this. The overall magnitude and trends of the charged-sphere prediction are clearly fairly good, although there is a $\approx 1\,\mathrm{MeV}$ over-estimate of the magnitude of the b term. As has been known for a long time (e.g. Bethe and Bacher 1936 [11]) such classical estimates of the Coulomb energy stored in a nucleus need to be modified to account for the effect of antisymmetrisation. Effectively, the average result of the Pauli principle is to keep the protons further apart than would be allowed classically, and this accounts for at least some of the discrepancy in the b coefficient in Fig. 1.

2.3 Energy Differences Between IAS

Total binding energy difference: Coulomb displacement energy (CDE)

The total binding energy difference between a particular state and its analogue state in another member of the isospin multiplet is referred to generically as the Coulomb displacement energy (CDE). For any two members of a multiplet of isospin T, transformed through exchange of p protons for neutrons, the CDE is given by

$$\mathrm{CDE}(T, T_z) = M_{T, T_z} - M_{T, T_z + p} + p\Delta_{nH} \tag{8}$$

where M is the atomic mass, Δ_{nH} is the neutron–hydrogen atomic mass difference and T_z is the isospin projection for the larger-Z isobar.

An example is shown in Fig. 2 for the $A = 21$ isobars with $T_z = \pm\frac{1}{2}, \frac{3}{2}$, where the ground-state binding energies form the usual Weizsäcker parabola. The $T = \frac{3}{2}$ isospin quadruplet shown includes the ground states of the two $T_z = \pm\frac{3}{2}$ nuclei, and the corresponding excited (isobaric analogue) states of the two nuclei in between. We have seen above, in the discussion of the IMME, that the binding energies of the IAS are also parabolic in Z – the *dashed line* in Fig. 2. The CDE indicated is the total binding energy difference between the two neighbouring members of the multiplet. Here, as is conventional, we associate the CDE specifically with the lowest energy set of IAS in a multiplet, whilst keeping in mind that bound excited-state sets of IAS will also exist in all probability in the multiplet.

In a refined calculation, the CDE is the most basic quantity one would want to reproduce theoretically. The CDE is, of course, directly related to the IMME coefficients: for any two IAS in a multiplet we have

$$\mathrm{CDE}(T, T_z) = -p(b + c[2T_z + p] - \Delta_{nH}) \tag{9}$$

where again we have exchanged p protons for neutrons and T_z is the isospin projection for the larger-Z isobar. The CDE has been the subject of much theoretical work – most notably the reviews of Nolen and Schiffer [12], Shlomo [13] and Auerbach [9]. In these reviews, CDE have been calculated for wide ranges of IAS for which experimental data exist for comparison. In the original work

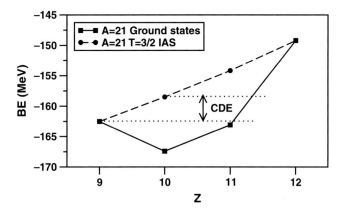

Fig. 2. The experimental binding energies of the $A = 21, T_z = \pm\frac{1}{2}, \pm\frac{3}{2}$ nuclei – both for the ground states and for the $T = \frac{3}{2}$ isobaric analogue states. The CDE between adjacent members of the $T = \frac{3}{2}$ quadruplet is indicated

of Nolen and Schiffer [12], a charge-symmetric and charge-independent interaction was assumed – and thus it is assumed initially that b and c originate entirely from the Coulomb interaction. The difference in Coulomb energy between adjacent members of the multiplets was computed using independent particle models. Here the dominant Coulomb shift was determined by computing the density distribution of the neutron excess in one member of the multiplet, and calculating the Coulomb shift when one of the neutrons is transformed into a proton. To this are added two further contributions – an exchange term (the result of the Pauli effect described above) and an *electromagnetic* spin–orbit term. The latter term turns out to be very significant in the interpretation of excited-states energy differences, and we will discuss this later. These three terms combined do not account fully for the experimental CDE, and so a number of corrections were computed. The largest corrections included the Coulomb distortion of analogue wavefunctions, isospin impurities in the core and intrashell interactions. When all the corrections were taken into account, there remained a consistent under-estimate of the CDE by around 7% on average – amounting to several hundred keV. This is the so-called "Nolen–Schiffer" anomaly. Much theoretical work on this anomaly has followed – especially in the comprehensive reviews of Shlomo [13] and Auerbach [9]. Here the calculations were revisited and refined, and further corrections were introduced – including effects associated with configuration mixing and polarisation of the core due to particle-vibration coupling.

Further studies have identified a number of phenomena that could account for the discrepancy. For example, it has been suggested [14, 15] that this anomaly could be principally associated with a charge-asymmetric component of the nucleon–nucleon interaction although Shlomo earlier showed [13] that this can only account for about half of the discrepancy in heavy nuclei, and that the effect of differences in neutron a proton radii could contribute significantly to the anomaly. Duflo and Zuker [16], studying the neutron skin, also show that a proper quantum treatment of the Coulomb and CSB interactions gives a very good description of the CDE, reducing the anomaly significantly. Whatever the source of the Nolen–Schiffer anomaly, it seemed from these studies that a detailed structural understanding of Coulomb effects at the level of less than, say, 100 keV was likely to be very difficult.

Energy differences between excited IAS

The differences in *excitation* energy between excited IAS states are generically termed Coulomb energy differences or CED. In determining CED between excited states, we effectively "normalise" the absolute binding energies of the ground states, thus the CED only reflect the *change in CDE* in relation to the ground state. The bulk of the CDE (e.g. the difference in bulk Coulomb energy) will simply cancel in this process. The CED between IAS in two members of a multiplet obtained through exchanging p protons for neutrons is given generically by

$$\text{CED}_{J,T} = E^*_{J,T,T_z} - E^*_{J,T,T_z+p}. \tag{10}$$

Here, we examine how the Coulomb energy (and other charge-dependent phenomena) varies as a function of excitation energy and angular momentum (spin) for a set of IAS. In particular, the IAS are expected to show slight differences with increasing spin associated with the influence on the Coulomb energy of the changing nucleon orbitals which contribute to the wave functions of the excited states. These measured CED are typically 100 keV or less, as we will show, and are remarkably sensitive to quite subtle nuclear structure phenomena which, with the aid of shell-model calculations, can be interpreted quantivately at the level of a *few tens of keV*. This is especially true when we restrict the study to excited structures in nuclei whose wave functions have major contributions from a single j shell – such as in the $f_{\frac{7}{2}}$ shell. In these cases, the CED can be reliably interpreted in terms of structural phenomena such as changes in the spatial correlations of pairs of valence protons and/or changes in radius/deformation as a function of spin. When the excitations involve significant changes in the single-particle contributions to the configurations, then larger CED (few hundred keV) can be observed, as we will see in Sect. 6. This, in turn, yields valuable information on the single-particle structure of the states and the nature of the excitations involved.

Recently, the study of CED between IAS has been pursued in considerable detail, yielding some remarkable results. Key experimental advances have been made in the last decade (which will be detailed in Sect. 3.1) and have afforded the opportunity to examine CED between many $T = \frac{1}{2}$ doublets and $T = 1$ triplets up to the highest accessible excitation energy and angular momentum. The most extensively studied cases are the $T_z = \pm\frac{1}{2}$ isobaric doublets (mirror pairs), where the $T = \frac{1}{2}$ states form a set of IAS. For a pair of mirror nuclei, the CED are specifically referred to as mirror energy differences (MED), defined, for any pair of mirror nuclei, as the difference in excitation energy as a function of spin:

$$\text{MED}_{J,T} = E^*_{J,T,T_z=-T} - E^*_{J,T,T_z=T} = -k\Delta b_J, \tag{11}$$

where we have assumed that the lowest isospin states are being studied – i.e. $T = |T_z|$. Here, again, k protons have been exchanged for neutrons, and in this specific case $k = 2T = 2|T_z|$. Here, we have used Eq. (5) to link the MED to the coefficients of the IMME, and we define Δb_J as the change in the b coefficient as a function of spin in relation to the ground state. We see that the MED give us *isovector* energy differences, the interpretation of which relies entirely on the concept of charge-symmetry.

The ability to study IAS of $T \geq 1$ as a function of spin has enabled a more detailed analysis of Coulomb (and other) phenomena in the context of the IMME. For example, if we consider an isobaric $T = 1$ triplet (i.e. $T = 1, T_z = 0, \pm 1$), then there is one other specific way of writing energy differences (solely for a set of IAS in a $T = 1$ triplet). These are the triplet energy differences (TED), defined as

$$\text{TED}_J = E^*_{J,T_z=-1} + E^*_{J,T_z=+1} - 2E^*_{J,T_z=0} = 2\Delta c_J. \tag{12}$$

Here we see that the TED depend only on the variation of the c coefficient with spin, hence these are *isotensor* energy differences which yield information on the Coulomb energy if charge-independence of the nuclear interaction is assumed (cf. charge-symmetry for MED). Conversely, deviations of the calculations from the experimental data will give evidence of violations of the charge-symmetry and/or charge-independence of the nuclear interaction. This turns out to be, however, quite difficult. For example, one essential ingredient in modelling Coulomb phenomena is the set of matrix elements that describe the Coulomb energy of a pair of protons as a function of their angular momentum – Coulomb matrix elements, CME. These either have to be modelled or extracted from the data – neither of which is particularly reliable. Nevertheless, a consistent picture of spin-dependent Coulomb phenomena (and other isospin non-conserving effects) is now emerging, which is discussed in the following sections.

3 Experimental and Theoretical Tools

3.1 Technical Advances

The ability to study these fascinating isospin-related effects requires detailed experimental spectroscopy of nuclei on – or more proton rich than – the line of $N = Z$. This is a considerable technical challenge for nuclei heavier than $A = 40$, where the line of stability and the line of $N = Z$ finally separate for ever. However, this has been achieved very successfully in the last decade or so due to the development of large Compton-suppressed gamma-ray spectrometers (e.g. see the review of Lee et al. [17]). Devices that have been used for the spectroscopy described here include the Gammasphere array [18] (based in the US at either the Lawrence Berkeley National Laboratory or the Argonne National Laboratory), the Euroball spectrometer [19, 20] (based at either IReS Strasbourg or Legnaro National Laboratory) and the GASP array [21] based at Legnaro National Laboratory. All three are arrays of hyper-pure germanium detectors (HpGe) – either single-crystal or composite detectors – with each detector surrounded by a BGO Compton suppression shield to improve the peak-to-background ratio. The arrays are based on the principle of maximising the total gamma-ray detection efficiency whilst maintaining a sufficient granularity to enable multiple gamma-ray coincidences to be recorded and to reduce the probability of more than one gamma ray hitting any crystal.

The power of these large arrays for this kind of spectroscopy has also been enhanced considerably by the use of efficient and highly selective ancillary detectors for identifying the final nucleus and using this identification to "tag" the observed gamma decays. This is essential for proton-rich nuclei, where cross sections in fusion-evaporation reactions are no higher than the millibarn level (i.e. roughly one reaction in 10^3 leading to the nucleus of interest) down towards the few *microbarn* level. Moreover, in general, only one or two (if any) gamma-ray transitions will have been identified previously in that nucleus. Thus, we have the two requirements of high-gamma-ray efficiency and clean reaction-channel identification to measure weak gamma-ray transitions and assign them to a particular nucleus.

There are several ways in which these large gamma-ray spectrometers can be used to identify the transitions associated with weak proton-rich reaction channels following fusion–evaporation reactions – usually involving $N = Z$ beam and target species. If no additional detectors are used, in some circumstances one can rely only on the power of the array to seek out the gamma rays of interest. For mirror nuclei, where there is by default some "approximate" information on the expected gamma-ray energies, this works well. For example, the yrast excited-state sequence in ^{53}Co was identified [22] this way. Using triple (or higher-fold) gamma-ray coincidences one can make very clean spectra by placing multiple conditions on the energies of some of the gamma rays observed. For example, Fig. 3(a) shows a spectrum after "double gating" on a number of gamma-ray transitions in the known yrast sequence of ^{53}Fe. The four strong peaks are all members of this yrast sequence, and the spectrum is extremely clean. Figure 3(b) shows the equivalent spectrum for ^{53}Co, produced using "analogue" gating conditions. The identicality of the spectra is obvious and the four analogue transitions in ^{53}Co are clear. The gating conditions for ^{53}Co were arrived at following an iterative procedure since the approximate energies of the transitions were known in advance from mirror-symmetry arguments. A triple-coincidence analysis confirmed the ordering of

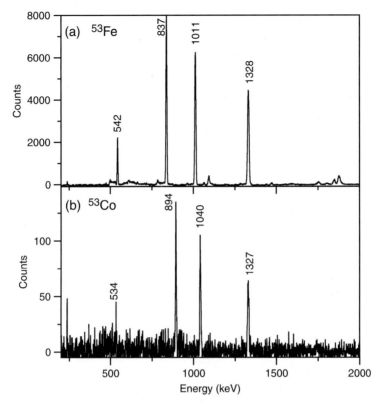

Fig. 3. (a) A spectrum of ^{53}Fe created by requiring double-coincidences between pairs of yrast transitions (see text). (b) A spectrum of ^{53}Co created in an identical manner to (a), by gating on the equivalent analogue transitions (See also Plate 6 in the Color Plate Section)

the gamma rays [22] and the level scheme of ^{53}Co was derived. A discussion of the resulting energy differences for this $A = 53$ mirror pair will be discussed later in Sect. 5.

The second method discussed here is the selection of weak reaction channels through A and Z determination of the recoiling nucleus. This requires a $0°$ ion-optical device to act as a mass separator downstream of the target, coupled to a device for determining Z. The example shown here is the identification of excited states in ^{48}Mn [23], for which the Argonne Fragment Mass Analyser (FMA) was used in conjunction with the Gammasphere array. The FMA has a combination of electric and magnetic dipoles which gives velocity selection to remove the beam particles and provides dispersion in A/Q – measured event-by-event from the horizontal position at the focal plane. Z-determination is achieved from energy-loss measurement in a gas-filled ionisation chamber after the focal plane.

The Z-identification in the ion chamber is essential for proton-rich spectroscopy. The reaction used for the ^{48}Mn experiment – ^{40}Ca(^{10}B,$2n$)^{48}Mn at a beam energy of 110 MeV – was ideal for this purpose as the recoiling ^{48}Mn nuclei were highly energetic (≈ 70 MeV) due to the inverse nature of the reaction. Thus, the different Z values in the ion chamber spectra were distinct and clean gates could be placed without the need for tricky background subtractions. The resulting gamma-ray spectra, requiring $A/Q \approx 3.0$ and Z identified in the ion chamber, are shown in Fig. 4. The $Z = 25$ (Mn) spectrum is shown in Fig. 4(a). This spectrum has also had additional requirements placed on recoil energy and time of flight to remove the charge-state ambiguities caused by low-level contamination in the beam and target. The gamma rays remaining here are all associated with ^{48}Mn, and a comparison with the ^{48}V spectrum in Fig. 4(b) shows the expected mirror symmetry. These spectra, and a subsequent gamma-ray coincidence analysis, resulted in the level scheme of ^{48}Mn shown in the left portion of Fig. 5, all of which were new. The energy differences for the main positive-parity structure of ^{48}Mn is discussed in Sect. 5.

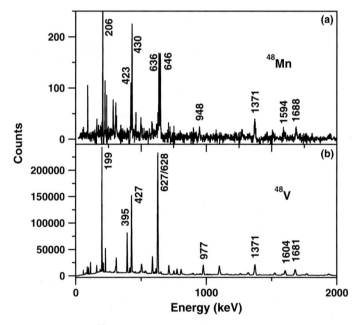

Fig. 4. Spectra from the ^{48}Mn experiment of [23]. (**a**) A spectrum requiring $A/Q = 3.0, Z = 25$ with additional requirements on the energy and time of flight of the recoiling nucleus – see text. Marked transitions are all assigned to ^{48}Mn. (**b**) A spectrum requiring $A/Q = 3.0, Z = 23$. The marked transitions are known to be in ^{48}V

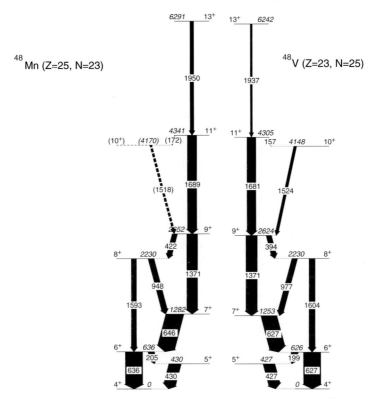

Fig. 5. (a) The level schemes of the yrast positive-parity states of ^{48}Mn (deduced in [23]) and of ^{48}V (taken from [40])

The final method to be discussed here also entails identification of the final nucleus, but through measuring the evaporated particles rather than the final residue to "tag" the observed gamma rays. To this end, detectors efficient in measuring evaporated protons, alphas and neutrons need to be employed. This technique is, in general, more flexible than recoil mass spectrometry, as many more channels can be studied. The disadvantage is that it is generally not as clean and, especially when neutron detection is required, can be less efficient overall. Charged-particle detection, for example, has been achieved using the ISIS array [24] and the EUCLIDES array [25], used in conjunction with the gamma-ray spectrometers GASP and Euroball, and the Microball array [26] used with Gammasphere. ISIS and EUCLIDES are 4π arrays of Si detector telescopes and the Microball array comprises a 4π array of CsI detectors. Additionally, neutron detection is essential, and arrays of neutron detectors are placed downstream of the target position and, in general, replace some of the forward-most detectors of the gamma-ray arrays. These are typically large-volume liquid scintillators coupled to photo-multiplier tubes. The Euroball Neutron Wall [27] comprised 50 such detectors, the Gammasphere

Neutron Shell [28] comprised 30 and the GASP n-Ring array [29] had six. This technique of combining gamma-ray, charged-particle and neutron detection has been very successfully applied to spectroscopy of isobaric analogue nuclei – such as for the $A = 50$ pair ^{50}Fe/^{50}Cr [30] and the $A = 45$ ^{45}V/^{45}Ti pair [31].

For all the studies described before, and for all the data presented in this work, the experiments have been performed using fusion–evaporation reactions. However, in the drive to access more exotic systems, a number of different reaction mechanisms are now being employed for in-beam gamma-ray spectroscopy for use with both stable and radioactive nuclear beams. These include fragmentation reactions, relativistic Coulomb excitation, transfer reactions and deep inelastic collisions. The first two of these, for example, are now widely employed at fragmentation facilities such as those available at GSI, GANIL and MSU, and use of these techniques is already affording rich opportunities for spectroscopy on the proton-rich side of the line of stability.

3.2 Shell-Model Developments

Parallel to the improvements in the experimental devices developed in the last decade, and discussed in Section 3.1, very important advances have been made in the theoretical techniques. The remarkable synergy between theoretical and experimental groups has allowed detailed studies of several physical properties of light- and medium-mass nuclei to be undertaken. Different interesting phenomena have been understood by means of shell-model calculations. Modern computational codes can deal with large bases and a large number of valence particles [32–34]. This means that not only nuclei near a closed shell, with single-particle behaviour, can be accounted for by the shell model, but collective phenomena, such as the rotation of a deformed nucleus, can be now reproduced with very good accuracy [35–37]. For this purpose the diagonalisation of the effective Hamiltonian has to be done in a large shell-model basis. The choice of the valence space is, however, limited by the capability of the computational procedure to deal with the dimension of the matrix to be diagonalised. Presently, exact large-scale shell-model calculations in the m-scheme, using the Lanczos method for the diagonalisation of the matrices, can cope with dimensions of the order of 10^9. This is the method used by the code ANTOINE, developed by the Strasbourg group [32, 33, 38]. Higher dimensions can be dealt with by quantum Monte Carlo techniques, such as the Monte Carlo Shell Model (MCSM) code developed by the theoretical nuclear group in Tokyo [34, 39].

A variety of effective interactions have been developed, and continue to be developed, to describe different mass regions in the table of isotopes. They are intimately related to the choice of the valence space, as they mock up the general Hamiltonian in the restricted basis. The $f_{\frac{7}{2}}$ shell, between the doubly magic nuclei ^{40}Ca and ^{56}Ni, constitutes a very special case, as it can be considered to be an isolated shell. This simplistic approximation allows straightforward predictions to be made [41]. However, it is clear that the $1f_{\frac{7}{2}}$ shell-model space is not sufficient to describe the spectroscopy of these nuclei with good accuracy – in particular the collective states – and that the rest of the fp orbitals, $2p_{\frac{3}{2}}$, $1f_{\frac{5}{2}}$ and $2p_{\frac{1}{2}}$, have to be taken into account in the calculations. In this respect, the most reliable interactions in the full fp valence space for the description of $f_{\frac{7}{2}}$-shell nuclei are FPD6 [42] and KB3G [43]. Recently, a new interaction, GXPF1 [44], has been introduced to describe nuclei in the whole fp main shell.

The USD interaction [45, 46], in the $d_{\frac{5}{2}}, s_{\frac{1}{2}}, d_{\frac{3}{2}}$ shell-model basis, gives a good description of the spectroscopy of positive-parity states of light sd-shell nuclei. An improved version, called USD05B, has been recently developed by B.A. Brown [47]. Beyond the middle of the shell, however, particle–hole excitations to the fp shell become important in the configuration of natural-parity states and are absolutely necessary for constructing negative-parity states. An exact calculation in the two main shells implies a large valence space and the dimensions of the matrices to be diagonalised become extremely large. Suitable truncations of the basis are therefore needed. For the upper sd shell, an effective interaction in the reduced valence space composed by the $s_{\frac{1}{2}} d_{\frac{3}{2}} f_{\frac{7}{2}} p_{\frac{3}{2}}$ orbitals, the $sdfp$ interaction, has been introduced by Caurier and collaborators [48]. This proves to give a good description, along the $N = Z$ line, of the spectroscopy of $A \sim 35$ nuclei such as ^{34}S [49] and ^{35}Cl [50], and for neutron-rich nuclei. Around $A = 30$, the closed shell at $N = Z = 14$ does not hold and excitations from the $d_{\frac{5}{2}}$ shell have to be considered. In these cases, the SDFP-M interaction [51] in the larger $d_{\frac{5}{2}} s_{\frac{1}{2}} d_{\frac{3}{2}} f_{\frac{7}{2}} p_{\frac{3}{2}}$ valence space can be used within the MCSM [34].

The reliability of shell-model calculations in describing the spectroscopy of medium-light nuclei, and in particular those in the $f_{\frac{7}{2}}$ shell, encouraged the extension of these calculations to the description of such small energy differences as the experimental MED and TED in isobaric multiplets. This constitutes a stringent test of the calculations due to the subtle details under examination.

In Section 4 we present some details on how the MED and TED are obtained in the shell-model framework. In Sects. 5 and 6 we compare the experimental data with the shell-model estimates. For these calculations we have used the code ANTOINE [32, 33]. The effective interactions used are the KB3G [43] for the $f_{\frac{7}{2}}$ shell, the $sdfp$ [48] for sd-shell nuclei, and the GXPF1 [44] for mirror nuclei above ^{56}Ni.

4 Description of Excitation Energy Differences

In the hypothesis of charge-symmetry and charge-independence of the nuclear force, differences in excitation energy between analogue states in mirror nuclei should be of purely electromagnetic origin. The Coulomb interaction only acts between protons, which induces an isospin dependence of the total interaction. The Coulomb field gives a contribution of the order of hundreds of MeV to the nuclear mass, and it is the Coulomb energy which plays the main role in the mass shifts between isobaric analogue states (CDE, see Sect. 2.3), which are of the order of tens of MeV. Other contributions to the CDE are the difference between proton and neutron masses and other minor effects of electromagnetic character. Compared with all these contributions to the CDE, isospin-breaking terms of the *nuclear* interaction are expected to be small [12].

When measuring the difference between *excited* states in isobaric multiplets, the large contributions due to the Coulomb field almost cancel out, as the ground states are normalised to zero excitation energy. Only small effects remain. In the $f_{\frac{7}{2}}$ shell, the measured energy differences between mirror nuclei (MED) amount to tens of keV and do not generally exceed 100 keV. Larger values (200–300 keV) have been encountered for some particular states in nuclei of the *sd* shell. For energy differences in $T = 1$ isobaric triplets in the $f_{\frac{7}{2}}$ shell, the measured TED values are smaller than 200 keV. Nevertheless, these rather small energy differences have demonstrated to act as a magnifying glass that highlights specific *nuclear* structure features. Moreover, if the Coulomb effects can be theoretically estimated, isospin-breaking effects due to the nuclear interaction could be revealed. In this section we describe the different contributions to the excitation energy differences, following the recent studies in the shell-model framework by Zuker and collaborators [16, 30, 52, 53].

4.1 Electromagnetic Effects

The Coulomb field or, more generally, the electromagnetic interaction is mainly responsible for differences in excitation energy between isobaric nuclei. This interaction yields several effects on the energy differences and, depending on the configuration of the states and on the mass region, some effects can be more evident than others. The possibility of having a rich quantity of good experimental data allows for a detailed study of these effects. In the next paragraphs we describe the several terms that contribute to the excitation energy differences. Following the formalism developed in [38, 52], the effective shell-model Hamiltonian is divided into a monopole plus a multipole component.

Multipole Coulomb term and nucleon alignment

The first steps in extending Coulomb energy differences between mirror nuclei to high-spin states were undertaken by Cameron et al. in nuclei in the $f_{\frac{7}{2}}$ shell at Daresbury [54]. They studied the mirror nuclei ^{49}Mn/^{49}Cr which are located in the middle of the shell and present stable deformation near the ground state. The MED of their ground-state rotational bands were measured up to the $J^{\pi} = \frac{19}{2}^{-}$ state. Small values were obtained for the MED at low

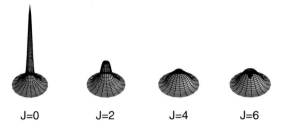

J=0   J=2   J=4   J=6

Fig. 6. A calculation of the probability distribution for the relative distance of two like-particles in the $f_{\frac{7}{2}}$ shell as a function of their coupled angular momentum. The calculations were undertaken in [55]. The centre of each plot corresponds to zero separation

spin, but at $J^\pi = \frac{17}{2}^-$ a substantial change was observed – the MED increased significantly for the highest two states. The enhancement of the MED was interpreted in terms of the *alignment* of nucleons along the rotational bands.

This results from the fact that the Coulomb interaction between two protons coupled in time-reversed orbits is larger than for any other coupling, as the spatial overlap of their orbits is maximum. This can be seen in Fig. 6, which demonstrates how the average separation of two like-particles in the $f_{\frac{7}{2}}$ shell increases with their coupled angular momentum [55]. Thus, when two protons coupled to $J = 0$ re-couple their angular momenta, the Coulomb energy decreases. In particular, when a pair of protons aligns to the maximum value $(2j-1)$ in a single j shell, the Coulomb energy between them reaches its minimum value as their spatial separation is largest. As the Coulomb interaction is repulsive, the effect of the alignment reduces the excitation energy of the nuclear state. Of course, the alignment of any pair of nucleons along a rotational band causes changes in the energy sequence, which is called back-bending as, in a spin vs. transition energy plot, the smooth behaviour is interrupted due to a decrease in the transition energy (see Fig. 7 for ^{48}Cr). This nuclear effect, however, will be equal in both mirror partners in the hypothesis of isospin symmetry of the nuclear interaction. On the other hand, only in the nucleus where the proton pair aligns will the Coulomb effect occur. Due to the isospin symmetry, in the other mirror partner, a pair of *neutrons* will align at the same state – without any Coulomb effect of course. Thus, just by looking at the experimental MED of a rotational mirror pair, one would be able to deduce which type of nucleons are aligning at the back-bend. A significant increase (decrease) of the MED would mean that a neutron (proton) pair is aligning in the proton-rich nucleus and, consequently, a proton (neutron) pair in its mirror partner.

The alignment process can be computed in the shell-model framework, by *counting* the number of proton pairs in a j orbital coupled to the maximum spin $(J = 2j - 1)$. In the $f_{\frac{7}{2}}$ shell, this can be performed by calculating the expectation value of the operator $A_\pi = [(a_\pi^+ \, a_\pi^+)^{J=6} \, (a_\pi \, a_\pi)^{J=6}]^0$ for each

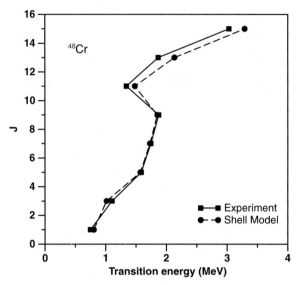

Fig. 7. Transition energies in the ground-state rotational band of ^{48}Cr ($\hbar\omega = \frac{E_\gamma}{2}$). Experimental data are taken from [36]; calculations are performed with the code ANTOINE [32, 33] and the effective interaction KB3G [43] in the full fp valence space

excited state of the rotational band. Doing this for both nuclei, one can then calculate the difference $\Delta A_\pi = A_\pi(Z_>) - A_\pi(Z_<)$ for the mirror pair, as a function of the angular momentum. If the alignment of a pair of protons in the nucleus with charge $Z_>$ – and, consequently, the alignment of a pair of neutrons in the $Z_<$ – occurs first, ΔA_π will increase, whilst it will decrease if the opposite happens. This was introduced by Poves and Sánchez-Solano [56], the results of which are presented in Fig. 8 for the mirror pair $A = 51$. The experimental MED$_J$ = $E_J^*(^{51}\text{Fe}) - E_J^*(^{51}\text{Mn})$ are shown in the first panel, while the lower panel shows the behaviour of $-\Delta A_\pi$. The similarity of the MED with the "quasi-alignment" $(-\Delta A_\pi)$ is impressive. Indeed, this pair of mirror nuclei constitutes a very particular case which can help to illustrate the effect. We take for example the nucleus ^{51}Fe, with six protons and five neutrons in the $f_{\frac{7}{2}}$ shell. As the odd neutron blocks the alignment of a neutron pair, the first alignment will be due to a pair of protons (to $J = 6$). Once this pair aligns, there are no more possibilities, within the $f_{\frac{7}{2}}$ shell, for proton re-coupling to increase the angular momentum of the nucleus, therefore, the neutrons will re-couple until the band termination is reached. The net decrease of the MED at $J^\pi = \frac{17}{2}^-$ in Fig. 8 corresponds to the alignment of two protons to the maximum spin $J = 6$ in ^{51}Fe and, consequently, a neutron alignment in its mirror, ^{51}Mn. At larger spins, the other type of nucleons start to re-couple and the curve increases until it reaches the band termination at $J^\pi = \frac{27}{2}^-$.

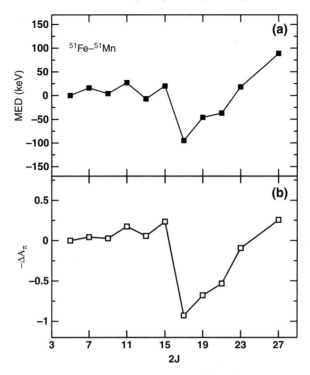

Fig. 8. MED and alignment for the mirror pair ^{51}Fe–^{51}Mn as a function of the angular momentum [56] (more up-to-data data on the MED is presented in Sect. 5). (a) The experimental MED. (b) The calculated "alignment" in the shell-model framework – see text for details

It is important to note here that the alignment process in nuclei can be computed in a straightforward way by *cranked* shell-model calculations [57, 58]. These calculations have been applied to several nuclei in the $f_{7/2}$ shell and a good qualitative description of the experimental Coulomb energy differences was achieved. While the cranked shell model calculates the alignment i_x along the rotational axis x, ΔA_π, obtained within the spherical shell model, indicates that there is an alignment but cannot distinguish between rotational alignment along the x axis (RAL), or deformation alignment (DAL) along the symmetry axis z. This subject was raised by Brandolini et al. [59] where it was suggested, for example, that the backbending in ^{50}Cr and ^{50}Fe is due to the crossing of the $K^\pi = 0^+$ ground state band with a $K^\pi = 10^+$ band – indicating a DAL process for the $A = 50$ mirror nuclei.

A better quantitative description was first obtained in the shell-model framework by G. Martínez-Pinedo and A. Poves [60], for the $T_z = \pm\frac{1}{2}$ mirror pairs of mass $A = 47$ and 49, where Coulomb matrix elements, calculated in the harmonic-oscillator basis for the whole fp main shell, were added to the KB3 [61] effective interaction. However, those calculations did not describe the experimental MED very well. To improve agreement with the data, Coulomb matrix elements had to be modified. In particular, those corresponding to the $f_{\frac{7}{2}}$ shell were replaced by the experimental values of mass $A = 42$ – a set of "effective" Coulomb matrix elements. With this interaction, the description of the experimental data for the mirror pairs $A = 47$ and 49 improved. However, further experimental data for heavier $T_z = \pm\frac{1}{2}$ and ± 1 mirror nuclei demonstrated the inadequacy of the "effective" Coulomb matrix elements for a quantitative reproduction of the MED values [30, 56]. It was then clear that the multipole Coulomb interaction was not enough to account for the experimental findings.

Monopole Coulomb contributions: the radial term

An important contribution to the comprehension of the origin of the MED was given by A. Zuker in the work described in [30]. It was pointed out that, in addition to the multipole Coulomb interaction between valence protons, there was a small but significant *monopole* Coulomb effect due to the change of the nuclear radius along the rotational band.

As discussed in the previous paragraph, the mechanism of generating angular momentum by aligning the valence-particle spins in a high-j orbit becomes energetically favoured with increasing rotational frequency. In $f_{\frac{7}{2}}$-shell nuclei, the occupation of orbits different from the $f_{\frac{7}{2}}$, important to create collective states near the ground state, decrease along the rotational bands producing changes of the nuclear radius. This affects the MED as valence protons in orbitals with smaller radii are *nearer* to the charged core and have *more* Coulomb energy.

Following [30, 52], the monopole Coulomb contribution to the MED can be deduced by considering the Coulomb energy of a uniformly charged sphere of radius R_C

$$E_C = \frac{3}{5}\frac{Z(Z-1)e^2}{R_C} \tag{13}$$

The difference between the energy of the ground states of $T_z = \pm\frac{k}{2}$ mirror nuclei $(Z_> = Z_< + k, Z = Z_>)$ is

$$\Delta E_C = E_C(Z_>) - E_C(Z_<) \simeq \frac{3}{5}\frac{k(2Z-k)e^2}{R_C}. \tag{14}$$

This energy difference amounts to tens of MeV and is the main ingredient in the evaluation of CDE (see Sect. 2.3 and [16]). When calculating the MED for each state of spin J as a function of the angular momentum, we refer the MED(J) values to the ground state. On doing so, the monopole effect of ΔE_C almost vanishes. A small contribution remains, however, due to the change in charge radius with the angular momentum, as discussed above. The monopole Coulomb radial contribution to the MED can thus be written:

$$\Delta_M < V_{Cr}(J) >= \Delta E_{\rm C}(J) - \Delta E_{\rm C}(0) = -\frac{3}{5}k(2Z-k)e^2\frac{\Delta R(J)}{R_{\rm C}^2}, \qquad (15)$$

where Δ_M is the MED and $\Delta R(J) = R_{\rm C}(J) - R_{\rm C}(0)$ and we assume, following Refs. [16, 30, 52], that it is *the same in both mirror nuclei*.

Nuclei that lie near the middle of the $f_{\frac{7}{2}}$ shell are well deformed at low spin. These states are more collective than the high-spin members of the rotational band and the wave functions have an important contribution of the $p_{\frac{3}{2}}$ orbit. In fact, it is the coupling between the $f_{\frac{7}{2}}$ and the $p_{\frac{3}{2}}$ orbits what gives rise to the quadrupole collectivity in this mass region [62]. With increasing angular momentum, the yrast bands evolve by progressively aligning the valence nucleons in the $f_{\frac{7}{2}}$ shell up to the band terminating state. On doing that, the occupation of the $p_{\frac{3}{2}}$ orbit decreases and the bands terminate in non-collective, high-spin states with all the valence particles in the $f_{\frac{7}{2}}$ shell. The role of the other orbits, $f_{\frac{5}{2}}$ and $p_{\frac{1}{2}}$, is less important, and does not change very much as a function of the angular momentum.

In the fp shell, p orbits have larger radius than f orbits and therefore, the Coulomb repulsion increases as the protons pass from the $p_{\frac{3}{2}}$ to the $f_{\frac{7}{2}}$ orbit. In other words, at high spin, when all nucleons are filling the $f_{\frac{7}{2}}$ shell, the monopole Coulomb contribution is larger than at low spin, where there is a significant $p_{\frac{3}{2}}$ contribution to the wavefunction. How to account for this effect in the shell-model framework will be discussed in Sect. 4.3, together with the other terms. It is important to note that the effect of the change of deformation in the MED was also introduced in [59] and calculated within the liquid drop model.

Single-particle corrections

The single-particle energies of protons and neutrons are modified by the monopole electromagnetic field in different ways [12, 16]. In [16], Duflo and Zuker show that the contribution of the monopole Coulomb interaction to the CDE can be expressed as the energy of a charged sphere (Eq. (13)) with single-particle corrections that account for shell effects. They affect the energy of the proton orbits proportionally to the square of the orbital momentum l in the harmonic-oscillator representation. The expression for the single-particle splittings for a proton in a main shell, with principal quantum number N, above closed shell $Z_{\rm cs}$ results [16]

$$E_{ll} = \frac{-4.5Z_{\rm cs}^{13/12}[2l(l+1) - N(N+3)]}{A^{1/3}(N+\frac{3}{2})}\text{keV}. \qquad (16)$$

The effect on the single-particle energies is sizable. In ^{41}Sc ($Z_{\rm cs}{=}20$, $N{=}3$), proton f orbits are lowered by $\sim$45 keV while the energy of p orbits is raised by $\sim$105 keV with respect to the neutron levels. The relative energy between the proton $f_{\frac{7}{2}}$ and $p_{\frac{3}{2}}$ orbitals is therefore increased by $\sim$150 keV with respect to the neutron energy difference.

Another interaction that affects the single-particle energies is the relativistic *electromagnetic* spin–orbit force (EMSO) [12, 63]. This interaction, analogous to the atomic case, results from the Larmor precession of the nucleons in the nuclear electric field due to their intrinsic magnetic moments and to the Thomas precession experienced by the protons because of their charge. The effect of the *nuclear* spin–orbit hamiltonian in the single-particle spectrum is very well known. It amounts to several MeV and acts on both protons and neutrons. The EMSO effect is almost two orders of magnitude smaller than the nuclear spin–orbit potential and has been in general ignored in MED calculations. However, as it acts differently on neutrons than on protons, its effect does not cancel when computing MED values and can become very important for some particular states. In [64] Trache and collaborators consider the EMSO interaction to calculate the single-particle energies in the $A = 57$, $T_z = \pm\frac{1}{2}$ mirror pair and in a recent work [53], Ekman et al. discuss the importance of the EMSO contribution to the large MED value observed at the yrast $J^\pi = \frac{13}{2}^-$ state in the mirror nuclei ^{35}Ar/^{35}Cl.

The general expression of the electromagnetic spin–orbit potential [12, 63] is:

$$V_{ls} = (g_s - g_l)\frac{1}{2m_N^2 c^2}\left(\frac{1}{r}\frac{dV_C}{dr}\right)\vec{l}\cdot\vec{s}, \qquad (17)$$

where g_s and g_l are the gyromagnetic factors, V_C is the Coulomb potential and m_N is the nucleon mass. The term proportional to g_s is the Larmor term. It can be deduced by considering the potential energy of a spin magnetic moment μ_s in an effective magnetic field due to its motion in the electric field generated by the protons in the nucleus. The second term in Eq. (17), proportional to g_l, is the relativistic Thomas term associated with the orbital magnetic moment μ_l, that vanishes in the neutron case.

A rough estimate of the energy shift produced by the relativistic electromagnetic spin–orbit term has been given by Nolen and Schiffer [12] assuming that V_C is generated by a uniformly charged sphere of radius R_C

$$E_{ls} \simeq (g_s - g_l)\frac{1}{2m_N^2 c^2}\left(-\frac{Ze^2}{R_C^3}\right)\langle\vec{l}\cdot\vec{s}\rangle. \qquad (18)$$

Using, for example, the free values of the gyromagnetic factors $g_s^\pi = 5.586$, $g_l^\pi = 1$ and $g_s^\nu = -3.828$, $g_l^\nu = 0$ for the proton and the neutron, respectively, it is easy to see that the energy shift will have different sign for a proton orbit than for a neutron one. The sign will also depend on the spin–orbit coupling, as $\langle\vec{l}\cdot\vec{s}\rangle = l/2$ when $j = l + s$ and $\langle\vec{l}\cdot\vec{s}\rangle = -(l + 1)/2$ when $j = l - s$. To illustrate the effect, we take two particular orbits, the $f_{\frac{7}{2}}$ and the $d_{\frac{3}{2}}$ which are involved in excited states of nuclei in the upper *sd* shell (see Sect. 6). The effect of the EMSO is to reduce the energy gap between the proton orbitals by ~120 keV and to increase it for neutrons by roughly the same amount. Therefore, in one nucleus the energy of a state whose configuration involves the excitation of one proton from the $d_{\frac{3}{2}}$ to the $f_{\frac{7}{2}}$ will be smaller than that of the analogue state in its mirror nucleus where a neutron undergoes the excitation. The MED for such states will reach large values. On the contrary, small MED will be obtained whenever the configuration of the state involves the excitation of one proton or one neutron with similar probabilities, as the effect is compensated.

4.2 The Isospin-Breaking "Nuclear" Term

The identification of the different electromagnetic terms that enter in the calculation of the MED and TED has allowed to improve the description of the data. Nevertheless, it was clear that certain renormalisations were needed to get a good fit. This was discussed in [30] where a single renormalisation seemed to be adequate to describe the data on MED so far available. However, this renormalisation of the Coulomb interaction could not account for new data on TED. A satisfactory solution was then proposed by Zuker, as described in [52]. Analysing the data for the $A = 42$ isobaric triplet, it was shown that the MED and TED values could not be reproduced by just considering the electromagnetic interaction and another isospin non-conserving term was thus called into play. We will not enter into technical details here but just outline the main points; a full description is given in [52].

Let us consider the yrast states $J = 0, 2, 4, 6; T = 1$ in the three isobaric nuclei ^{42}Ti, ^{42}Sc and ^{42}Ca and assume that they have essentially $f_{\frac{7}{2}}^2$ configurations. Therefore, a two-body effective interaction in the $f_{\frac{7}{2}}$ shell can be obtained from the experimental data, using Eqs. (11) and (12). In particular, the MED values account for the isovector term of the interaction whilst the TED give the isotensor component as follows,

$$\mathrm{MED}_J(A = 42, T = 1) = V^{(1)}_{\mathrm{CM}, f_{\frac{7}{2}}}(J) + V^{(1)}_{B, f_{\frac{7}{2}}}(J)$$

$$\mathrm{TED}_J(A = 42, T = 1) = V^{(2)}_{\mathrm{CM}, f_{\frac{7}{2}}}(J) + V^{(2)}_{B, f_{\frac{7}{2}}}(J), \qquad (19)$$

where, in addition to the Coulomb term, an isospin non-conserving term V_B is considered. It is reasonable to neglect changes of deformation and single-particle effects along this yrast sequence and therefore, only the multipole part of the Coulomb interaction will contribute (V_{CM}). The Coulomb term can be calculated for two protons in the $f_{\frac{7}{2}}$ shell (Table 1) and, as in [52], we choose the harmonic-oscillator basis for the calculation of the matrix elements, $V^{(1)}_{\mathrm{CM}, f_{\frac{7}{2}}}(J) = V^{(2)}_{\mathrm{CM}, f_{\frac{7}{2}}}(J) = V^{\mathrm{ho}}_{\mathrm{CM}, f_{\frac{7}{2}}}(J)$. By subtracting these terms from the MED and TED data in Eq. (19), in the hypothesis of isospin symmetry and independence, respectively, the contribution of $V^{(1)}_{B, f_{\frac{7}{2}}}$ and $V^{(2)}_{B, f_{\frac{7}{2}}}$ for all J values should be negligible. The numbers, reported in Table 2 [52] demonstrate that these values are not small, but of the same order of magnitude of the Coulomb matrix elements.

Table 1. Coulomb matrix elements for two protons in the $f_{\frac{7}{2}}$ shell. All values are in keV

	$J = 0$	$J = 2$	$J = 4$	$J = 6$
ho	399	341	309	304
Exp.	391	421	315	245
Fit (47–49)	330	393	331	247

   This means that the (ISB) interaction cannot be ignored in the calculation of the excitation energy differences. The key point relates to the relative values of the matrix elements for the different J couplings. The maximum value is obtained at $J = 0$ for the isotensor matrix elements while for the isovector components the peak is obtained for two $f_{\frac{7}{2}}$ nucleons coupled to $J = 2$. This effect was already noted in early studies performed by Brown and Sherr [65] but the origin of this charge-dependent interaction is still an open question. This was considered of nuclear nature in [52], however, contributions from the renormalisation of the Coulomb interaction could also contribute. This is discussed further in Sect. 5.

   The empirical ISB matrix elements of Table 2, obtained in the $f_{\frac{7}{2}}$ orbit, have to be generalised to the whole fp shell to be used in the shell-model calculations of the MED and TED for other masses. The extension to the main shell is, however, not straightforward. As shown in [52], a multiplicative prescription, where the interaction in the fp shell consists of just $f_{\frac{7}{2}}$ matrix elements scaled by an overall factor, can be viable in some cases. The extremely simple ansatz proposed in [52] consists of constructing a ISB hamiltonian in the fp shell by just taking one $f_{\frac{7}{2}}^2$ matrix element with a strength determined by the data for $A = 42$,

Table 2. $f_{\frac{7}{2}}$ Coulomb matrix elements calculated in the harmonic-oscillator basis $V^{ho}_{CM,f_{\frac{7}{2}}}$; and the isospin non-conserving isovector ($V^{(1)}_{B,f_{\frac{7}{2}}}$) and isotensor ($V^{(2)}_{B,f_{\frac{7}{2}}}$) terms deduced from the experimental data in $A = 42$. The centroids $V_{centr} = \frac{\sum_J V_J \times (2J+1)}{\sum_J 2J+1}$ have been subtracted from the matrix elements. All values are in keV

	$J = 0$	$J = 2$	$J = 4$	$J = 6$
$V^{ho}_{CM,f_{\frac{7}{2}}}$	83	25	−6	−12
$V^{(1)}_{B,f_{\frac{7}{2}}} = \mathrm{MED}(A = 42) - V^{ho}_{CM,f_{\frac{7}{2}}}$	5	93	17	−48
$V^{(2)}_{B,f_{\frac{7}{2}}} = \mathrm{TED}(A = 42) - V^{ho}_{CM,f_{\frac{7}{2}}}$	117	81	3	−43

$$V_{B,fp}^{(1)} = \beta_1 V_{B,f_{\frac{7}{2}}}(J = 2)$$

$$V_{B,fp}^{(2)} = \beta_2 V_{B,f_{\frac{7}{2}}}(J = 0), \qquad (20)$$

where $V_{B,f_{\frac{7}{2}}}(J)$ are matrix elements with unit value. The choice of $V_{B,f_{\frac{7}{2}}}(J = 2)$ for the isovector and $V_{B,f_{\frac{7}{2}}}(J = 0)$ for the isotensor components is based on the leading terms in Table 2. In [52], the strengths used were $\beta_1 = \beta_2 = 100$ keV. The results of this simple ansatz are very successful, as the same interaction can be used to reproduce with good accuracy the data on MED and TED for all the $f_{\frac{7}{2}}$-shell isobaric multiplets measured so far (see Sect. 5).

4.3 Calculation of MED and TED

Although MED are extremely sensitive to the details of isospin non-conserving effects, in performing shell-model calculations, they can be treated in first-order perturbation theory. To appreciate the weight of the single effects on the MED and TED, it is useful to perform a diagonalisation of the nuclear effective interaction and to calculate the contribution of the different terms by means of the expectation values,

$$\text{MED}(J) = \Delta_M < V_{\text{Cr}}^J + V_{\text{CM}}^J + V_B^{(1,J)} + V_{ll}^J + V_{ls}^J > \qquad (21)$$

$$\text{TED}(J) = \Delta_T < V_{\text{CM}}^J + V_B^{(2,J)} >, \qquad (22)$$

where Δ_M means the difference between the mirror nuclei (Eq. (11)) and Δ_T stands for the difference in the triplet (Eq. (12)).

Changes in the charge radius can be accounted for, within the shell model, by considering the evolution of the occupation numbers of the different orbits as a function of the angular momentum along the yrast bands [30]. As stated above, for nuclei in the $f_{\frac{7}{2}}$ shell, it is the relative occupation of the $p_{\frac{3}{2}}$ orbit which determines the main changes of radii. The assumption of equal radii of the mirror partners means that a calculation of the *average* of proton and neutron occupation numbers of the $p_{\frac{3}{2}}$, m_π and m_ν, respectively, is required: $\frac{m_\pi + m_\nu}{2}$. The contribution of the monopole Coulomb radial term to the MED of mirror nuclei with $|T_z| = k/2$ can be parameterised as,

$$\Delta_M < V_{\text{Cr}}^J > = k \, \alpha_r \left(\frac{m_\pi(0) + m_\nu(0)}{2} - \frac{m_\pi(J) + m_\nu(J)}{2} \right), \qquad (23)$$

where the constant α_r can be deduced from the single-particle relative energies in mass $A = 41$, as discussed below. It is important to note, however, that the radial term in the calculation of the MED is not a single-particle effect and therefore it cannot be accounted for by setting different single-particle energies for protons and neutrons in the shell-model calculation. The radial contribution vanishes in the calculation of the TED [52].

Monopole Coulomb single-particle contributions to the MED are proportional to the *difference* between the proton and the neutron occupation numbers. This difference is small in most of the $T_z = \pm\frac{1}{2}$ and $T_z = \pm 1$ mirror nuclei studied in the $f_{\frac{7}{2}}$ shell. On the other hand, for nuclei in the upper sd shell, the promotion of nucleons to the fp shell becomes important already at low spin and excitation energy. In these cases, the difference of occupation numbers between protons and neutrons are significant and single-particle Coulomb effects can become important.

The energy shifts between the different orbits E_{ll} and E_{ls} can be calculated using Eqs. (16) and (18). Once the relative energies are known, the value of α_r in Eq. (23) can be estimated. As an example, we calculate the shift between the $p_{\frac{3}{2}}$ and the $f_{\frac{7}{2}}$ orbits due to the monopole Coulomb radial effect. The electromagnetic spin–orbit interaction lowers the proton $f_{\frac{7}{2}}$ orbit by about 120 keV with respect to the corresponding neutron orbit, while the effect on the proton $p_{\frac{3}{2}}$ single-particle level is a decrease of about 40 keV. Adding this effect to that of the E_{ll} shift, we find that the energy difference between the $p_{\frac{3}{2}}$ and the $f_{\frac{7}{2}}$ single-particle orbits is $\sim +200$ keV larger for protons than for neutrons. Experimentally, by comparing the spectrum of ^{41}Sc and ^{41}Ca , the resulting energy difference is ~ -200 keV. This means that the effect due to the radial term is of the order of $\alpha_r \sim 400$ keV.

5 Isobaric Multiplets in the $f_{\frac{7}{2}}$ Shell

In Sect. 4 we have seen how the state-of-the-art shell-model calculations can be employed to model Coulomb (and other isospin non-conserving) contributions to energy shifts between excited IAS. In this section we will discuss some of the experimentally observed shifts in the $f_{\frac{7}{2}}$ shell, and demonstrate how they provide an extremely rigorous test of the model predictions on a state-by-state basis.

We begin the discussion of experimental results with MED in the $f_{\frac{7}{2}}$ shell. In all cases shown, a comparison of the analogue states up to the $f_{\frac{7}{2}}$-shell band termination has been achieved. The band-terminating state has the maximum angular momentum allowed assuming a pure $f_{\frac{7}{2}}$ valence configuration. This terminating angular momentum is largest for nuclei in the centre of the shell, reducing as the closed shells are approached. The diagrams show experimental MED as a function of spin, as defined by Eq. (11). Error bars are not shown on the data points as they are generally smaller than the symbols as plotted. Each experimental curve is accompanied by a prediction from the full fp shell model as described in Sect. 4.3. The shell-model predictions are broken down into four separate components corresponding to the different terms of Eq. (21). In the diagrams, CM is the Coulomb multipole contribution describing the changing Coulomb effect of different angular momentum couplings for pairs of protons. The multipole term VB takes account of the isospin non-conserving phenomena (Sect. 4.2). This will be discussed later in this section.

The Cr term is the monopole radial term and takes account of changes in radius/deformation along the yrast line as a function of spin. Finally, the $E_{ll} + E_{ls}$ term is the sum of the two single-particle contributions described at the end of Sect. 4.1. The calculations were performed with a consistent set of parameters (e.g. the coefficients of the VB and Cr terms), the justification for which can be found in [66]. Coulomb matrix elements for the CM term were taken from the harmonic oscillator. The two single-particle contributions, E_{ll} and E_{ls}, do not contribute significantly to the MED for yrast states in $f_{\frac{7}{2}}$-shell nuclei (except for states above the band termination) and, although they are included in the calculations, are not discussed in detail here. Their effect is more significant in the sd and upper fp shells – see Sect. 6. The cases chosen illustrate effectively the three other main effects in turn: re-coupling of angular momenta of pairs of protons, changes in nuclear radius with increasing spin and the effect of the additional isospin non-conserving contribution for $J = 2$.

We start the discussion of the odd-A mirrors in the shell with the $A = 49$ and 47 mirror pairs, $^{49}_{25}\mathrm{Mn}_{24}/^{49}_{24}\mathrm{Cr}_{25}$ and $^{47}_{24}\mathrm{Cr}_{23}/^{47}_{23}\mathrm{V}_{24}$ – those closest to the centre of the shell. The data shown here comes from the latest study of these pairs [67, 68]. For the four $f_{\frac{7}{2}}$ protons and three $f_{\frac{7}{2}}$ neutron holes in $^{49}\mathrm{Cr}$, the maximum valence angular momentum is determined by considering the maximum projections allowed by the Pauli principle – i.e. $\Omega_\pi = \frac{7}{2} + \frac{5}{2} + \frac{3}{2} + \frac{1}{2}$ and $\Omega_\nu = \frac{7}{2} + \frac{5}{2} + \frac{3}{2}$. Thus $J_{\max} = \frac{31}{2}$ and the band-terminating state is therefore $J^\pi = \frac{31}{2}^-$. The MED for both pairs, up the band-termination, are shown in Fig. 9.

The experimental MED for $A = 49$ in Fig. 9(a) shows a smooth variation with increasing J, the largest effect of which is the large rise in the MED at around $J^\pi = \frac{19}{2}^-$. The effect is well understood (see Sect. 4.1) since nuclei near the centre of the shell demonstrate some collectivity (e.g. [69]). At around $J^\pi = \frac{19}{2}^-$ an alignment of a pair of protons occurs in $^{49}\mathrm{Cr}$ – the alignment of neutrons being blocked by the unpaired $f_{\frac{7}{2}}$ neutron. As the protons align from $J = 0$ to the maximum allowed $J = 6$, there is a reduction in their spatial overlap, with a simultaneous reduction in the Coulomb energy. Mirror symmetry dictates that the alignment in $^{49}\mathrm{Mn}$ must be a neutron alignment, but here there is no Coulomb effect of course. Thus we have a difference in variation of Coulomb energy with J, and the peak in the MED occurs. To generate further spin, as the band-terminating state is approached, the alignment of the other particle type is required (recall that all particles are maximally aligned at the band-termination), and the effect reverses leading to a reduction in the MED at high spin. The major changes in the MED are understood this way.

The $A = 47$ nuclei are the *cross-conjugate* nuclei of the $A = 49$ pair. Cross conjugacy is a symmetry of a single-j shell model – i.e. where the valence space is restricted to one single shell-model level. In this simplistic model, cross-conjugate partners are those defined by a simultaneous exchange of particles

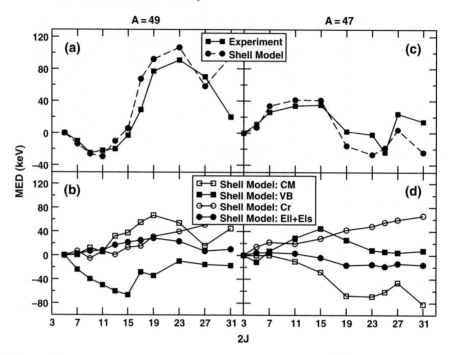

Fig. 9. The experimental and predicted MED for the $T_z = \pm\frac{1}{2}$ mirror pairs with $A = 49$ (*left*) and 47 (*right*). The experimental MED values are shown in (**a**) and (**c**) along with the predictions of the shell model – see text for details. In (**b**) and (**d**) are shown the four contributions to the shell-model prediction that sum to give the total calculated MED in (**a**) and (**c**), respectively. CM is the Coulomb multipole term, VB is the isospin non-conserving contribution, Cr is the radial monopole effect and "$ll + ls$" is the sum of the two single-particle corrections described in Sect. 4.1

for holes and protons for neutrons, and the wave functions of the predicted states of cross-conjugate partners are identical. For example, ^{49}Cr (four proton particles and three neutron holes in $f_{\frac{7}{2}}$) is the cross conjugate of ^{47}V. Of course, this is only an approximate symmetry in reality, although it retains some validity in the $f_{\frac{7}{2}}$ shell which is somewhat isolated from other shell-model orbitals and where the wave functions of yrast low-energy states are dominated by $f_{\frac{7}{2}}$ configurations. The upshot of this is that if cross-conjugate symmetry is valid, the multipole Coulomb effects associated with changes in angular momentum coupling should be essentially identical but have the opposite sign in $A = 47$ than for their cross-conjugate partner pair with $A = 49$. Figure 9(c) shows that this is approximately the case. The alignment effect at $J^\pi = \frac{19}{2}^-$ in $A = 47$ is less marked, but clearly present, as is the reversal of the effect as the band-termination is approached.

The agreement with the shell model is impressive in both cases, and its clear from Fig. 9(b) and (d) that all components of the model are required to reproduce the trends of the experimental data. The Coulomb multipole term, as expected, reflects the alignment effect described before, with the expected effect at around $J^\pi = \frac{19}{2}^-$ reversing towards the band-termination. The other terms, however, are also important and will be discussed in more detail in the context of other mirror pairs in the discussion to follow.

The multipole effect of the re-coupling of proton angular momenta (CM in the model prediction) is seen strikingly in the next examples, shown in Fig. 10, beginning with the $A = 50$ $T = 1, T_z = \pm 1$ mirror pair $^{50}_{26}\text{Fe}_{24}/^{50}_{24}\text{Cr}_{26}$ [30]. With ten particles in the $f_{\frac{7}{2}}$ shell, the band-termination is at $J^\pi = 14^+$. To date, ^{50}Fe has been observed up to $J^\pi = 11^+$ and the MED for yrast structures of $^{50}\text{Fe}/^{50}\text{Cr}$ up to this state are shown in Fig. 10(a) and (b). Again, we consider the effect of alignment of pairs of particles. Without the blocking effect present in the odd-A $T_z = \pm\frac{1}{2}$ mirror nuclei, both types of particle alignment are possible in even–even mirrors, and the trend of the MED will be sensitive to the details of which alignment occurs first (i.e. at the lowest J). In this case, calculations [58] indicate that, in ^{50}Fe, the neutrons should align first (i.e. *protons* in ^{50}Cr), followed by the alignment of the other kind of particle at $J^\pi = 10^+$ in each case. This can be deduced from the CM curve in Fig. 10(b). However, the contributions from the monopole Coulomb and the ISB VB terms are equally important and give together a good description of the data.

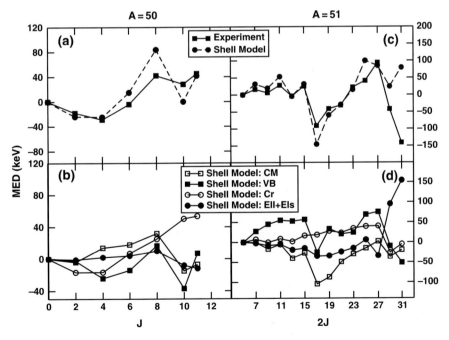

Fig. 10. The experimental and predicted MED for the $T_z = \pm 1, A = 50$ mirror pair (*left*) and the $T_z = \pm\frac{1}{2}$ mirror pair with $A = 51$ (*right*). See caption to Fig. 9 for details

Considering $A = 51$ mirror nuclei, the level schemes of the yrast negative-parity structures (up to the $J^\pi = \frac{27}{2}^-$ band-termination are shown in Fig. 11. The mirror symmetry of the level energies and the intensities of the gamma-ray transitions (indicated by the width of the *arrows*) is obvious. The MED for these states [56, 70] and for some core-excited states beyond the $f_{\frac{7}{2}}$

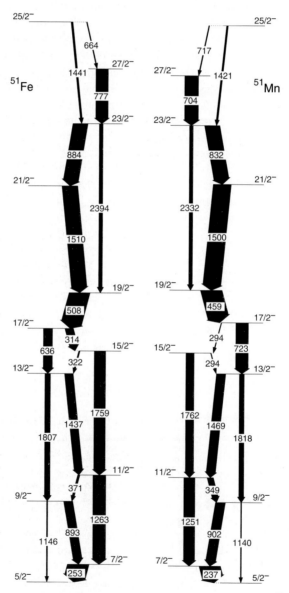

Fig. 11. Partial level schemes of the yrast negative-parity states of ^{51}Fe (data taken from [70]) and of ^{51}Mn (data taken from [72])

band-termination [71] are shown in Fig. 10(c). The most striking feature is the large dip in the MED at $J^\pi = \frac{17}{2}^-$. The explanation is again straightforward in terms of angular momentum alignment. ^{51}Fe has two $f_{\frac{7}{2}}$ proton holes and three $f_{\frac{7}{2}}$ neutron holes. The blocking effect requires that the pair of proton holes align first. At $J^\pi = \frac{17}{2}^-$, a favoured configuration is formed, similar to the $J^\pi = \frac{5}{2}^-$ ground state, but with the proton–hole pair re-coupled from $J = 0$ to 6. The same state in ^{51}Mn is formed from a neutron–hole alignment, resulting in the sharp dip in the MED. Again, to generate further angular momentum, alignment of the other particle type is required, and the MED effect reverses. Figure 10(d) again shows that the CM term reflects this alignment very clearly although, as with the other cases discussed, the other components to the calculation are also important.

The next contribution to be discussed is the radial term (Cr in the model). This effect is particularly striking in the odd–odd $T = 1$ mirror pair, ^{48}Mn/^{48}V [23]. The MED have been established from the $J^\pi = 4^+$ ground state to $J^\pi = 13^+$ (the band-termination being at $J^\pi = 15^+$). The fact that these nuclei are "mid-shell" has an important effect here. In a pure $f_{\frac{7}{2}}$ valence space, then both ^{48}Mn and ^{48}V have three active valence protons (holes and particles, respectively) and three active valence neutrons (particles and holes, respectively). As a result the multipole effects (associated with angular momentum re-coupling of the valence particles) in each nucleus must be identical. Thus, even though a pure $f_{\frac{7}{2}}$ structure is unrealistic, we still expect the (usually dominant) multipole contributions to the MED variations to be much reduced. This is seen clearly in the multipole components of the model (CM and VB – see Fig. 12(b)) which are small and, in fact, of the opposite sign to the trend of the data in Fig. 12(a). The monopole effect we consider here (Cr) is clearly the remaining dominant component.

The Cr term steadily increases with spin, understood as follows. The monopole term depends on the changes of $p_{\frac{3}{2}}$ occupancies and, near the middle of the $f_{\frac{7}{2}}$ shell, one expects significant admixtures from the $p_{\frac{3}{2}}$ orbit near the ground state, which decrease at high spin due to alignments. The occupancy of the $p_{\frac{3}{2}}$ orbit decreases steadily with spin, which causes the reduction of the effective nuclear radius as J increases. This has the effect of increasing the Coulomb energy for both members of the pair, but by more for the $T_z = +1$ member of the pair due to the larger Z. A rise in the MED is therefore observed. We have already seen from the data in Figs. 9 and 10 that this term is essential for a good description of the data. In this case, however, Fig. 12(b) shows that it is the major component of the calculation, and that the rise in the MED is almost entirely associated with the monopole radial effect.

The final effect we discuss in this section is the so-called "$J = 2$ anomaly" – included in the model through the inclusion of the isospin non-conserving term, VB. Much evidence has built up that the multipole contribution to the MED cannot be reproduced entirely through Coulomb effects alone, if a sensibly behaved set of Coulomb matrix elements is used (such as those from

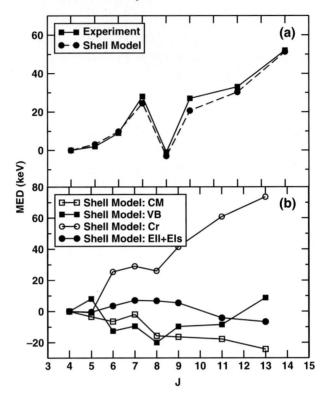

Fig. 12. The experimental and predicted MED for the odd–odd $T_z = \pm 1, A = 48$ mirror pair. See caption to Fig. 9 for details

the harmonic oscillator). It is found repeatedly that an additional isospin non-conserving component at $J = 2$ needs to be included, and this is accounted in the model through the VB term. An inspection of the data discussed (e.g. in Figs. 9 and 10) reveals how important this can be, particularly at low spins. It is seen very clearly in two further cases – the $A = 53$ mirror pair ^{53}Co/^{53}Fe [22] and the $T = 1$; $A = 54$ mirrors ^{54}Ni/^{54}Fe [73].

In these cases, the structures are relatively simple, with just a few holes in a ^{56}Ni core. For the $A = 53$ pair, one might consider that the yrast sequences up to the band-termination have simple three-hole configurations: $\nu(f_{\frac{7}{2}})^{-2}\pi(f_{\frac{7}{2}})^{-1}$ for ^{53}Co and $\nu(f_{\frac{7}{2}})^{-1}\pi(f_{\frac{7}{2}})^{-2}$ for ^{53}Fe. For ^{53}Fe, in progressing from the ground state to the band-termination, one requires a gradual re-coupling of the proton–hole pair from $J = 0$ to the maximum angular momentum of $J = 6$ (a neutron alignment for ^{53}Co, of course). Thus we expect a simple MED rise across the whole spin range associated with the angular momentum re-coupling of one pair. This is seen clearly in Fig. 13(a). However, it is also clear from Fig. 13(b) that for good agreement with the experimental trend at low spins, the inclusion of the VB term is essential.

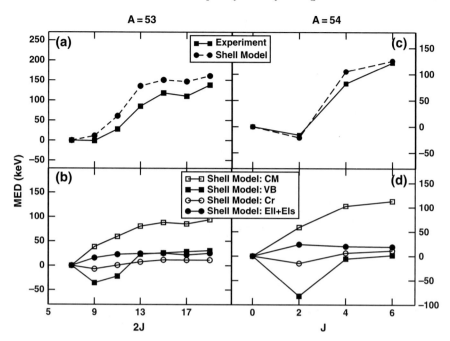

Fig. 13. The experimental and predicted MED for the $T_z = \pm\frac{1}{2}$ mirror pair with $A = 53$ (*left*) and the $T_z = \pm 1, A = 54$ mirror pair (*right*). See caption to Fig. 9 for details

This is even more striking in $A = 54$. Here, the structure might be considered to be even simpler – two neutron holes in ^{56}Ni for ^{54}Ni, and two proton holes in ^{54}Fe. Thus we expect an MED trend for the $J^\pi = 0^+, 2^+, 4^+, 6^+$ states that reflects the re-coupling of the proton–hole pair for ^{54}Fe and neutron–hole pair for ^{54}Ni. This should give a smooth rise in the MED. In fact, the data in Fig. 13(c) reveals a dip at $J = 2$ before the expected rise appears. This cannot be accounted for in the model (or intuitively) without the inclusion of the isospin non-conserving term – see Fig. 13(d).

Interestingly, the $A = 42$ mirror pair, representing two $f_{\frac{7}{2}}$ particles outside a closed shell, rather than two holes, also shows the same effect. Although one must consider cross-shell excitations in this analysis (i.e the structures are not that simple) it is nevertheless intriguing that the $J = 2$ effect is consistent at both ends of the shell. Moreover, the effect is clearly important in the deformed region in the middle of the shell. The anomaly is clearly present in a consistent fashion across the entire shell. It is also interesting to note that the effect was hinted at some years ago by Brown and Sherr [65], although in a study relating to CDE, not excitation energies. In their shell-model study, displacement energies across the $f_{\frac{7}{2}}$ shell were calculated, and were then fitted to the data to extract a single set of two-body isovector matrix elements $(pp - nn)$. Evidence of the anomaly is present in the form of a non-zero and positive $J = 2$ isovector component, which remains once Coulomb effects had been subtracted.

The fact that two-body Coulomb matrix elements (CME) require this additional isovector component at $J = 2$ needs to be explained. One possibility implies a non-Coulomb isospin-breaking effect (i.e. CSB, as we are discussing isovector energy differences) – or, more specifically, a spin-dependence of such an effect. This was indeed suggested [52], although a theoretical basis for such an effect would need to be established to proceed further with the discussion. A second possibility implies that the anomaly is associated with spin-dependent interactions with the core (i.e. configuration mixing). Indeed, re-normalisation of matrix elements to account for different core-interactions is a common feature of shell-model calculations. This was suggested in [30] but no renormalisation that could account for both the MED *and* TED could be found. The origin is still not clear, though the effect is certainly present across the shell and is now an essential ingredient in any realistic calculation in this region.

6 Isobaric Multiplets in the *sd* and *fp* Shells

The successful studies of mirror and isobaric multiplets in the $f_{\frac{7}{2}}$ shell presented in Sect. 5 have encouraged the extension of these investigations to other mass regions. This is important in order to check the limits of validity of isospin symmetry for different masses, to identify the nature of the symmetry breaking, to look for new Coulomb effects and to explore the experimental evidence of ISB terms of the nuclear interaction. The production of $N \sim Z$ nuclei of the *sd* and upper *fp* shells at high spin is, however, a difficult task. For nuclei heavier than ^{56}Ni, the stability line bends toward the neutron-rich side of the nuclide chart and the cross section for $T_z < 0$ nuclei using stable beams and targets decreases very rapidly. In the *sd* shell, when producing mirror nuclei in fusion–evaporation reactions, the rather low relative impact parameter prevents the formation of a high-spin residue. So far, the mirror nuclei studied in the *sd* and *fp* shells do not exhibit the collective behaviour of those in the $f_{\frac{7}{2}}$ shell. On the other hand, they put in clear evidence the electromagnetic single-particle effects. The shell-model description of these nuclei is presently not as accurate as in the $f_{\frac{7}{2}}$ shell. This is partly due to the large dimensions of the matrices to be diagonalised which, by imposing truncations of the valence space, introduce inaccuracies. The lack of appropriate and reliable residual interactions that, in some cases, have to take into account more than one main shell, also precludes a good description of the data. As we will see in this section, MED provide an optimum tool in order to disentangle the configurations of the excited states and, therefore, they constitute a stringent test of the calculations, putting constraints on the model space and effective interactions.

In the sd shell, the experimental information on MED has been for many years rather scarce and limited to low-spin states. Recently, interesting results have been obtained at high spin in the $T_z = \pm\frac{1}{2}$ mirror pairs of mass $A = 31, 35, 39$. Large MED values, of the order of 300 keV, have been observed in a few non-natural-parity states, which have been explained in terms of the electromagnetic spin–orbit interaction [53, 74–76]. These states present particular configurations where a single nucleon (a proton or a neutron) is excited from the sd to the fp shell. In these cases, as discussed in Sect. 4.1, the EMSO interaction, that acts differently on protons and neutrons, induces changes to the single-particle energies. In particular, if a $d_{\frac{3}{2}}$ proton is promoted to the $f_{\frac{7}{2}}$ orbit the gain in energy with respect to the excitation of a neutron amounts to 200–250 keV. This does not produce a difference in the MED if the negative-parity state is formed by exciting, with similar probabilities, a proton and a neutron, as the effect of the EMSO compensates. On the contrary, if only one type of nucleon is excited in one of the mirror nuclei – which implies that a nucleon of the other type is excited in the mirror partner – the effect of the EMSO is large. To illustrate this, we report in Fig. 14(a) and (b) the MED values for the negative-parity yrast states in the $T_z = \pm\frac{1}{2}$, $A = 35$ [76] and $A = 39$ [75] mirror pairs, respectively. Very large values of MED are obtained for all the high-spin yrast states and the experimental values are very similar.

We have calculated with the shell model the multipole Coulomb CM contribution to the MED for both mirror pairs. The nuclear wave functions were obtained using the $sdfp$ interaction [48] in the $s_{\frac{1}{2}} d_{\frac{3}{2}} f_{\frac{7}{2}} p_{\frac{3}{2}}$ valence space. As usual, Coulomb matrix elements were calculated in the harmonic-oscillator basis. The two curves are shown in Fig. 14. Interestingly, the multipole Coulomb contribution follows in both cases the qualitative behaviour of the data. For the $A = 35$ mirror nuclei the agreement is also quantitatively good. The predictions fail completely in the $J^\pi = \frac{9}{2}^-$ states. This could be due to the fact that excitations from the $d_{\frac{5}{2}}$ orbit may play a role in the configuration of this state, a possibility not allowed in the present calculations.

The relative importance of the single-particle contributions, E_{ls} and E_{ll} (Eqs. (18) and (16)), are shown in Fig. 14. It is clear that the electromagnetic spin–orbit interaction plays a very important role in both cases. Its contribution is comparable to that of the multipole Coulomb term CM. On the other hand, the orbital E_{ll} contribution does not significantly change the results in the present cases. Its effect on the proton single-particle energies is to further reduce the gap between the $f_{\frac{7}{2}}$ and the $d_{\frac{3}{2}}$ by ~ 30 keV. For the $A = 39$ mirror pair, the calculated curve, including the multipole Coulomb and the single-particle contributions, reproduces the experimental MED with very good accuracy. In the case of mass $A = 35$, where the multipole Coulomb contribution already gave a good quantitative description of the data, the addition of the single-particle term results in an over-prediction of the MED (absolute) values.

Shell-model calculations indicate that states that give rise to large MED values have a dominating configuration with a pure single-particle excitation

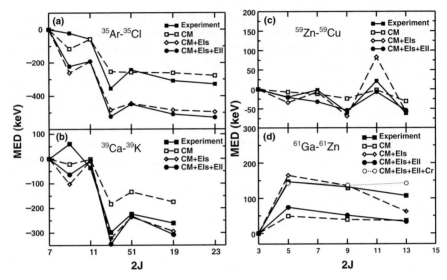

Fig. 14. Experimental MED for the negative-parity yrast states of the $T_z = \pm\frac{1}{2}$ mirror nuclei with $A = 35$ (**a**) [53, 76], $A = 39$ (**b**) [75], $A = 59$ (**c**) [77, 78] and $A = 61$ (**d**) [79], compared with shell-model calculations. The MED here are determined relative to the excitation energy of the $J^\pi = \frac{7}{2}^-$ state in (**a**) and (**b**), and relative to the ground state in (**c**) and (**d**). The theoretical values include: the multipole Coulomb contribution CM (*open squares*), the multipole Coulomb with single-particle energies corrected by the electromagnetic spin orbit effect E_{ls} (*diamonds*), CM plus both single-particle corrections E_{ls} and E_{ll} (*full circles*), finally, the *open circles* show the curve where the radial Cr term is also added

from the sd to the fp shell. In particular, the proton-rich nucleus of the mirror pair excites a proton, while the neutron-rich excites a neutron, or, when three nucleons are excited to the fp shell, the $T_z = -\frac{1}{2}$ nucleus excites two protons plus a neutron and vice versa for the $T_z = +\frac{1}{2}$. Although we do not report the MED for positive-parity states, which are small and limited to low spins, it is important to note that the shell-model results, which include the terms discussed above, reproduce them very well.

In the last few years experimental data on mirror nuclei above the doubly magic ^{56}Ni have become available. Recently, excited states up to $J^\pi = \frac{13}{2}^-$ have been observed in the $T_z = \pm\frac{1}{2}$ mirror pairs ^{59}Zn/^{59}Cu [77] and ^{61}Ga/^{61}Zn [79]. In this mass region, excitations from the $p_{\frac{3}{2}}$ to the $f_{\frac{5}{2}}$ play an important role in generating angular momentum. The EMSO single-particle energy shifts reduce the gap between these two orbits for neutrons and increase it for protons. The net gain in the promotion of a neutron from the $p_{\frac{3}{2}}$ to the $f_{\frac{5}{2}}$ orbit, with respect to a proton excitation, amounts to ~ 190 keV. When the shift due to the E_{ll} term – that acts only on protons – is considered, the gain reduces to just ~ 40 keV. This indicates that, while in the $A \sim 35$

mass region E_{ll} plays a minor role, its effect is considerable in $A \sim 60$. We report in Fig. 14(c) and (d) the MED values for the $A = 59$ and 61, $T_z = \pm\frac{1}{2}$ mirror pairs, which are compared with the theoretical curves, in the same way as in the previous cases.

The shell-model calculations have been performed using the GXPF1 interaction [44] in the fp valence space allowing for four nucleons to be excited from the $f_{\frac{7}{2}}$ shell to the upper three orbits (a $t = 4$ truncation). This interaction does not describe very well the level scheme of the $A = 59$ mirror nuclei but gives a good description of the energy levels for $A = 61$. As can be seen in Fig. 14(c) and (d), in both cases the E_{ll} and E_{ls} single-particle contributions are important, producing opposite effects on the MED. The overall theoretical description for $A = 59$ is in good agreement with the data, but they are underestimated for $A = 61$.

The role of the ISB term, V_B, in the form deduced in Sect. 4.2 from the $A = 42$ spectra, vanishes in these calculations, as the $f_{\frac{7}{2}}$ shell is almost not active. The other term we have not yet considered in the calculation of the MED is the monopole Coulomb radial contribution, V_{Cr}. This term depends on the average of proton plus neutron occupation numbers of the different shells. Assuming that changes in radii are associated to changes in the occupation numbers of p orbits, the radial term in Eq. (21) can be calculated. This requires a very good description of the spectroscopy and therefore we calculate the Cr contribution for the $A = 61$ mirror pair. Using the same parameterisation of the $f_{\frac{7}{2}}$ shell for α_r, the contribution of the radial term together with the multipole Coulomb and the single particle E_{ls} and E_{ll}, bring the theoretical values in very good agreement with the experimental MED (see Fig. 14(d)).

An important conclusion can be drawn from the analysis of the MED in the sd and fp nuclei: In all of the studied mirror nuclei, the CM contribution reproduces the trend of the experimental MED curves. In most of the cases it underestimates the absolute values by a factor of 2–3; a smoother behaviour is also predicted. In addition, important contributions arise from the EMSO E_{ls} and the orbital E_{ll}, since states with single-particle configurations are considered. Although the effective interactions are not completely reliable and the valence spaces have to be truncated to cope with the present computational capabilities, the experimental MED give us valuable information about the configuration of the states, as they distinguish clearly between pure single-particle and mixed configurations.

7 Conclusions and Outlook

In these lectures, we have presented tests of isospin symmetry in nuclei near $N = Z$ through analysis of energy differences between isobaric multiplets. We have concentrated much of the discussion on the $f_{\frac{7}{2}}$ shell, where the Coulomb energy differences can be followed up to high-spin states and interpreted by means of state-of-the-art shell-model calculations. We have also discussed how these ideas can be developed and extended into the sd and upper-fp shells. These CED turn out to give extremely valuable and precise information on nuclear structure – providing a level of consistency in detail that might seem surprising, given the long-standing issues regarding modelling of Coulomb effects in nuclei (e.g. the Nolen–Schiffer anomaly). These studies have now developed into a mature field, and the systematic investigation of energy differences between analogue states is starting to yield some fascinating questions – such as the $J = 2$ anomaly.

In this contribution, we have concentrated on energy differences as a test of isospin symmetry. As a result, some other important aspects of isospin symmetry of isobaric multiplets have not been discussed. For example, subtle information on the isospin degree of freedom can be derived through the study of the T_z-dependence of electromagnetic transition matrix elements (see, for example, [80, 81] for the $f_{\frac{7}{2}}$ shell). Mirror nuclei also provide an ideal laboratory for measurement of effective charges, as lifetimes of analogue states can be accurately determined – see, for example, the work of Du Reitz et al. [81], where information on isoscalar and isovector effective charges has been deduced from the lifetimes in the $A = 51$ mirror pair ^{51}Fe/^{51}Mn. Another key result relates to $E1$ decays in mirror nuclei. Because $E1$ decays are purely isovector in origin, in the limit of good isospin symmetry, $\Delta T = 0$ $E1$ transition strengths in mirror nuclei should be identical – i.e. have identical strengths. However, it now appears that $E1$ transitions have shown some anomalous results. For example, in the $A = 35$ [53, 75, 82], $A = 31$ [74, 83] and $A = 45$ [31] mirror nuclei, strong $E1$ decays have been observed from certain states in one member of the mirror pair, which are either absent, or highly hindered, in the other. This breakdown of the selection rule has been interpreted in terms of isospin mixing [53], although we need to await results of lifetime measurements to address this issue further.

The development and availability of the first generation of radioactive beam facilities has allowed for further access to exotic nuclei. These, and the planned next generation of ISOL and fragmentation facilities, will open up unprecedented access to proton-rich nuclei. As we proceed towards the spectroscopic study of proton-rich nuclei of both larger isospin and heavier mass, one may expect other effects to come into play. For example, the assumption has been made so far that the wave functions of the analogue states are essentially identical. When the analogue states of interest in the proton-rich member of the multiplet are weakly bound, this is no longer expected to be the case, and some part of the energy difference observed will be due to the different spatial distributions of the analogue wave functions. This shift, the Thomas–Ehrman shift [84, 85], will become more significant as the proton-rich states become more weakly bound.

Exploration of the isospin degree of freedom is certain to be one of the key nuclear-structure objectives of the new generation of radioactive beam facilities.

Acknowledgement

The authors are grateful to A.P. Zuker for helpful and illuminating discussions. The results presented have been obtained in collaboration with many colleagues and through fruitful and interesting discussions. The authors are particularly grateful to (alphabetically) J.A Cameron, C. Chandler, E. Caurier, F. Della Vedova, J. Ekman, A. Gadea, N. Mărginean, G. Martínez-Pinedo, D.R. Napoli, F. Nowacki, C.D. O'Leary, A. Poves, D. Rudolph, J.P. Schiffer, P. Van Isacker, D.D. Warner and S.J. Williams.

References

1. J. Jänecke in *Isospin in Nuclear Physics*, Ed. D.H. Wilkinson, North Holland, Amsterdam, 1969
2. R. Machleidt and H. Muther, Phys. Rev. C **63** 034005 (2001)
3. G.Q. Li and R. Machleidt, Phys. Rev. C **58** 1393 (1998)
4. D.E. González-Trotter et al., Phys. Rev. Lett. **83** 3788 (1999)
5. E.P. Wigner, Proceedings of the Robert A Welch Conference on Chemical Research (Robert A Welch Foundation, Houston, Texas) **1** (1957) 67
6. W. Benenson and E. Kashy, Rev. Mod. Phys. **51** 527 (1979)
7. W.E. Ormand, Phys. Rev. C **55** 2407 (1997)
8. W.E. Ormand and B.A. Brown, Nucl. Phys. A **491** 1 (1989)
9. N. Auerbach, Phys. Rep. **98** 273 (1983)
10. J. Britz, A. Pape and M.S. Anthony, At. Nuc. Dat. Tab. **69** 125 (1998)
11. H.A. Bethe and R.F. Bacher, Rev. Mod. Phys. **8** 82 (1936)
12. J.A. Nolen and J.P. Schiffer, Ann. Rev. Nucl. Sci **19** 471 (1969)
13. S. Shlomo Rep. Prog. Phys. **41** 66 (1978)
14. G.A. Miller, Chin. J. Phys. **32** 1075 (1994)
15. M.H. Shahnas et al., Phys. Rev. C **50** 2346 (1994)
16. J. Duflo and A.P. Zuker, Phys. Rev. C **66** 051304(R) (2002)
17. I.Y. Lee, M.A. Delaplanque and K. Vetter, Rep. Prog. Phys. **66** 1095 (2003)
18. I.Y. Lee, Nucl. Phys. A **520** 641c (1990)
19. F.A. Beck, Prog. Part. Nucl. Phys **28** 443 (1992)
20. J. Simpson, Z. Phys. A **358** 139 (1997)
21. C. Rossi Alvarez, Nucl. Phys. News **3** 3 (1993)
22. S.J. Williams et al., Phys. Rev. C **68** 011301 (2003)
23. M.A. Bentley et al., Phys. Rev. Lett. **97** 132501 (2006)
24. E. Farnea et al., Nucl. Instr. Meth. A **421** 87 (1997)
25. A. Gadea *et al.*, Legnaro National Laboratory Ann. Rep. 1999, INFN(REP) 160/00 p.151
26. D. Sarantites, Nucl. Instr. Meth. A **381** 418 (1996)
27. O. Skeppstedt et al., Nucl. Instr. Meth. A **421** 531 (1999)
28. D. Sarantites, Nucl. Instr. Meth. A **530** 473 (2004)
29. C. Rossi Alvarez et al., Legnaro National Laboratory Ann. Rep. 1999, INFN (REP) 160/00 p.155
30. S.M. Lenzi et al., Phys. Rev. Lett. **87** 122501 (2001)
31. M.A. Bentley et al., Phys. Rev. C **73** 024304 (2006)
32. E. Caurier, Code ANTOINE, Strasbourg, 1989-2005;
33. E. Caurier and F. Nowacki, Acta Phys. Pol. B **30** 705 (1999)
34. T. Otsuka, M. Honma, and T. Mizusaki, Phys. Rev. Lett. **81** 1588 (1998)
35. E. Caurier et al., Phys. Lett. B **75** 2466 (1995)
36. S.M. Lenzi et al., Z. Phys. A **354** 117 (1996)
37. F. Brandolini et al., Nucl. Phys. A **642** 387 (1998)

38. E. Caurier, G. Martínez-Pinedo, F. Nowacki, A.Poves and A.P. Zuker, Rev. Mod. Phys. **77** 427 (2005)
39. T. Otsuka, M. Honma, T. Mizusaki and N. Shimitzu, Prog. Part. Nucl. Phys **47** 319 (2001)
40. F. Brandolini et al., Phys. Rev. C **66** 024304 (2002)
41. W. Kutschera, B.A. Brown and K. Ogawa, Riv. Nuovo Cimento **31** 1 (1978)
42. W.A. Richter et al., Nucl. Phys. A **523** 325 (1991)
43. A. Poves et al., Nucl. Phys. A **694** 157 (2001)
44. M. Honma et al., Phys. Rev. C **65** 061301 (2002)
45. B.H. Wildenthal, Prog. Part. Nucl. Phys **11** 5 (1984)
46. B.A. Brown and B.H. Wildenthal, Ann. Rev. Nucl. Part. Sci. **38** 29 (1988)
47. B.A. Brown and W.A. Richter, Jr. Phys.: Conference Series 20 (2005) 145; and submitted to publication.
48. E. Caurier, K. Langanke, G. Martínez-Pinedo, F. Nowacki and P. Vogel, Phys. Lett. B **522** 240 (2001)
49. P. Mason et al., Phys. Rev. C **71** 014316 (2005)
50. F. Della Vedova et al., *AIP Conf. Proc.* 764, 205 (2005)
51. Y. Utsuno, T. Otsuka, T. Glasmacher, T. Mizusaki and M. Honma, Phys. Rev. C **70** 044307 (2004)
52. A.P. Zuker, S.M. Lenzi, G. Martinez-Pinedo and A. Poves, Phys. Rev. Lett. **89** 142502 (2002)
53. J. Ekman et al., Phys. Rev. Lett. **92** 132502 (2004)
54. J.A. Cameron et al., Phys. Lett. B **235** 239 (1990)
55. D.D. Warner, M.A.Bentley and P. Van Isacker, Nature Phys. **2** 311 (2006)
56. M.A. Bentley et al., Phys. Rev. C **62** 051303 (2000)
57. J.A. Sheikh, P. Van Isacker, D.D. Warner and J.A. Cameron, Phys. Lett. B **252** 314 (1990)
58. J.A. Sheikh, D.D. Warner and P. Van Isacker, Phys. Lett. B **443** 16 (1998)
59. F. Brandolini et al., Pgys. Rec. C. 66 021302(R) 2002
60. E. Caurier, A.P. Zuker, A. Poves and G. Martinez-Pinedo, Phys. Rev. C **50** 225 (1994)
61. M.A. Bentley et al., Erratum, Phys. Lett. B **451** 445 (1999)
62. A. Poves, J. Phys. G **25** 589 (1999)
63. D.R. Inglis, Phys. Rev. **82** 181 (1951)
64. L. Trache et al., Phys. Rev. C **54** 2361 (1996)
65. B.A. Brown and R. Sherr, Nucl. Phys. A **322** 1979 (61)
66. M.A. Bentley and S.M. Lenzi. Prog. Part. Nucl. Phys **59** 497 (2006)
67. C. O'Leary et al., Phys. Rev. Lett. **79** 4349 (1997)
68. M.A. Bentley, et al., Phys. Lett. B **437** 243 (1998)
69. J.A. Cameron et al., Phys. Rev. C **58** 808 (1998)
70. J. Ekman et al., Eur. Phys. J. A **9** 13 (2000)
71. J. Ekman et al., Phys. Rev. C **70** 057305 (2004)
72. J. Ekman et al., Phys. Rev. C **70** 014306 (2004)
73. A. Gadea et al., Phys. Rev. Lett. **97** 152501 (2006)
74. D.G. Jenkins et al., Phys. Rev. C **72** 031303 (2005)
75. T. Anderson et al., Eur. Phys. J. A **6** 5 (1999)
76. F. Della Vedova et al., Phys. Rev. C **75** 034317 (2007)
77. C. Andreoiu et al., Eur. Phys. J. A **15** 459 (2002)
78. C. Andreoiu et al., Eur. Phys. J. A **14** 317 (2002)
79. L.L. Andersson et al., Phys. Rev. C **71** 011303 (2005)
80. D. Tonev et al., Phys. Rev. C **65** 034314 (2002)
81. R. du Rietz et al. Phys. Rev. Lett. **93** 222501 (2004)
82. F. Della Vedova et al., Legnaro National Laboratory Ann. Rep. 2004, INFN (REP) 204/2004 p. 7
83. F. Della Vedova et al., Legnaro National Laboratory Ann. Rep. 2003, INFN (REP) 202/2004 p.3
84. R.G. Thomas. Phys. Rev. **88** 1109 (1952)
85. J.B. Ehrman. Phys. Rev. **81** 412 (1951)

Beta Decay of Exotic Nuclei

B. Rubio[a] and W. Gelletly[b]

a. IFIC, CSIC–Un. Valencia, Apdo. 22085, E–46071 Valencia, Spain
b. Physics Department, University of Surrey, Guildford, Surrey GU2 7H, UK

Abstract In this chapter β-decay is discussed as a tool for studying the structure of atomic nuclei. An attempt is made to give a simple account of the topic so that the student can understand what they find in the literature about β-decay. The quantities of spectroscopic interest are defined and the student is shown how they can be derived from experiments. In the study of exotic nuclei often the first thing we learn about a nucleus is how it β-decays. A series of examples of β-decay studies of exotic nuclei is presented with the aim of both illustrating the most up-to-date techniques and showing the student the breadth of physics that can be addressed

1 Introduction

Our starting point for this chapter on the β–decay of atomic nuclei is the chart of the nuclides shown in Fig. 7. Often called the Segré chart the version we see here shows the stable nuclear species as *black squares* plotted as a function of the proton number (Z) and neutron number (N).

There are only some 283 such nuclear species which are stable or sufficiently long–lived to be found on Earth. We see them stretching initially along the $N = Z$ line but then moving steadily to the neutron–rich side of the chart because of the increasingly disruptive effect of the Coulomb force. The chart also indicates how many protons (neutrons) the nuclear ground state can hold and still be bound. Our present knowledge of atomic nuclei suggests that some 6,000–7,000 distinct nuclear species live long enough to be created and studied. The limits are set by the drip lines for protons and neutrons and by the heaviest elements that can exist [1, 2].

In essence we have two main ways of studying nuclear properties, namely in reactions and in radioactive decay. The study of the prompt radiation from reactions and the delayed radiations from decay complement each other. There are many different types of reaction, each with different properties, but they have some features in common. To illustrate the contrasting features of reaction and decay studies we have chosen the case of the fusion–evaporation

Rubio, B., Gelletly, W.: *Beta Decay of Exotic Nuclei.* Lect. Notes Phys. **764**, 99–151 (2009)
DOI 10.1007/978-3-540-85839-3_4   © Springer-Verlag Berlin Heidelberg 2009

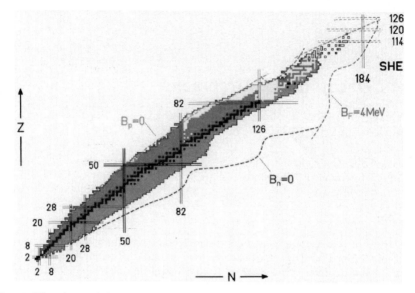

Fig. 1. The chart of the nuclides showing the neutron and proton drip lines which are defined by demanding that the respective binding energies, B_n and B_p, be zero. The neutron and proton numbers for the closed shells are indicated by the *horizontal* and *vertical lines*. The figure also shows the line where the fission barrier B_F goes below 4 MeV and a prediction of the possible shell structure for the super–heavy elements (SHE) (See also Plate 7 in the Color Plate Section)

reaction, which was already presented and explained in [3]. Figure 2 shows schematically how such a reaction proceeds. From the name it will be no surprise to the student that the two nuclei fuse together to form a compound nucleus that lasts for a long time compared with the time taken for the projectile nucleus to "cross" the target nucleus, i.e. 10^{-22} s. The compound nucleus is now a "hot" (high temperature), rapidly rotating (50–80 $\hbar$), charged liquid drop. Not surprisingly the system decays in a way which reduces the temperature, namely it emits particles. This process stops when the excitation energy of the system lies below the separation energy for either a proton or a neutron. The system is now much cooler but still rotating rapidly and further decay occurs by the emission of γ–rays. This means that, typically, a long cascade of γ–rays is emitted, leading, through a series of excited states, to the ground state. The ground state then decays by β–emission on a much longer timescale.

Later in this chapter, when we show examples of β–decay studies, the reader should remember that, in general, many things have happened prior to the formation of the β–decaying state but most of the time the decay experiments will not be sensitive to them. This is particularly true in experiments where the nuclear species of interest has been physically and/or chemically separated from all the other nuclei produced.

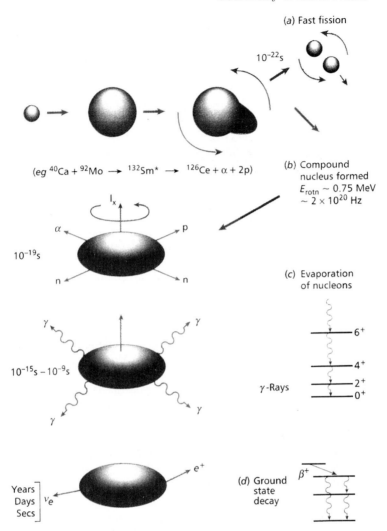

Fig. 2. Schematic view of what happens in a fusion–evaporation reaction. In general it is only the ground state, shown on the *lower right*, which decays by emitting a β–particle. In some cases two long–lived, low–lying, β–decaying states may be populated

There are, of course, many possible different types of reaction. In some cases, a compound nucleus is formed and in others it is not. In some, a large amount of angular momentum is brought into the system and in others not. However, they have common features including the fact that the radiation and particles are generally emitted over a timescale very short compared with the subsequent radioactive decay of the ground state.

Typically we can alter the amount of angular momentum injected and the nuclear temperature by varying either the projectile/target nucleus or the bombarding energy. Often this allows us to emphasise some aspect of nuclear properties we wish to study. In this sense reactions are a flexible tool. They suffer, however, from the disadvantage that often many reaction channels are open, in the case of the fusion–evaporation reaction this is dictated by the number and character of the particles evaporated. This greatly reduces the sensitivity of the measurements of any prompt radiation. As we shall see later, very similar considerations apply to other types of reaction such as spallation, fragmentation or fission where we may have several hundred reaction channels open. Decay studies, in contrast, benefit from the much longer timescale. In general, this allows one to separate the particular nuclear species of interest from any others produced and the resulting selectivity greatly enhances the sensitivity of the measurements. If separated from other species present during its production it can also be studied with only a very small number of atoms. This sensitivity also means that it is a prolific source of applications, a forensic tool that scientists have used in a wide variety of contexts. The corresponding disadvantage is that nuclear decay is essentially God–given and can only be altered under certain unusual conditions [4–6].

Thus radioactive decay is more limited as a tool for studying nuclei but it is often the first means of identifying a new nuclear species and hence also the first harbinger of a knowledge of its properties. Most nuclear ground states decay by β–decay, either β$^-$ decay on the neutron–rich side of stability or β$^+$/EC decay on the proton–rich side. Examination of Fig. 7 shows us that studies of radioactive decay can provide information about nuclear properties over a wide range of N and Z. Since this process is governed by the weak interaction, which is a slow process, the half lives are relatively long, ranging from tens of ms to 10^{15} years. At and near the proton drip line proton-decay [7] is also important and as we gradually establish the means to study more and more neutron–rich nuclei we may encounter neutron–decay near the neutron drip line [8]. In heavy elements the nuclei may also be unstable to α–decay or spontaneous fission. In these processes the strong interaction dominates and the decays will be much faster than β–decay in principle, except where they are delayed by the Coulomb barrier. Consequently they are often in competition with β–decay for large values of Z.

It is the study of β–decay which is the focus of this chapter and we now turn our attention to this topic. The first question is why such studies are important. As indicated above one main reason is that it is often the primary source of information about a newly identified nucleus. Once we have demonstrated that a particular nucleus exists, even a rough measurement of the half life or its decay Q_β–value, relatively simple quantities to measure, can provide important clues to its properties. Later when we are able to undertake a more complete study we are tapping into a much richer vein of information about nuclear structure. This is another reason why β–decay studies are important and in this chapter we will encounter examples to illustrate this.

In addition, a knowledge of β–decay is important for our understanding of the creation of the heavier elements in explosive stellar processes. There are many other applications of β–decay such as Positron Emission Tomography or γ–radiography, and a knowledge of β–decay properties is essential for the proper design of nuclear reactors and for associated questions of shielding and safety. In operation 7–8% of the total heat in a reactor is generated in the radioactive decay of the fission products. Following shutdown this heating remains and we need a good knowledge of many of the decays involved in order to be able to calculate the subsequent decay heating as a function of time.

Our aim in this chapter will be to describe how β–decay studies provide a practical tool to unravel the complexities of nuclear structure rather than a primer to aid our theoretical understanding of the weak interaction and β–decay, in particular. Naturally, however, we will provide enough theoretical information to underpin our "rude mechanical" approach. The student who seeks a more profound understanding of the background is referred to the lecture by Severijns [9] and to other articles [10, 11].

2 Beta Decay and Nuclear Structure

One advantage of studying β–decay is that we now have a good understanding of the process. It was not always so. In spectroscopic terms β–decay was difficult to understand initially because of the continuous nature of the β–spectrum, which is in stark contrast with the discrete line spectra of α– and γ–decay, which are readily understood in terms of transitions between the discrete quantum states in nuclei. Pauli's explanation [12] in terms of the neutrino and hence a three–body process clarified this. Soon afterwards, the present basis of our understanding was laid by Fermi's theory of β–decay [13].

In this chapter we are concerned not with the history of the subject but with extracting nuclear structure information from β–decay studies as a tool to understand nuclear structure. In simple terms we can imagine the β–decay process as involving the transformation of a proton into a neutron or a neutron into a proton. In an intuitive way we can immediately see that the probability of this process will depend inter alia on the relation between the wave–functions of the initial and final nuclear states, just as in any transition between states in a quantised system.

This intuitive idea bears little relation to what a student first encounters when she seeks information about β–decay. In modern times she will open the Table of Isotopes [14] (or insert a query into a nuclear data base such as [15]), where she will find a radioactive decay scheme similar to the one shown in Fig. 3. Here we see a summary of our knowledge of the β–decay of ^{132}Sn to levels in ^{132}Sb at the time of printing. The so–called parent state, which is, in general but not exclusively, the ground state of the decaying nucleus, is here the ground state of ^{132}Sn. Figure 3 shows its spin and parity, 0^+, and its half life, 39.7 s. It also shows the states in the daughter nucleus

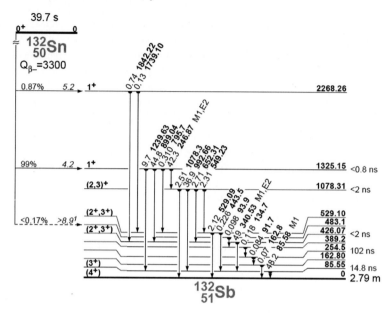

Fig. 3. Partial decay scheme for the ground state of ^{132}Sn. See Ref. [14] for details

^{132}Sb which are known to be populated directly or indirectly in the β–decay. It gives their excitation energies relative to the ground state of ^{132}Sb, their spins and parities and the ways in which they γ–decay. To the left of each state we see tabulated the percentage of the total direct β–decay feeding to each state and the log ft value or comparative half life as it is sometimes called. We will return to this latter quantity later. We also see the Q_β–value, which is the difference between the nuclear masses of the two ground states. It is our intention, in this chapter, to explain why these quantities are measured and their significance in terms of the study of β–decay and nuclear structure.

Our starting point is that the description of a β–decay process is not as simple as the transformation of a proton into a neutron or vice versa. More properly the initial state is a particular state, usually the ground state, in the parent nucleus but the final state consists of a state in the daughter nucleus together with an emitted β–particle and a neutrino. However, if we temporarily ignore the complication that we are dealing with transitions from one nuclear state to another and think only in terms of the transformation of an individual nucleon, we can characterise the main types of β–decay as described below.

Nuclei to the right of the Valley of Stability, mentioned in the introduction to Fig. 7, decay with the emission of an electron and an antineutrino

$$\beta^- : n \rightarrow p + e^- + \bar{\nu}. \tag{1}$$

The neutron–deficient nuclei, left of the valley of stability, decay by emitting a positron and a neutrino

$$\beta^+ : p \rightarrow n + e^+ + \nu. \tag{2}$$

There is a third process called electron capture (EC) which competes with positron emission. In this case an atomic electron is absorbed by the nucleus with the emission of a neutrino. The end result of this process is to populate states in the same daughter nucleus as in positron decay.

Again if it were a single proton that interacts with the atomic electron we can write this process as

$$EC : p + e^- \rightarrow n + \nu. \tag{3}$$

The reader should note that this process can only occur if the atomic electrons are present. If the atom is not dressed with electrons, as in a storage ring filled with fully stripped ions [5, 6] or in a stellar plasma, EC cannot occur and the nuclear lifetime will be altered.

The processes described above occur in real nuclei where the equations can be written as follows:

$$^A_Z X_N \rightarrow {}^A_{Z+1} X^*_{N-1} + e^- + \bar{\nu}_e \tag{4}$$

$$^A_Z X_N \rightarrow {}^A_{Z-1} X^*_{N+1} + e^+ + \nu_e \tag{5}$$

$$^A_Z X_N + e^- \rightarrow {}^A_{Z-1} X^*_{N+1} + \nu_e + X_{ray} \tag{6}$$

where $^A_Z X_N$ represents a nucleus with chemical symbol X, proton number Z, neutron number N and mass number A, and X_{ray} represents a characteristic X–ray from the daughter element emitted following EC. The student who is alert will realise that Auger emission competes with X–ray emission and so we may have an Auger electron instead of the X–ray.

The electron (positron) and the antineutrino (neutrino) in the final state influence the β–decay transition rate in three ways, namely

a) For a given energy released in the decay there is a density of possible final states for both the electron and antineutrino since they share the energy.
b) The β–particle will "feel" the Coulomb field created by the protons in the nucleus. In other words the wave–functions of the electron (positron) are enhanced (suppressed) close to the nucleus.
c) The possible angular momentum and parity in the final state. The product of the electron and neutrino wave–functions has parity $(-1)^L$, where L is the orbital angular momentum carried away by the electron.

The first two effects (a) and (b) can be calculated and the combination of the two is usually called the Fermi Integral.

Naturally the transition rate to a particular state in the final nucleus is also dependent on the change in nuclear structure, which is embodied in the matrix element, which we will call M_{fi}. We define this quantity as

$$M_{\text{fi}} = < \psi_{\text{f}}^* |V| \psi_{\text{i}} >, \tag{7}$$

where ψ_{f} and ψ_{i} describe the final and initial states, respectively, and V represents the operator responsible for β–decay. However, we know that for Quantum systems we cannot measure matrix elements directly; they have to be deduced from the quantity that can be measured, namely the transition probability. In Nuclear Physics the transition probabilities are denoted by B, where B is the square of the matrix element.

From the point of view of nuclear physics it is interesting to isolate this part of the transition rate and this is done using the ft value, where t is the partial half life for the transition to a particular state in the daughter nucleus and f, the Fermi Integral, takes account of the effect of the neutrino and electron wave–functions. Using the ft value allows us to compare β–decay probabilities in different nuclei. As we will see later it turns out that

$$1/ft \; \alpha \; M_{\text{fi}} \tag{8}$$

In practice one finds that the partial half lives, and hence the ft values, take a very wide range of values. As a result one has to have recourse to the logarithm of the ft value (see Fig. 4). Thus it is log ft which is tabulated in the decay scheme of Fig. 3.

Historically, this wide range of observed ft values led to an empirical classification of the transitions according to the log ft value. The fastest transitions, with log $ft < 6.0$, were called *allowed transitions* and slower transitions with larger log ft values were called *forbidden*. This is one of the many examples in Physics where the original nomenclature, introduced empirically without an understanding of the underlying phenomenon, is retained although it may be misleading if taken literally. Forbidden transitions are not in reality forbidden but actually occur and are simply slower than allowed transitions; it turns out for good reason. It is worth noting that our present knowledge of the log ft values of *allowed transitions* as tabulated in Fig. 4 shows that they may have much larger values than 6.0 under some circumstances.

The introduction of the theory of β–decay by Fermi [13] and its early modification by Gamow and Teller [16] provided an explanation of the observed large variation in the measured comparative half lives. It is not our intention to repeat the many descriptions of Fermi's theory which can be found in numerous textbooks [17, 18]. The students will find these readily for themselves. Instead we will adumbrate the essence of the theory in order to make clear the essential features for an understanding of how it explains the phenomenon of β–decay.

Fermi begins with his *Golden Rule* derived from perturbation theory, which is familiar to all undergraduates, for the calculation of transition rates (λ) in quantum systems

$$\lambda = \frac{2\pi}{\hbar} \times |\langle \psi_f \varphi_e \varphi_\nu |V| \psi_i \rangle|^2 \times \rho\left(E_f\right), \tag{9}$$

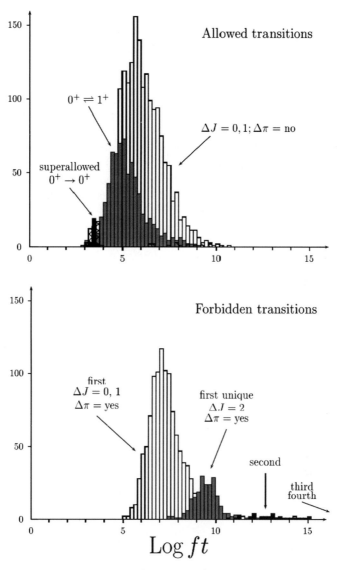

Fig. 4. Numbers of known allowed (*upper panel*) and forbidden (*lower panel*) transitions of different types (see text) plotted as a function of the log *ft* values [19]

where $\rho(E_f)$ is the density of states available to the electron and neutrino.

In 1934 Fermi did not know the form of the interaction V. Intuitively he chose it by analogy with the well–known electromagnetic interaction. He found five operators with the correct mathematical form which satisfied special relativity. These operators have different transformation properties. They are vector (V), axial vector (A), scalar (S), pseudoscalar (P) and tensor (T) in form. It is not our purpose here to follow the ups and downs of the experimental attempts to determine which of these operators is favoured by Nature. It took more than 20 years of experiments examining the symmetries involved in the decays and the spatial properties of the decay products before this question was settled.

Initially Fermi's theory used the V or S forms of the operator but at a very early stage it was necessary to introduce the A or T forms as well. The former govern the so–called Fermi decays and the latter Gamow–Teller (GT) decays, named after the physicists who introduced the corresponding idea. Eventually, it was shown that parity [20] and charge conjugation [21] are not conserved in β–decay although the combined CP operation is conserved. This fixes the V–A combination of operators for the Fermi and GT decays. Any student who is particularly interested in the vicissitudes of the theory in this period can avail themselves of the articles in [10]. Some of the same material is covered in [9] and [11].

Returning to Fermi's theory after this short digression we find that he made the assumption that the interaction takes place at a point, the centre of the nucleus. This is equivalent to assuming that the electron and neutrino carry away zero orbital angular momentum ($L = 0$).

In general terms there is ample justification for the assumption. Let us consider a decay in which a 1 MeV electron is emitted. In a medium-heavy nucleus with a radius of approximately $6\,F$ the maximum orbital angular momentum for this electron is 1.4 MeV/c. In terms of $h/2\pi$ this is equal to ~ 0.04 so the probability of a 1 MeV electron having $1\,h/2\pi$ orbital angular momentum at the surface is small compared with $l = 0$. In other words Fermi's assumption is a good one. If further justification were needed for Fermi's assumption we can find it in the modern theory of the electroweak interaction, where phenomena such as β–decay are explained in terms of the exchange of $W^{+/-}$ bosons with a mass of $\sim 80\,\text{GeV}/c^2$. The masses of these exchange particles are so large that the interaction is effectively at a point.

Looked at another way we can write the wavefunctions for the electron and neutrino as

$$\varphi_e(\vec{r}) = \frac{1}{\sqrt{A}} e^{i\vec{p}.\vec{r}/\hbar} \tag{10}$$

and

$$\varphi_\nu(\vec{r}) = \frac{1}{\sqrt{A}} e^{i\vec{q}.\vec{r}/\hbar}, \tag{11}$$

where A represents a spherical volume in the momentum space. We can expand these expressions as follows

$$\varphi_e(\vec{r}) = \frac{1}{\sqrt{A}} \left(1 + i\vec{p}.\vec{r}/\hbar - (\vec{p}.\vec{r}/\hbar)^2 + \ldots\ldots\right) \tag{12}$$

Since $\vec{p}.\vec{r}/\hbar \leq 0.04$, Fermi's approximation is well justified and we can approximate

$$\varphi_e(\vec{r}) = \frac{1}{\sqrt{A}} \tag{13}$$

over the whole nuclear volume. This is the *allowed approximation*.

By definition the only change in the nuclear angular momentum must come from the spins of the particles since

$$I_i = I_f + L + S, \tag{14}$$

where I_i is the spin of the parent nucleus, I_f is the spin of the state populated in the daughter nucleus, L is the orbital angular momentum carried away by the leptons (effectively the electron) and S is the vector sum of the electron and neutrino intrinsic spins, both of which are $s = 1/2$. In the allowed approximation $L = 0$ by definition and $S = 0$ or 1. As a result $\Delta I = 0, +/-1$ in an allowed transition.

In the case with $S = 0$, the electron and neutrino have their spins antiparallel. This is a Fermi decay, called thus because it corresponds to Fermi's original assumption. The other possibility, where we have the spins parallel to each other and $S = 1$, means that the change in the nuclear angular momentum can be either $0, +/-1$. Such transitions are GT transitions. From Eq. (14) it follows that in this case 0^+–0^+ is not possible.

If we turn now to the so–called *forbidden transitions*, where $L \neq 0$ the first thing to note is that the name is very misleading as we indicated earlier. Such transitions are not "forbidden" but merely suppressed relative to the allowed transitions (see Fig. 4). This does not come as a surprise given our discussion above of the small probability of an electron of 1 MeV in a medium–heavy nucleus having $L = 1$ compared with $L = 0$. The probability will be even smaller if $L = 2$. The forbidden transitions are further classified by their degree of "forbiddenness", which corresponds to the value of L. For $L = 1$ we have *first forbidden transitions*, $L = 2$ corresponds to *second forbidden ones* and so on. Since the parity of the electron plus neutrino wavefunctions is given by $(-1)^L$ first forbidden transitions must involve a change in parity of the nuclear states involved. Here again we can have both Fermi and GT transitions.

For the former, $S = 0$ means that

$$\Delta I = |\, I_f - I_i \,| = 0, +/-1 \tag{15}$$

with $0 \to 0$ not possible because of angular momentum coupling and a change in parity. For the latter $L = 1$ and with $S = 1$ this means

$$\Delta I = |\, I_f - I_i \,| = 0, +/-1, +/-2, \tag{16}$$

again with $0 \to 0$ not possible and $\Delta \pi = $ yes.

All of these transitions are suppressed in rate compared with the allowed transitions. As we see from the simple rules listed above most first forbidden transitions can be a mixture of Fermi and GT transitions. The exception is for $\Delta I = +/-2$ when only GT transitions are possible. This leads to a further quirk in the nomenclature with these transitions said to be *unique first forbidden transitions.* This is an example of a more general situation for a given L, where transitions with $\Delta I = L + 1$ are only possible via GT transitions and such transitions are called "unique".

As we said earlier the most probable case is when L is equal to 0, which means that transitions involving $L = 1$ or more are slower in principle. In other words, in a normal decay, where allowed transitions are possible, they will dominate, and the forbidden ones will be too weak and therefore "invisible" in experiments. However, there are cases, for instance, for small Q_β–values, where only the forbidden decay is possible and then we can observe it and quantify it. The reader should note that as we move away from the line of stability the Q_β–value will increase with the consequence that the average electron energy emitted in β–decay will also increase. The corollary of our earlier simple estimate of the average value of L carried away by an electron of 1 MeV is that the relative probability of the so–called forbidden transitions must increase with the increase in energy released. This is likely to be of importance for our understanding of phenomena such as the astrophysical r–process where the reaction pathways pass through a network of nuclei with large Q_β–values far from stability.

Closer attention to Fig. 4 indicates that the allowed GT transitions and the first forbidden transitions have some overlap in terms of log *ft* values. In contrast with the Fermi case, where only one state in the daughter nucleus is populated, in the GT case it can happen that many final states are possible. As a consequence the total strength can be fragmented between many individual levels with relatively large log *ft* values.

There is yet another classification of allowed and forbidden decays which can be readily understood in terms of the isospin formalism [22, 23]. If we assume that nuclear forces are charge-independent we can define the proton and the neutron as two states of the same particle, the nucleon, characterised by the isospin $T = {}^1/_2$ and third component $T_Z = -1/2$ for the proton and $T_Z = +1/2$ for the neutron. For a nucleus, a system with several protons and neutrons, the total isospin can be constructed as the sum of the isospins of the nucleons following the same coupling rules as ordinary vectors. The corresponding value of $T_Z = (N - Z)/2$. The only unhindered, allowed Fermi decays are the so–called *super–allowed* decays, those with $\Delta T = 0$. In other words, only T_Z changes. Such transitions occur between isobaric analogue states (IAS). Consequently, in β–decay they only occur when the IAS of the parent state lies within the Q_β–window in the daughter nucleus.

In the isospin formalism the operator responsible for the decay takes a very simple form. In a Fermi decay only the isospin operator τ acts, either increasing or lowering the third component of the isospin; in a GT decay the isospin operator is also active, but in addition the spin operator σ can flip the spin of the nucleon.

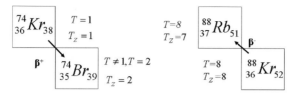

Fig. 5. Examples of β^+ and β^- decays indicating where Fermi transitions are possible (see text)

As one can see in the typical examples of β^+ and β^- decay shown in Fig. 5, there is another consideration regarding the Fermi transitions. In the β^+ decays of $N > Z$ nuclei the absolute value of T_z always increases by one unit, consequently it is impossible to keep the T quantum number unaltered unless one is at negative values of T_z. Thus Fermi decays in nuclei decaying by β^+ emission are very often isospin–forbidden. If we again look at Fig. 5 we see that this is not a problem in the nuclei unstable to β^- emission. Here, however, since the final states usually have T values one unit higher than the ground state, the IAS is, in general, at high-excitation energy and consequently outside the Q_β–window. The reader should note that we have assumed here that the ground state and low–lying levels in the nucleus have the lowest isospin possible. This is, in general, a good assumption apart from the nuclei near $N = Z$.

Turning to GT transitions we face a different situation. As for Fermi transitions one can imagine that the operator transforms a neutron to a proton or vice versa but at the same time the nucleon can have its spin "flipped". If nuclear forces were spin–independent as well as charge–independent we would have a very similar situation to the Fermi decays with the transitions going to the IAS. However, nuclear forces are strongly spin-dependent, as is evidenced from the success of the Shell Model (see [17, 18]). As a result the states are mixed into nuclear states over a wide range of energy centred at the energy of the expected "resonance" state. Moreover, we know now that the residual nucleon–nucleon interaction, and more particularly the $\sigma\sigma\tau\tau$ term, because of its repulsive character, moves the strength from the few MeV, zeroth–order excitation energy of the daughter particle–hole excitations to a resonance peak at typically 15–20 MeV excitation and consequently outside the Q_β–window. As in the case of the Fermi transitions, there is a difference between the β^- and β^+ decays. In general, all that we have said above applies to the β^- case but the β^+ transitions are either very suppressed or forbidden. This is because the allowed orbitals (the place where "the proton which is transformed into a neutron" could "go" following the selection rules) are often occupied on the neutron side. However, this is not always the case as we shall see later. For instance, sometimes a transition between a state involving a proton with high-orbital angular momentum can proceed to its spin–orbit partner state on the neutron side ($J \uparrow \rightarrow J \downarrow$).

Near the line of stability, where the Q_β–values are small, decay schemes are simpler and more amenable to experimental study because there are fewer states that can be populated. At the same time they carry less information. The present thrust of research in this area is to study transition probabilities in β–decays far from stability where the Q_β–values are large, a wider range of states can be populated and we have better access to a large fraction of the GT strength. Although we have been studying β–decay for a long time we have only really scratched the surface in terms of studying exotic nuclei. For a variety of reasons experimenters have not devoted the same effort to develop instrumentation for such studies as they have to develop silicon (Si) and germanium (Ge) detector arrays for use in studying prompt radiation from reactions. As a result only a limited number of cases have been studied in detail but they reveal what one might expect to learn in the future. Some good examples are the studies of allowed decays to the GT resonance in nuclei (a) just below ^{100}Sn [24–26] and (b) in the rare–earth region [27–29], measurements in $A \sim 70$–80 nuclei that have allowed the shapes of the parent ground state to be deduced [30–32] or measurements of super–allowed Fermi decays of importance as a test of our understanding of the weak interaction [33–35].

In Sect. 1 we suggested that one could imagine the β–decay process as the transformation of a proton into a neutron or vice versa. It turns out that there is a type of nuclear reaction, called a charge exchange (CE) reaction, which can be described in just the same way. The simplest examples are the (p,n) or (^{3}He, t) reactions, where a neutron is changed into a proton or the (n,p) reaction, where a proton is changed into a neutron. Under some specific experimental circumstances, namely if they are carried out at zero degrees and at the appropriate projectile energy (about 200 MeV), these reactions are dominated by $L = 0$ transfer and can be described in terms of the Fermi, $\tau_{+/-}$ (isospin) or the GT operator, $\sigma\tau_{+/-}$ (spin–isospin), just as in the case of the β–decay process.

As the reader might well imagine, these reactions and the corresponding β–decays are intimately related. At the same time there are essential differences between them, which we should point out straight away. In general, CE and β–decay are measured on different nuclei. Beta decay starts with an unstable nucleus whilst at present, because of experimental limitations, CE is usually carried out on a stable target nucleus. The immediate corollary is that CE will provide information near the valley of stability whilst β$^-$ decay, as emphasised earlier, will carry more information if we go far from stability where the Q_β–values are large.

One very important advantage of the CE reactions is that they are not restricted to an energy window as in β–decay. The other advantage is that one can study the (p,n) and the (n,p) reactions on the same target. This is important because there is a model-independent rule called the Ikeda sum rule [36, 37] that relates the GT transition probability $B(GT)$ in these two processes, namely

$$B(GT)^- - B(GT)^+ = 3(N - Z). \tag{17}$$

In other words, if we can measure the total $B(GT)$ strength on the "β^-" and "β^+" sides of the same nucleus, the difference between these two quantities should be equal to six times the third component of the isospin of the nucleus concerned.

There are cases where $B(GT)^+ = 0$ and therefore one expects to see $B(GT)^- = 3(N - Z)$. Our example of the decay of ^{132}Sn (see Fig. 3) is just such a case. To understand this one has to look at the representation of the ground state of this doubly magic nucleus from the point of view of the shell model. This is shown in Fig. 6. It is not possible in this case for any of the protons to be transformed into a neutron. In contrast for all of the neutrons marked in the figure there are one or two possibilities for them to be transformed into a proton. As a result, if one were able to measure all the $B(GT)^-$ in the β^- decay of ^{132}Sn unrestricted by the energy window, one should obtain 96 units. We will see later, that the $B(GT)^-$ is in this case severely cut by the energy window. For that we will have to wait until we learn how to deduce the $B(GT)$ value from the experimental observables.

It came as a surprise when CE experiments were first carried out that, although the Ikeda sum rule is model-independent, only a part of its value is obtained. This is a long-standing problem that has puzzled us for more than 20 years. The interested reader is referred to [38–41] for a detailed view of this problem. It seems to be related to problems on the experimental as well as the theoretical side. It is important in this context for the student to know that CE reactions are experimentally difficult to study because of ambiguities in the background and also that the extraction of the $B(GT)$ from the experimental cross-section requires a normalisation factor which is

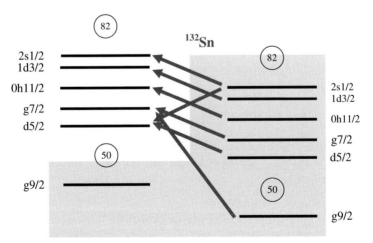

Fig. 6. Schematic view of the proton and neutron orbitals near the surface for ^{132}Sn. The *arrows* indicate the β–transitions that are possible in the β–decay of ^{132}Sn

not trivial to determine. In contrast, in β–decay the relationship between the $B(\text{GT})$ and the experimental quantities is very well defined, as we will see in Sect. 3, but we will always face the problem of the limitation imposed by the energy window accessible in the decay.

In only a very few cases has one been able to compare the experimentally measured B_{GT} in β–decay with the Ikeda sum rule. As it happens it seems that a normalising factor has to be applied in this case also. The justification is that the axial–vector constant, extracted from the experiments on the decay of the neutron has a different value when the decay occurs inside the nuclear medium. Some times this is also addressed as the influence of the sub–nucleonic degrees of freedom.

As indicated above this is a long–standing problem that cannot really progress until new experimental input becomes available. This is among the things one can expect from the new experiments with radioactive beams, where for the first time one will be able to carry out a β–decay measurement on a given nucleus, and a CE reaction on the same radioactive nucleus. The β–decay will provide the normalisation factor without ambiguities, and the CE reaction will provide the full $B(\text{GT})$ without energy restrictions. However, one needs improvements in beam intensities and developments in experimental technique before this happens.

3 Experimental Considerations

3.1 Measuring the Quantities of Interest

In our brief outline of the Fermi theory we put the emphasis on the matrix element $\mathbf{M_{fi}}$ as the quantity carrying information about nuclear structure. However, as mentioned before we have to extract this from the quantity that can be measured, namely the transition probability. For Fermi and GT transitions we define it as

$$B(\text{F}) = | < \psi_f{}^* | \tau | \psi_\text{I} > |^2,$$

and

$$B(\text{GT}) = | < \psi_f{}^* | \sigma\tau | \psi_\text{I} > |^2.$$

(18)

To proceed further we have to relate these theoretical expressions to the quantities we can measure in an experiment, namely the ft value mentioned earlier. For the general case of a mixed Fermi and GT transition we define the ft value as:

$$ft = \frac{k}{g_V^2 B(\text{F}) + g_A^2 B(\text{GT})},$$

(19)

where $k = \frac{2 \ln 2 \pi^3 \hbar^7}{m_e^5 c^4}$ and the constants g_V and g_A are the weak interaction vector and axial–vector coupling constants, respectively.

If we consider a pure Fermi transition then the GT part of this expression is zero. Now as we mentioned earlier the Fermi transitions are very simple and since only the third component of the isospin changes we can best treat them in the isospin formalism, where the initial state is defined as $|T, T_z>$, and the $\tau_{+/-}$ operator converts this state into the state with $|T, T_z + / - 1>$.

In isospin space the algebra of isospin is the same as angular momentum algebra so we can write

$$< T,\ T_z + / - 1|\tau_{+/-}|T,\ T_z >= [(T - / + T_z)(T + / - T_z + 1)]^{1/2}. \qquad (20)$$

A typical example would be the β–decay of ^{14}O, with $T = 1$ and $T_z = -1$. Thus for the super–allowed transitions we have

$$< 1,0|\tau_+|1,\ -1 >= (2)^{1/2} \text{ and } B_F = 2 \text{ and in this case } ft = k/2g_V{}^2. \qquad (21)$$

As it happens one can assume in general that $B(F) = |N-Z|$, the constant $g_V{}^2$ can be extracted from measurements of super–allowed transitions in $N = Z$ and $N - Z = -2$, $T = 1$ parent states [35].

Using this value we can write for the ft value of a GT transition,

$$\frac{1}{ft} = \frac{1}{6,147 \pm 7} \left(\frac{g_A}{g_V}\right)^2 B(\text{GT})_{i \to f} \qquad (22)$$

The ratio of g_A/g_V is -1.266 [42] and is derived from measurements on the decay of the free neutron. The reader should note that the above expression describes the ft value for the transition to a particular state in the daughter nucleus and t is the partial half life to this particular state.

However, for large Q_β–values where decay will occur to regions of high-level density, the information is most conveniently expressed by a strength function [43]

$$S_\beta (E_x) = \frac{1}{6,147 \pm 7} \left(\frac{g_A}{g_V}\right)^2 \sum_{E_f \in \Delta E} \frac{1}{\Delta E} B(\text{GT})_{i \to f}, \qquad (23)$$

where the average transition probability in the energy interval between E_x and $E_x + \Delta E$ is used in the summation. In other words, we are anticipating that in experiment we will measure the transition probability in energy bins of this size.

The ft value and the strength function are related to the experimental observables by the expressions

$$\frac{1}{ft} = \frac{I_\beta(E_f)}{f(Q_\beta - E_f, Z)T_{1/2}}, \qquad (24)$$

and

$$S_\beta(E_x) = \frac{\sum_{E_f \in \Delta E} \frac{1}{\Delta E} I_\beta(E_f)}{f(Q_\beta - E_x, Z)T_{1/2}}, \qquad (25)$$

where I_β is the direct β–feeding either to a state of energy E_f in the first case or the average to all the levels contained inside the interval ΔE in the second, f (which depends on the excitation energy, Q_β–value and Z) is the Fermi Integral and $T_{1/2}$ is the β–decay half life of the parent nucleus.

From Eq. (25), if we recall that the Fermi integral depends linearly on $(Q_\beta - E_x)^2$ one can see that even a small amount of feeding at high-excitation energy close to the Q_β–value will carry a significant part of the strength. The term $f(Q_\beta - E, Z)$ in Eq. (25) depends on the energy available to the electron and the antineutrino which is just the difference between Q_β, the total energy available in the decay, and the excitation energy of the level populated in the daughter nucleus, and Z.

We met the Q_β–value first in Fig. 3, where it was one of the quantities tabulated in the decay scheme for ^{132}Sn taken from [14]. In the simplest terms it is defined just as the difference in total energy between the initial and the final system. More formally we should write in terms of nuclear masses

$$\left.\begin{array}{l} \text{For } \beta^- \text{ decay } N(^A_Z X_N) - N(^A_{Z+1} X_{N-1} + e^-), \\[2mm] \text{For } \beta^+ \text{ decay } N(^A_Z X_N) - N(^A_{Z-1} X_{N+1} + e^+), \\[2mm] \text{For EC decay } \quad N(^A_Z X_N + e^-) - (^A_{Z-1} X_{N+1}), \end{array}\right\} \qquad (26)$$

where N represents a nuclear mass and the other symbols have the obvious meanings. This is fine but in reality we are usually dealing with tabulations of atomic masses rather than fully stripped ions and we must refine our definitions so that we have,

$$\begin{array}{l} \text{For } \beta^- \text{ decay } \left[M(^A_Z X_N) - M(^A_{Z+1} X_{N-1})\right] c^2, \\[2mm] \text{For } \beta^+ \text{ decay } \left[M(^A_Z X_N) - M(^A_{Z-1} X_{N+1})\right] c^2 - 2m_e c^2, \qquad (27) \\[2mm] \text{For EC decay } \left[M(^A_Z X_N) - M(^A_{Z-1} X_{N+1})\right] c^2 - B_e. \end{array}$$

The most recent tabulation of masses is given in [44], where we find the values we require in the useful form of mass excesses. If we use these quantities instead of the masses themselves in Eq. (25) then we get the Q_β–value directly in keV.

The question now arises of how to obtain the I_β feeding to each state. At first sight this seems straightforward, one has only to measure the β–decay to each individual state, but the β–spectra are continuous and in complex decay schemes difficult to disentangle. As a result one cannot derive the β–feeding to individual states in the daughter nucleus from measurements of the β–particles themselves except in a few simple cases. The most popular solution up to now is to use the intensities of the β–delayed γ–rays. This is normally done using high-resolution Ge detectors. In order to assign transitions to a particular decay, one uses all of the information available which includes previous knowledge of the transitions known in the daughter nucleus, if available, the time behaviour of the observed γ–rays (see later), coincidences with the

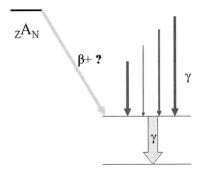

$_zA_N$

$\beta+$?

γ

γ

Fig. 7. Feeding of a typical level in the daughter nucleus in β decay (see text)

daughter X–rays, etc. The next step is to establish the decay scheme from the coincidence relationships between the γ–rays assigned to the decay. Then, having established the level scheme, one can determine the balance of intensity, feeding and de–exciting each level. If the experiment is properly done, the difference in intensity for a given level must come from direct β–feeding (see Fig. 7).

Let us do this for our example for the decay of ^{132}Sn using the information shown in Fig. 3. If we consider the level at 1,325 keV all of the γ–intensity de–exciting the level sums up to 97% of the parent nucleus decays and since there is no γ–feeding, this level should be all due to direct β–feeding. However, to our surprise we see that instead of 97%, the number quoted is 99%. The reason is, of course, that not all of the de–excitation of the level proceeds by γ–emission. Part of the decay goes via the competing process of internal conversion, which in the case of the low-energy transition of 246.9 keV energy adds the extra 2%. As we will discuss later in Sect. 3.1 there is an inherent problem in the technique of deducing β–feeding from high-resolution measurements, but if we assume for the moment that the measurement is correct, we can attempt to derive the log ft value to this level, and this is done in the following way.

As input we need the parent β–half life, which for the decay of ^{132}Cd is given as 39.7 s (remember to take the β–branching ratio into account in any case where there is a competing decay mode such as α–decay, fission or proton-decay), the β–decay energy $(Q_\beta - E_F) = 3,300 - 1,324 = 1,976$ keV, and the direct β–feeding to this level which is $I_\beta = 99\%$. The partial half life to this level is simply

$$t = \frac{39.7}{0.99} = 40.1 \, \text{s}. \tag{28}$$

We obtain the required value of log f from the tabulated values for β^- decay and $Z = 51$ in Table 1 of [45]. We find for a β–decay energy of 1,976 keV a value of log $f^- = 2.62$ and, since log $t = 1.61$, we have log $ft = $ log $f +$log $t = 4.23$, the value given for this level in Fig. 3. As an alternative to this simple procedure one can use the website (http://www.nndc.bnl.gov/logft/) to find exactly the same number.

What does that mean in terms of $B(\text{GT})$ units? If we now use the value $g_A/g_V = -1.262$ [42] in Eq. (24) we obtain

$$B(\text{GT}) = \frac{1}{10^{4.2}} \times \frac{6,147}{(1.262)^2} = 0.24. \qquad (29)$$

This quantity is to be compared with the expected total $B(\text{GT})^- = 96$ derived from the Ikeda sum rule in Sect. 2. As anticipated earlier we see only a small fraction of the expected total $B(\text{GT})^-$ strength, namely 1/400th, within the energy window accessible in the β–decay.

As mentioned above, the procedure we have described for obtaining the β–feeding is typical for determining this quantity. There is, however, an inherent deficiency in such measurements with semiconductor Ge detectors. Figure 7 can be used to illustrate the problem. The level in the daughter nucleus is fed directly in β–decay and it is fed indirectly by electromagnetic transitions from higher-lying levels. However, Ge semiconductor detectors, which have moderately good energy resolution, have modest efficiencies. At present even the best arrays of such detectors have efficiencies of about 20% for γ–rays of 1,332 keV energy [3]. To make matters worse their detection efficiency is strongly dependent on energy. If the β–feeding to a level is deduced from the difference between the γ–ray intensity feeding the level and the intensity de-populating it, and many weak transitions are unobserved, then their strength can add up to a sufficiently large number that we get quite the wrong number for the β–feeding. Thus, if we use detection techniques where the detection efficiency is much less than one, we cannot reliably extract the β–feeding or *ft* values simply from the γ–ray intensity balances. As we move away from stability and Q_β–values increase, one expects greater fragmentation of the β–feeding because of the rapid increase in level density with excitation energy. One consequence is that, in general, the average γ–ray intensity will also be reduced and more γ–rays will not be observed. In addition, there will be more feeding of levels at higher energies, which will be de–excited by higher energy γ–rays, on average, for which the detection efficiency is lower. Thus we can expect this difficulty to get worse. This problem was recognised some time ago [46] and was named the *Pandemonium* effect by Hardy et al. after the city where Lucifer reigned in Milton's epic poem *Paradise Lost*, a place where one might expect chaos to reign. The reader should note they may encounter the expression "apparent log ft" in the literature which addresses exactly this problem.

Can this problem be overcome? One solution is to adopt a quite different approach to the measurements. Total absorption gamma spectroscopy (TAGS) [47, 48] offers just such an approach and will be explained in Sect. 3.2.

3.2 Total Absorption Spectroscopy

In this method of determining the β–strength one still detects the secondary γ–rays but one aims to measure the population of the levels directly rather than indirectly as described earlier. In the ideal case TAGS involves a γ–ray detector with 100% detection efficiency. In the spectrum from such a detector one will detect for each and every β–decay the summed energy of all the γ–rays in the resulting cascade de–exciting the level that is fed initially. It will be readily obvious that the main difficulty in applying the technique is the creation of a spectrometer with 100% detection efficiency.

One question that immediately springs to mind is "If it is important to obtain reliable and accurate β–strength distributions and TAGS provides the remedy to the problem outlined earlier, why is it not in widespread use?" The answer lies partly in the difficulty of making a spectrometer with sufficiently high efficiency, partly in the complexity of the analysis of the data collected in such experiments and partly the lack of a detailed study of the assumptions underlying the analysis methods and the associated systematic uncertainties. As we will find out in the following, all of these questions have now been addressed.

We are not concerned in this lecture with the history of how the technique of TAGS has developed. The interested reader is referred to [47, 48]. The critical point is that the early measurements [49–52], although carried out competently, simply used detectors which were too small and hence were too far from having the ideal efficiency. They may be characterised as partial TAGS experiments. In their introduction of the fictitious nucleus *Pandemonium*, Hardy et al. exposed the difficulties in measuring β–strength functions. In particular, it is essential for the success of the method that the spectrometer has as high an efficiency as possible. At the same time, it is imperative that ways are found to analyse the data. There are now two spectrometers [47, 53] which have the required characteristics and Tain and Cano–Ott [54–56] have developed a suite of analysis programmes that can be used to analyse the data from them.

In terms of analysing the data we should remind our reader that the relationship between the β–feeding $I(E_j)$ and the data d_i measured in channel i in the total absorption spectrometer is given by

$$d_i = \sum_{j=1}^{j_{\max}} R_{ij} I_j, \qquad (30)$$

where R_{ij}, the response function, is the probability that feeding at an energy E_j produces a count in channel i. In order to determine the response function we need to know how the spectrometer responds to individual quanta and β–particles as a function of energy and also have a knowledge of the branching ratios for the electromagnetic transitions de–exciting the levels [55]. From Eq. (30) we see that if we want to determine the β–feeding we must solve this inverse problem. This is not a trivial exercise because Eq. (30) falls into the class of so–called "ill–posed" problems and their solution is neither trivial nor straightforward. Tain and Cano–Ott [54–56] have devoted considerable effort to examining how to optimise the solutions and these authors make recommendations on how the analysis should be carried out. For details the reader is referred to their papers. In a nutshell they examined the suitability of three different de–convolution algorithms for extracting the correct intensity distribution from the data.

It is necessary in the analysis either to have a knowledge of the branching ratios of the levels populated or to make some assumption about them in cases where we do not know the level scheme. Tain and Cano–Ott also examined how much the results of the analysis depended on assumptions about the branching ratios and were able to show that in the cases they studied the results are insensitive to the initial assumptions. However, in other cases [57] it is important to have a solid knowledge of the level scheme at least up to some reasonable excitation energy. In recent years, considerable progress has been made in the Monte Carlo simulation of the response to individual single quanta and this is also an ingredient in the analysis procedure developed by Tain and Cano–Ott. Taken overall these authors have put the analysis of TAGS data on a sound footing although the methods must be applied with due care and attention to the individual case under study.

The most successful TAGS instrument built to date was installed at the GSI on–line mass separator [53]. This instrument, called the GSI–TAS, involved a larger single NaI detector than any that had been used previously [52]; it is of cylindrical shape with dimensions 35.6×35.6 cm, with a central well which could be filled with a matching plug of NaI. The activity from the mass separator was implanted on to a transport tape which was used both to allow a freshly prepared source to be carried to the counting position and to carry away residual daughter activity after a preset counting time. This is a standard procedure used in many experiments at mass separators (see Sect. 4.1). The set up included a small Ge detector and ancillary Si detectors placed inside the well, close to the source position, to allow coincidence measurements of γ–rays in the large NaI with X–rays and β– or α–particles, respectively. A whole series of measurements have been made with this device including measurements of the β–decays of spherical, rare–earth nuclei [27, 28] and neutron–deficient nuclei just below ^{100}Sn [24–26]. In addition to their work on the solution of the "inverse, ill–posed" problem the Valencia group looked at the effect of the non–linearity of the light output in the NaI scintillator [58] and of the pile–up in the electronic circuitry [59] and showed that these effects could be taken into account satisfactorily.

More recently, the present authors were involved in building and installing a new total absorption spectrometer *Lucrecia* at the CERN–ISOLDE mass separator. There were two main aims for the use of this spectrometer, namely to take advantage of the wide range of separated nuclear species available from the ISOLDE separators and to be able to arrange that the separated activity can be deposited directly at the centre of the spectrometer, thus eliminating the delay in carrying the sources from an external point of implantation into the spectrometer. In this mode the tape is used to carry away the daughter activities rather than to refresh the sources under study. The system was designed in such a way that it could still be operated as at GSI with the sources implanted externally then carried to the counting position in the centre of the detector. The *Lucrecia* spectrometer consists of an even bigger single crystal of NaI (38 × 38 cm) with a 7.5 cm through hole, symmetrically placed, at right angles to the axis of the cylinder. From one side the tape and the beam, depending on the half life of interest, enter the crystal. From the other, a number of ancillary detectors can be placed close to the counting position so that one can measure the total absorption spectrum in coincidence with β–particles, positrons, X–rays and γ–rays. In the experiments with *Lucrecia* this involved a 2 mm thick plastic detector for detecting β–particles placed close to the source and a Ge telescope, consisting of a 1 cm thick planar detector backed by a 5 cm thick, co–axial Ge detector.

Detectors as large as the GSI–TAS and *Lucrecia* are a much better approximation to the ideal detector than those used earlier. This certainly reduces the uncertainties in the analysis. However, it also means that they are more efficient in terms of detecting background radiation. In both cases an effort has been made to minimise the background and hence improve the sensitivity. At ISOLDE, where the activities are produced in the fission or fragmentation of heavy targets with a beam of 1.4 GeV protons from the PS–Booster, one might expect a significant background from both γ–rays and neutrons in the experimental hall. To minimise the overall background the spectrometer system was placed inside an 11-ton shield, made up of successive layers of boron–loaded polyethylene (10 cm thick), lead (5.1 cm), copper (1.5 cm) and aluminium (2 cm). The resulting spectra show that this is effective with the main background being due to ^{40}K, which is present as a natural contaminant in the crystal.

Figure 8 makes a comparison of the performance of the two detector systems. On the left it shows views of the two central NaI detectors. The GSI–TAS was mounted with the central axis in the vertical direction. *Lucrecia* on the other hand is mounted horizontally, with the through hole pointing along the beamline delivering the radioactive sources. In the photograph we see the through hole with the Ge telescope withdrawn from it on the far side from the direction in which the beam enters. On the right–hand side of Fig. 8 we see the total and photopeak efficiencies for the two spectrometers. The empty and filled points represent the results for the GSI–TAS and the *Lucrecia* TAGS, respectively. The latter is the larger of the two detectors and instinctively one feels it should have the higher efficiency but one must remember that it was designed to allow the radioactive beam to enter directly and a through hole was used to allow the insertion of the ancillary detectors. In addition, although a matching plug detector is available to fill the hole, it is not used in general since the space is used for the β– and γ– detectors. In contrast, in the GSI–TAS an effort was made to design a special Ge detector that would allow the NaI plug to fill all the space except for that reserved for the cold finger. As a result the efficiency of the Lucrecia spectrometer is lower than that of the GSI–TAS. However, we should remember that these curves represent

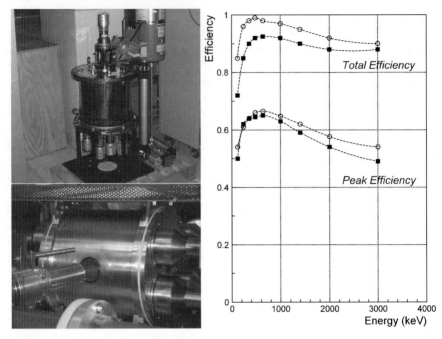

Fig. 8. Photographs of the GSI–TAS [53] (*upper left panel*) and Lucrecia [47] (*lower left panel*) and corresponding total and photopeak efficiency curves for these two detectors (*right panel*). The empty and filled points represent the results for the GSI–TAS and the Lucrecia TAGS, respectively

the efficiency for detecting a single γ–ray of a given energy. In practice, the de–excitation of a given nuclear level generally involves more than one γ–ray and in the case of positron emission two 511 keV quanta are also produced. Thus the probability that at least one of the γ–rays leaves some energy in the crystal is very high and consequently the difference in total efficiency is less dramatic than shown in Fig. 8. In other respects, the two spectrometers are comparable with the energy resolutions and intrinsic backgrounds being very similar. We will see examples of the use of both these spectrometers in Sect. 4 below.

4 Some Illustrative Examples

In this section we will discuss the results of a number of different β–decay studies. They have been chosen both to illustrate some of the topics dealt with earlier and show the breadth of the science that can be addressed in such decay studies.

4.1 The ISOL Method and ^{130}Cd Decay

Our first example is of an experiment that makes use of what one might call the classical "ISOL" technique [60] with the γ–rays being detected using high-resolution Ge detectors. In this method the nuclear species of interest is produced in a reaction, which can be anything from neutron capture in a thermal reactor to spallation by a high-energy proton beam (~ 1 GeV) or fission induced by fast neutrons. The nuclear species produced are transformed into ions and later into a beam of ions of low energy, typically 60 keV, and then separated in mass with an analysing magnet. In some cases chemical properties are used to isolate the species of interest by Z as well as A. The overall aim in doing this is to produce a beam of high purity, the highest possible intensity and good optical quality which we can transport as far away as possible from the production site, where there is a large background from reactions, to a well–shielded experimental setup.

In our case of ^{130}Cd the nuclei were produced at the ISOLDE facility at CERN. In this case the selectivity for producing ^{130}Cd was enhanced by using two techniques which have been recently added to the standard ISOL mass separation repertoire. The first of these involves the suppression of spallation products in the primary reaction process which is then dominated by fission. This reduces the production of proton–rich isobaric elements, which are not the subject of study here. In essence the idea is to use neutrons to induce fission in an actinide target. The neutrons are produced using the 1 GeV protons from the CERN PS–Booster by bombarding a W or Ta target. The neutrons produced are then used to induce fission in a UC$_X$ target situated near the neutron converter [61]. The fission process leads to the production of the neutron–rich nuclei of interest. The second innovation was to use laser ionisation techniques [62] in order to selectively ionise the Cd atoms. Although laser ionisation enhances the ionisation of the species of interest, Cd in this case, other elements are still ionised. However, by recording data with "laser on" and "laser off" we can distinguish the lines corresponding to the decay of interest. This was essential to differentiate between the Cd decay lines and the In decay lines of the daughter, which is the most abundant contaminant produced. In these experiments the high-resolution mass separator (HRS) was used with $M/\Delta M \sim 4300$, the third essential element in obtaining the required selectivity in the experiment. The purified, mass–separated beam containing ^{130}Cd was implanted into the surface of a moving tape to form a source which was transported from the collection point to the measuring position in front of the detectors. In this experiment measurements were made either with four large HpGe detectors for γ–ray singles and coincidence measurements or with a ΔE–E_β–telescope replacing one of the Ge detectors so that β–γ coincidences could be measured.

As explained in Sect. 3, one set of quantities we would like to extract from the data is the β–feeding to each state. Here for ^{130}Cd the intensities of the β–delayed γ–rays were used; the most common procedure. Firstly the decay scheme is established using the coincidence relationships between the observed γ–rays assigned to this decay and measured energy sums. Having established the decay scheme it is then possible to determine the direct feeding to each level from the difference in the summed intensities of γ–rays feeding the level and those de–exciting it. To be sure that one has established the fraction of decays feeding each level we need two other pieces of information. We need the amount of direct feeding to the ground state of the daughter, with which no γ–rays are associated and we need to be sure that there is no feeding from other states, which we have missed or which is too weak to measure. The former is, in general, difficult to measure although there are cases where it is possible. In [63] this was not possible and the ground state feeding had to be calculated using the Gross Theory of β–decay [64]. As far as the latter feeding is concerned other γ–rays were assigned to ^{130}Cd decay but they are not shown in the partial decay scheme of [63]. In particular, a cluster of seven levels with excitation energies about 4.4 MeV was thought to be fed in the decay. Their combined β–feeding was ∼ 3.5%. Following these considerations we are left with the percentage β–feeding to each level shown in the extreme left–hand column in Fig. 9. The reader should note that the figure shows only a partial level scheme and, if they check, they will find that the β–feeding to the excited states, plus the β–decay to levels which decay by neutrons (3.6%, see below) does not add up to 100%. The remaining 12.9% is the fraction of the feeding the authors estimated they have detected but were unable to locate in the level scheme.

To go further we also need the half life of the parent nucleus. This has been measured several times for ^{130}Cd. The value quoted in Fig. 9 was measured in an earlier experiment [65, 66] by the same authors. The activity was again produced at CERN–ISOLDE. In a small percentage of decays ^{130}Cd emits β–particles followed by a neutron. They used a 4π neutron detector to record the delayed neutrons and measured their intensity as a function of time after the production of a fresh source. The result of 162(7) ms is given in Fig. 9. In this case and before all the selectivity techniques described above were fully developed; this was the best way to avoid contamination from isobars closer to the valley of stability. The reason is that there are no neutrons emitted in the decay of possible contaminants in the same isobaric chain. In the following we will explain how to extract the half life from the observation of the time behaviour of the radioactivity, whether it takes the form of β–particles, γ–rays, neutrons, protons, α–particles, etc. In many studies the γ–rays are used because they are characteristic of the decay (see the example in Sect. 4.2). Once one has identified the radiation associated with a given decay one can extract the half life by recording the number of counts in the γ–ray peaks as a function of time. In [63] this was not the best method because of the elaborate subtractions of spectra needed to obtain a "clean" spectrum.

To understand how we extract the half life we need only to remember the radioactive decay law

$$N = N_0 \, \exp(-\lambda t), \tag{31a}$$

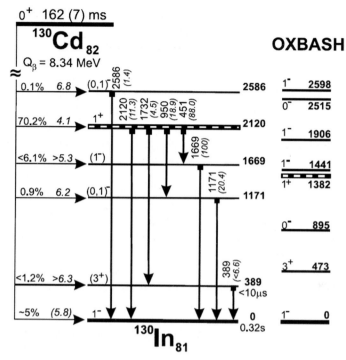

Fig. 9. Partial decay scheme for ^{130}Cd [63]. Reprinted with permission from I. Dillman et al., Phys. Rev. Lett. **91**, 162503 (2003). Copyright (2003) by the American Physical Society

where N and N_0 are the numbers of nuclei of the decaying species at times t and $t = 0$, respectively, and λ is the radioactive decay constant, characteristic of the decay in question. If we recast this familiar equation in the differential form

$$-(dN/dt)/N = \lambda, \qquad (31b)$$

we see that the probability of decay is constant in time. Using Eq. (31b) and setting $t = 0$ we can rewrite Eq. (31a) in the form

$$dN/dt = (dN/dt)_{t=0} \exp(-\lambda t), \qquad (31c)$$

where the subscript $t = 0$ indicates the rate of decay at time zero. In the case of a β–decaying nucleus this gives us the rate of decay of the parent nucleus. There is, of course, a one–to–one correspondence between the number of nuclei in the parent which decay and the number of nuclei in states in the daughter nucleus which are created. Since the percentage feeding to individual levels in a β–decay is fixed, then for any prompt γ–rays which follow the decay, we have the same expression for the rate of γ–emission in the daughter nucleus,

$$dN\gamma/dt = (dN\gamma/dt)_{t=0} \exp(-\lambda t), \qquad (31d)$$

where the subscript now indicates the number of γ–rays emitted. Thus measuring the rate of emission of the β–delayed γ–rays as a function of time allows us to determine the half life of the parent decay. The reader should note that we have assumed that the lifetimes of the γ–decaying levels are short compared with the lifetime of the β–decay involved.

The radioactive decay law also leads to the definition of half life as the time in which the number of nuclei is reduced by half or the equivalent, namely that the rate of decay of the nucleus under study has reduced by a factor of 2. Again it is worth noting that if one has identified a γ–ray as belonging to a nucleus under study it is a relatively easy quantity to measure "on–line" and hence check that you are dealing with the activity you expected. It is also useful in identifying that a series of γ–rays comes from the same decay.

In practice, if we are to determine a precise value for the half life from measurements of γ–ray intensities there are a number of practical considerations of importance. Thus it is important, amongst other things, how one determines peak intensities in the γ–ray spectra, how one determines the background (often a delicate matter in γ–ray spectroscopy) and one needs spectra which are well determined statistically. One must also take due account of changes in the electronic deadtime and pulse pile–up as the source decays as well as correcting for the effect of the way in which the counts are binned with time.

As mentioned before one must take account of any short–lived activity feeding the species of interest. This is avoided if one has chemical separation or if the parent activity is produced with considerably less yield. Daughter activity normally present in the sources does not affect the situation in principle since the γ–rays from the daughter will normally have different energies from those of the parent.

After this digression, let us return to the case of ^{130}Cd where, as mentioned, the half life was measured by detecting the delayed neutrons as a function of time, since in this particular case they are not emitted from ^{130}In, the main contaminant in the experiment. In addition to the $T_{1/2}$ we also need the Q_β value if we are to determine the log ft values. In [63] this was measured directly using a version of the so–called Fermi–Kurie plot. In the allowed approximation one can readily show [17, 18] that

$$(Q_\beta - E) \infty \text{ sqrt.} N(p)/p^2 F(Z, p), \tag{32}$$

where $N(p)$ is the number of electrons with momentum p and F is the Fermi function. If we plot these quantities against one another we should obtain a straight line which has an end point equal to $(Q_\beta - E)$. One has to be very careful, however, when we try to extract the $N(p)$ from the experimental data since they are folded with the instrumental response function. Dillmann et al. did this kind of analysis for the β–particles in coincidence with all of the transitions de–exciting the 2,120 keV level shown in Fig. 9. The coincidence condition guaranteed that the β–particles were indeed from the ^{130}Cd decay. Another word of caution with regard to this kind of analysis is the possible influence of β–particles in the spectra populating the 2,120 keV level indirectly via γ–rays from higher lying levels. In other words, the "pandemonium" effect might deform the spectrum at lower energies. This is probably the reason why the authors of the article used only the higher part of the spectrum in their analysis. The resulting end point energy added to the level energy in their work gave a value of $Q_\beta = 8,344(\pm^{165}_{157})$ keV. The short format of the published letter did not allow the authors to explain in detail how the final uncertainties were calculated, but one would hope that they included all of these effects. This is the first direct measurement of this Q–value and it is of considerable significance in terms of shell quenching at $N = 82$ (see discussion in [63]).

The question of how well this value agrees with the predictions of mass formulae and the tabulated mass values of [44] is an interesting one but is not germane to our present discussion and we put it to one side. Returning to our decay scheme of Fig. 9 we now have all the information required to determine the ft values shown in the second column. The 1^- spin and parity of the ^{130}In ground state are taken from earlier studies. The fast transition to the 2,120 keV level with $\log ft = 4.1$ is clearly an allowed GT transition and this fixes the spin and parity of this level as 1^+. The $\log ft$ values for the feeding to the 1,171, 1,669 and 2,586 keV levels all suggest non–unique, first forbidden transitions (see Fig. 4). The spins and parities assigned to these levels are tentative (they are shown in parentheses) but are based on this assumption. This leaves the isomeric level at 389 keV, which had been seen in fragmentation. The spin and parity for this level do not derive from [54] and are based on the lifetime of the level. The observed limit on the $\log ft$ is consistent with this spin and parity.

It is interesting to compare the experimental results obtained for the decay of ^{130}Cd with shell model predictions. Figure 10 shows schematically the single particle levels appropriate to a discussion of ^{130}Cd, which we can see in the simplest terms as having a structure dictated by having two holes, associated with two neutrons, in the doubly closed ^{132}Sn core (shown in Fig. 6). A significant number of GT transitions is possible if we consider all of the neutron levels in the $N = 50$–82 major shell. However, only one of them, namely the $\nu g_{7/2}$–$\nu g_{9/2}$, is expected at low-excitation energy in ^{130}In because the final $(\pi g^{-1}_{9/2}\, \nu g^{-1}_{7/2})_{1+}$ state is a two–hole state, whereas all other transitions will lead to a four–particle excitation. As we can see in Fig. 9 we find only one transition with a $\log ft$ corresponding to an allowed transition, namely the transition to the state at 2,120 keV. This state is thus identified as the $(\pi g^{-1}_{9/2}\, \nu g^{-1}_{7/2})_{1+}$ state. On the other hand, the only way to produce a spin and parity of 1^- for the ground state of ^{130}In is by the coupling $(\pi g^{-1}_{9/2}\, \nu h_{11/2}^{-1})$. Thus the excitation energy of the 2,120 keV level is determined by the difference in the single particle energies of the $\nu g_{7/2}$ and $\nu h_{11/2}$ orbitals and the difference in residual interaction between the $(\pi g_{9/2}\, \nu g_{7/2})$ and $(\pi g_{9/2}\, \nu h_{11/2})$ nucleon–nucleon residual interactions. This difference can be calculated in the framework of the shell model and on the right-hand side of Fig. 9 we see the authors' attempt to do this with the OXBASH code [67].

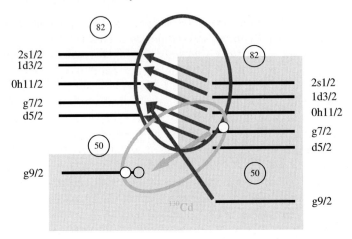

Fig. 10. Schematic view of the proton and neutron orbitals near the surface for ^{130}Cd. The *arrows* indicate the β–transitions that are possible in the β–decay of ^{130}Cd

The fact that this excited state lies at much higher excitation energy in experiment than in the calculation was a surprise and means that the residual interaction is larger than anticipated. If this result was repeated in all of the neighbouring nuclei then the β–decay half lives would be longer than previously thought.

The motivation for Dillmann et al. to study this decay lies in its significance as a waiting point nucleus in the astrophysical r–process [68]. The reader will find an up–to–date description of the r–process in [69]. Here we simply sketch out the basic ideas. Most of the heavy nuclei beyond ^{56}Fe are thought to be produced in reactions induced by neutrons, the process of element building in stars by charged particle–induced reactions having been brought to an end by the fact that such reactions become endothermic. Examination of the measured solar abundances of the elements reveals two peaks related to each of the closed neutron shells in heavy nuclei. In simple terms the two peaks are thought to relate to the so–called s– and r–processes, where s and r stand for slow and rapid neutron capture, respectively. In both processes there is competition between neutron capture and β–decay and in the r–process competition with photo–disintegration as well. In a situation where the neutron flux is low, say 10^8 neutrons per cm^3, β–decay usually occurs before another neutron can be captured. Since we are on the neutron–rich side of stability and are concerned with β$^-$ decay this increases the nuclear charge by one unit before another neutron capture occurs. Accordingly, in such a moderate flux we will get a series of neutron captures followed by β–decays and we will slowly climb up the Segre chart, remaining always close to stability. This is the s–process.

In contrast, if the neutron flux is high, say 10^{20} neutrons per cm^3 or more, we have a rather different situation. Now three processes are important, namely neutron capture and the inverse process of photo-disintegration (the (γ, n) reaction) and β–decay. In explosive processes, where the neutron flux is high, the flux of γ–rays will be very high as well. Starting with stable species we will have, for a given isotopic chain, an abundance determined by the competition between the (n, γ) and (γ, n) reactions. In general the former will "win" until we reach a point where the neutron–separation energy has dropped to the point where we have an equilibrium. This happens at the point where we reach the neutron magic numbers since the capture cross–section is small and the neutron–separation energy is also small and hence the (γ, n) cross–section is larger. Now the nuclei in question last long enough for β–decay to occur back to the line of stability producing a peak in the abundance curve related to the neutron magic number in question. Thus these closed shell nuclei are referred to as "waiting point" nuclei.

In Fig. 11 we see the observed solar system elemental abundances in the region of interest in the case of ^{130}Cd. We also see two calculated curves called "short" half lives and "long" half lives, respectively. The former is based on results available prior to the measurements of Dillmann et al. [63] and the latter was calculated by them assuming that all of the half lives of the $N = 82$ nuclei are longer than previously thought, using the residual interaction deduced from ^{130}Cd (see Fig. 11). The latter curve is clearly in better agreement with the observations in this $A \sim 130$ mass region. If the

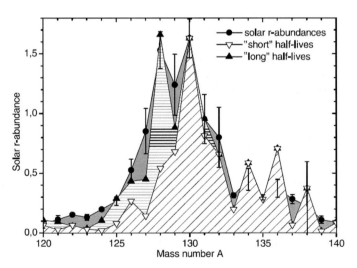

Fig. 11. The Solar system abundances observed for the r–process are shown by the *filled circles*. The other two sets of points labelled "short" and "long" half lives were calculated based on the knowledge of ^{130}Cd β–decay before and after the measurements reported in [63]. Reprinted with permission from I. Dillman et al., Phys. Rev. Lett. **91**, 162503 (2003). Copyright (2003) by the American Physical Society

"long" half–life assumption is confirmed, this is an important result. It also emphasises how important it is to measure such quantities since we do not have reliable nuclear models with which to calculate them. In general we are confident of the basic idea that the heavy elements are created in the s– and r–processes. There is still uncertainty about where these processes occur and whether they occur in one type of site only or in more than one. Opinion is that the s–process occurs in the He–burning shell of pulsating red giant stars, where $T \sim 30\,\mathrm{keV}$ and the neutron flux is 10^8 neutrons per cm^3. More recently opinion favours Asymptotic Giant Branch stars of low mass in particular. In the r–process the neutron flux is much higher and a variety of possible sites have been proposed. Here the favoured site is a core collapse supernova but it may be that it also occurs in neutron star mergers or similar events. There is evidence that more than one type of site is required if the observed abundances are to be reproduced. If good models of the various possible processes are to be produced then reliable measurements of the properties of exotic nuclei such as ^{130}Cd are essential to underpin them.

4.2 Fragmentation and Beta Decay of Exotic Nuclei

In the previous example, the experiments used were what we called earlier the classical "ISOL" technique [60]. As the reader of this reference will find there are limitations to this technique with two obvious drawbacks in particular. Firstly there are significant delays in time in extracting a nucleus produced by the primary beam from the target/ion source. For short–lived isotopes this will lead to a considerable loss in yield because they will have decayed before they reach the experimental apparatus. Secondly, in some cases the chemistry may be such that the technique cannot be used since the ions are not released from the target. This is true, for example, for refractory elements. There are, however, steadily fewer cases of this type as techniques are developed to overcome the difficulties [62, 70].

An alternative is to use the fragmentation of high energy heavy ions and their separation "in flight". This minimises or eliminates these problems although it introduces new limitations of its own. The technique is discussed in detail in [71]. In essence it involves the peripheral interaction of a high energy heavy projectile nucleus, of energy $20\,\mathrm{MeV/u}$ to $2\,\mathrm{GeV/u}$ or higher, with a light target nucleus. In the reaction some nucleons are removed from the nucleus and essentially all of the products are focussed into a narrow forward cone because of the high-initial velocity. There are recoil effects on the heavy product which open the cone. They are due to the removal of the nucleons and the deflection due to the Coulomb field. But these effects are small. Similar considerations apply to fission fragments produced in fission at high energy in inverse kinematics. In both cases these products can then be separated in–flight using a fragment recoil separator [72–75]. Figure 12 shows such a separator schematically. Typically, they can work in either monochromatic or achromatic mode. In this brief outline we will confine ourselves to the achro-

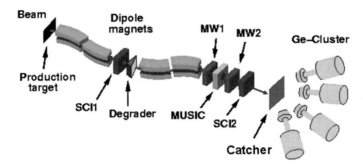

Fig. 12. A schematic view of a typical fragment recoil separator. (In reality the fragment separator at GSI [72]). The mode of operation of such a separator is described in the text. Here the cocktail of ions from the separator is stopped in a catcher, a passive piece of material. In the experiments described in Sect. 4.2 the catcher is replaced by an "active" stopper in the form of a stack of double–sided Si strip detectors (DSSD)

matic mode, where the horizontal position and angle of a particle does not depend on its momentum as explained in [71]. In the first section of the device magnetic elements are used to select the ions by the ratio of momentum–to–charge and at the same time reject the primary beam particles. The latter is important because scattered beam particles may produce background at the end of the spectrometer. There is then an energy loss degrader or "wedge", so–called because of its physical shape, then a second magnetic system which is used to select the ions we want to study by their charge–to–momentum ratio.

In principle such a separator works on the basis of a few, relatively simple physical ideas. If we neglect relativistic effects for the moment then we know from our elementary physics that a charged particle with mass m, velocity v and charge q moving in a magnetic field B experiences a force $\mathbf{F} = q.\mathbf{v}_\otimes\mathbf{B}$. For the case of constant field B perpendicular to the velocity of the ions we can relate its momentum, charge and the constant radius ρ of its trajectory by

$$mv/q = B\rho. \tag{33}$$

We also know that when a charged particle passes through material it loses energy at a rate proportional to $(q/v)^2$ and so its momentum loss in such a process is proportional to (q/v). In high-energy fragmentation and fission many nuclear species with the same momentum–to–charge ratio are produced. In the first half of a typical fragment recoil separator the ions are separated by m/q by a series of dipole magnets. If we concern ourselves only with fully stripped ions they are dispersed in the x–direction in $m/q = A/Z$ when they arrive at the wedge. The ambiguities are removed by the wedge because the momentum loss is proportional to Z/v. Since the ions all have approximately the same velocity when they enter the wedge, ions with the same A/Z but different Z will come out of it with different momenta and have different trajectories in the second part of the spectrometer because they will have different $B\rho$ values. As a result the different ions will be dispersed in the X direction when they arrive at the final focal plane of the spectrometer. If we measure $B\rho$, the time–of–flight (TOF) and the energy loss in a final detector we can identify the individual ions in A and Z. How do we do this?

From measurements of the x and y positions of the ions after the wedge and at the final focal plane together with the measurent of the field B in the magnets one obtains $B\rho$ which is proportional to vA/Z (see Eq. (33)). The velocity v can be extracted from the TOF in the second half of the spectrometer and the Z from the energy loss in an ancillary detector, leading us finally to the determination of A and Z. This simplified picture ignores many of the details such as relativistic effects and the fact that the ions may be in different charge states, which must be taken into account in a real experiment.

If the energies are relatively low, as at the NSCL, MSU, or if one keeps the dispersion in $B\rho$ small, for example by making the ions pass through narrow slits, then the TOF is enough to clearly separate the ions. We see such an example in Fig. 13. It is taken from the next experimental example we want to explain in this chapter. The data were recorded with the A1900 fragment separator at NSCL, MSU by Hosmer et al. [76] using a primary beam of 140 MeV/u ^{86}Kr.

In this particular experiment the aim was to study the β–decay of ^{78}Ni by first implanting the selected ions into a Si detector system and then observing

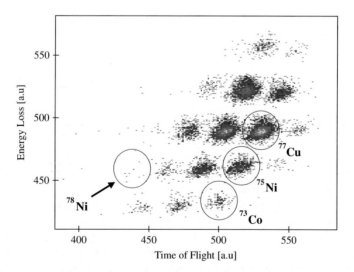

Fig. 13. Particle identification plot using energy loss versus time–of–flight for 140 MeV/nucleon ^{86}Kr on a Be target studied at NSCL [76]. Reprinted with permission from P. Hosmer et al., Phys. Rev. Lett. **94**, 112501 (2005). Copyright (2005) by the American Physical Society

the subsequent β–decay. In practice, three double–sided Si strip detectors (DSSD) were used at the final focal plane. The energy loss of the heavy ions was measured in the first two DSSDs. An Al degrader was located in between to allow the adjustment of the final energy in order to implant the nuclei of interest in the last DSSD which had 985 μm thickness and was segmented with 40, 1mm strips in the horizontal on one side and 40, 1 mm strips in the vertical on the other, resulting in 1,600 effective pixels. This detector was used to measure the signal from the implants as well as the energy loss (ΔE) of the β–particles produced in the subsequent decay of the implanted ion. Because of the large difference in energy range of these two signals, one of the order of a few GeV and the other of hundreds of keV, two different electronic chains were used with different amplifications to provide suitable signals for processing and recording.

The implanted ion generates a pulse in part of the DSSD which can be identified in x and y and hence as occurring in a defined pixel. In the subsequent β–decay a pulse will be produced in the same pixel. Such correlated signals are the markers for events of interest. In essence a histogram of the number of events with differences in absolute time between the implant events and their correlated β–particle pulses in the same pixel provides a radioactive decay curve (see Eq. (29)).

If, in addition, the DSSDs are surrounded by an array of γ–ray detectors it is then possible to record correlated signals from implants, β–particles and the subsequent delayed γ–rays.

This is the essence of the methods which are being developed and are already being used to exploit fragmentation and high-energy fission reactions to study β–decay.

As indicated earlier such processes have the advantage that all chemical species are accessible, since they do not depend on chemical properties, as are very short–lived species since the only time constraint here is that they can survive transit through the separator, which is typically of the order of 100 ns. The transit time then sets a lower bound on the half life that can be studied. Accordingly this technique, in general terms, is presently the method of choice if we are to study β–decays as far away from stability as possible. As one might suspect, however, there are difficulties hidden in this idealised explanation. For instance, one must find a means of identifying the centre of gravity of the energy deposited by the implants since the incoming ion will induce charges in a number of neighbouring pixels as well as the one that it hits. In addition a limit in β–decay lifetime is set to such measurements by random coincidences. The rate of the latter depends on the degree of pixellation, the rate of implantation, the half lives of the implanted ions, and the time interval during which we want to measure the nucleus of interest, which is normally related to its half life.

In Sect. 4.1, we presented the case of a measurement at an ISOL facility of the half life of a nucleus with $N = 82$, which was of importance for our understanding of the astrophysical r–process. Here we present a similar case, namely the experimental study of the half life of a nucleus with $N = 50$ but even further away from stability. Indeed the ^{78}Ni doubly-magic nucleus is so far away from the line of stability that there is little hope of producing it with enough yield to survive the long process of separation at an ISOL facility. As we can see from Fig. 13 the total number of ^{78}Ni ions produced in the experiment is quite small, indeed there are too few to determine the half life from a radioactive decay curve. Instead the authors had to have recourse to a maximum–likelihood analysis which has been used in other cases [77, 78] where the statistics are poor. As the reader who peruses [76] will find the authors had to struggle with the uncertainties in the fraction of decays in which β–delayed neutrons occur, the half lives of daughter and granddaughter activities and even the probability that one of the small number of ions might be misidentified. The end result is a half-life value for ^{78}Ni of $110(\pm^{100}_{60})$ ms. Although it is determined with limited precision this number is an important test of nuclear structure models of very neutron–rich nuclei and was the last of the important $N = 50$ waiting point nuclei in the astrophysical r–process to be measured [79]. In this regard it is thought to be especially important in terms of the overall delay that nuclei in this mass region impose on the flow of the r–process towards heavier nuclei. In contrast with our previous example, the experimental half lives of ^{78}Ni and other nuclei measured in the same experiment are shorter than predicted with current models. This example again highlights the need to measure such properties for exotic nuclei.

The example given above is a particularly simple one since it does not exploit the full power of the technique to study the β–decay of exotic nuclei. Another example, again taken from studies at NSCL (MSU) illustrates how one can exploit correlations between the implants and the subsequent β–particles and γ–rays emitted. Although it is not aimed at answering a key physics question in the way the previous example does, it is an excellent example of how important it is to carry out detailed spectroscopy if one is to establish the properties of nuclei properly. Prior to this experiment there was some confusion about the half life of the nucleus ^{60}Mn. This was largely due to the fact that two states in ^{60}Mn decay by β–emission. In addition the long half life of the ^{60}Mn ground state and the apparent direct feeding of the ^{60}Fe ground state, giving a log ft of 6.7, would make it a candidate for isospin–forbidden β–decay. To clarify the situation Liddick et al. [80] produced the ^{60}Mn ground state via the decay of the ground state of the parent ^{60}Cr, which only populates the state with lower spin in ^{60}Mn. Exactly the same techniques and setup were used as described above but with the addition of 12 Ge detectors from the SEGA array [81] surrounding the β–detector in order to detect the β–delayed γ–rays. The same primary beam and target were used as in the previous example and this time, fragment–β and fragment–β–γ correlations were recorded. Figure 14 shows the delayed γ spectrum for β–decay events occurring within 1 s of the implantation of a ^{60}Cr ion. Since both ^{60}Cr and its daughter ^{60}Mn were found to have half lives less than 1 s their β–delayed γ–decays should appear in Fig. 14.

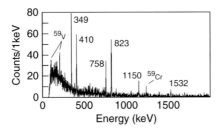

Fig. 14. The spectrum of delayed γ–rays from sources of ^{60}Cr as seen in [80]. The spectrum contains γ–rays from both ^{60}Cr and its daughter ^{60}Mn. Gamma rays from a few contaminants are marked. Reprinted with permission from S.N. Liddick et al. Phys. Rev. C **73**, 044322 (2006). Copyright (2006) by the American Physical Society

Figure 15 shows the decay curves associated with (a) fragment–β, (b) fragment–β–349 keV γ and (c) fragment–β–823 keV γ correlations. The assignment of the observed γ–rays to the mother or daughter nucleus was accomplished using these and other such decay curves. The reader will recall that the radiation from a single decaying state will follow the radioactive decay law. If, however, we are interested in the decay of a daughter or granddaughter activity then the radiation will follow a composite curve given by the Bateman equations [82]. Since the ^{60}Mn is produced in the decay of ^{60}Cr the radiation from its decay follows a more complex curve. The decay curve of Fig. 15, part (a) was fitted with a function that included the exponential decay of ^{60}Cr, the growth and decay of ^{60}Mn and a linear background. The 823 keV γ–ray had been assigned in an earlier experiment to the de–excitation of the first excited 2$^+$ state in ^{60}Fe, the daughter of ^{60}Mn. The decay curve in Fig. 15, part (c), which involves the triple correlation between the implant, the β–particle and this γ–ray clearly exhibits the growth and decay of the ^{60}Mn. Decay curves gated on the 1,150 and 1,532 keV γ–rays show the same shape. The other γ–rays have half lives consistent with the ^{60}Cr parent decay and were interpreted as transitions de–exciting levels in ^{60}Mn.

All of this information was used to construct the level schemes presented in Fig. 16 with the half life of ^{60}Mn being deduced from the fitting of the curve displayed in Fig. 15(c). The newly derived half life of 0.28(2) s immediately called into question the 0$^+$ spin and parity assignment of the ^{60}Mn since it did not fit for an isospin–hindered, 0$^+$ → 0$^+$ Fermi transition as proposed earlier, since such transitions have longer half lives. All those known have a log $ft > 6.5$. This work led to a change in the assignment of the spin and parity of both the ground state and the 272 keV isomeric state in ^{60}Mn which decays to the ground state by a $\Delta J = 3$ transition. The new assignment of 1$^+$ to the ground state is then happily consistent with an allowed GT decay.

We have already indicated the main advantages of using fragmentation. They also apply to high-energy fission carried out in inverse kinematics.

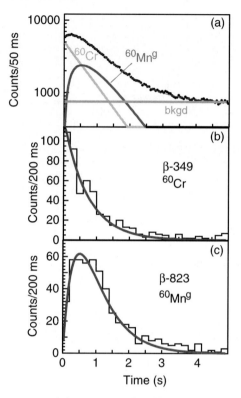

Fig. 15. Decay curves for (**a**) fragment–β, (**b**) fragment–β–349 keV γ and (**c**) fragment–β – 823 keV γ correlations reported in [80]. Panels (**b**) and (**c**) show the pure decay of ^{60}Cr and the growth and decay of its daughter ^{60}Mn respectively. Reprinted with permission from S.N. Liddick et al., Phys. Rev. C **73**, 044322 (2006). Copyright (2006) by the American Physical Society

The two examples outlined in this section show first that useful information can be derived following the production of only a few ions and second that even far from stability one can make β–decay spectroscopy studies of some precision.

4.3 The GT Resonance Observed in the Decay of ^{150}Ho

This experimental investigation illustrates very well many of the features of β–decay studies we have described above. It not only shows the GT resonance very clearly within the β–window but also shows why one should study β–decay with both the TAGS technique and in high resolution. Figure 17 gives a schematic view of the single particle orbitals available above the ^{146}Gd double–closed shell, or at least those that are relevant to the present discussion. The ^{150}Ho nucleus has two, low–lying, β–decaying states with spins and parities 2^- and 9^+, respectively. Figure 17 shows the expected configuration of the

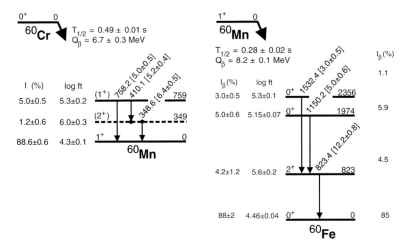

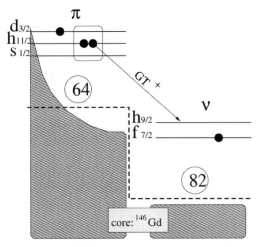

Fig. 16. Partial decay schemes for ^{60}Cr and its daughter ^{60}Mn established in [80]. Reprinted with permission from S.N. Liddick et al., Phys. Rev. C **73**, 044322 (2006). Copyright (2006) by the American Physical Society

2^- isomer with the odd neutron in the $f_{7/2}$ shell and a proton in the $d_{3/2}$ orbit. The reader should remember that the pairs of protons in the $h_{11/2}$ orbits, which couple to 0^+, can scatter between all three orbits ($s_{1/2}$, $h_{11/2}$ and $d_{3/2}$).

One reason for the interest in this case and in the decays of neighbouring nuclei is that, unusually, most of the GT strength lies within the Q_β-window.

Fig. 17. Schematic view of the single particle orbits available above the ^{146}Gd doubly closed shell relevant to the present discussion

It is not a very common situation in β^+ decay, as we explained earlier, and, in this case, it occurs because the proton is transformed in the transition into a neutron in the spin–orbit partner orbital, which lies within the Q_β-window. Here the transition is from the $\pi h_{11/2}$ to $\nu h_{9/2}$ level since the decay of the unpaired proton (in the $d_{3/2}$ orbit) is forbidden. A similar case open to study is the $\pi g_{9/2}$ to $\nu g_{7/2}$ transition which occurs in the $N \sim Z$ nuclei close to ^{100}Sn.

The four–particle states with spins and parities 1^-, 2^- and 3^- populated in the ^{150}Dy daughter nucleus have the configuration $[(\pi d_{3/2}\nu f_{7/2})(\pi h_{11/2}\nu h_{9/2})]$. A simple approximation to the excitation energy of these states is just twice the pairing gap for protons plus twice the pairing gap for neutrons plus the neutron $h_{9/2}$ single particle energy, i.e. at ~ 5 MeV excitation energy. This is well within the Q_β-window of ~ 7 MeV for this decay. It should be noted that this decay is closely connected to the decay of ^{148}Dy, a simpler case since it has just the single proton pair outside the ^{146}Gd core. The two cases must clearly be much the same and should have a comparable log ft namely 3.95(3) [27]. It will, of course, be slightly different because of the presence of the $d_{3/2}$ proton which will modify the probability of the proton pair occupying the $\pi h_{11/2}$ orbital. With this minor caveat we can say that we expect the ^{150}Ho 2^- state to decay strongly to levels at ~ 5 MeV with a log ft of about 3.9.

An important feature of this work is that it has been studied in detail in two ways [28, 29]; with the GSI–TAS described earlier and with a highly efficient Ge array called the "Cluster Cube". This array consisted of six EUROBALL cluster detectors [83] in a highly compact geometry, with four of the detectors 10.2 cm from the source and the other two at a distance of 11.3 cm. The photopeak efficiency of the array at 1,332 keV was 10.2(0.5)%. As we shall see this means that one can compare the two methods directly. Because the direct production of ^{150}Ho in a heavy ion reaction would inevitably favour the 9^+ isomeric state (see Fig. 2 and the related discussion), the 2^- state was produced as the daughter activity of ^{150}Er (see the discussion of ^{60}Mn decay in Sect. 4.2). This results in clean production of the ^{150}Ho via the decay of the 0^+ ground state of ^{150}Er to 1^+ states in ^{150}Ho which decay to the 2^- ground state. The details can be found in [28] and [29]. Our concern here is with the results.

In Fig. 8 we see the measured β–strength as a function of excitation energy derived from measurements both with the GSI–TAS and Cluster Cube. To give our reader a feeling for the quality of the CLUSTER CUBE results we should mention that 1,064 γ–ray lines were observed, which were arranged into a decay scheme with 295 levels in ^{150}Dy. Compare this with the much more difficult study of ^{130}Cd discussed earlier, where the nuclear species involved is much harder to produce. In the ^{150}Ho case, on the assumption that the β–decays are allowed GT transitions, it was possible to assign spins and parities to most of them. In this study an analysis, based on the shell model, provides a prediction of the distribution of $B(GT)$ strength between the 1^-, 2^- and

3^- states of 3.6:4.0:7.4 normalised to 15 arbitrary units. This is in excellent agreement with the measured ratios of 3.4:4.2:7.4 (now in units of $g_A^2/4\pi$) derived from the CLUSTER CUBE measurements.

In Fig. 8 one can see clearly the most distinctive feature of the spectra from both types of spectrometer, namely the very strong β–feeding to a narrow interval in energy near 4.4 MeV excitation with a width of about 240 keV. This is the $\pi h_{11/2}$ to $\nu h_{9/2}$ transition we anticipated seeing earlier. This is the peak of the GT resonance, more or less at the energy anticipated. The two spectra have the same shape, which gives confidence in the analysis techniques used for the TAGS spectrum. There is, however, a clear loss in sensitivity in the CLUSTER CUBE spectrum at higher energies. Quantitatively we can say that the total $B(GT)$ up to the highest observed level at 5.9 MeV is 0.267 corresponding to $\log ft = 4.16$ when derived from the CLUSTER CUBE spectrum. This compares with values of 0.455 and 3.93 obtained for these quantities from the TAGS up to the same energy. If we take the total $B(GT)$ up to the Q_β-window then we miss, in total, 50% of the $B(GT)$ in this very high quality Ge measurement. On the other hand, the individual levels and γ–transitions can only be disentangled in the spectra from the Ge detector array. This is shown qualitatively in Fig. 9, where the region of the resonance measured with the GSI–TAS is compared with the spectra from some coincidence gates showing γ–rays de–exciting levels in the same region.

What can we conclude from these studies? Firstly, it demonstrates very clearly and beautifully the population of the GT resonance in β–decay within the Q_β–window. Secondly, it demonstrates the clear need for the use of both techniques in such cases. The TAGS measurements are essential because it is the only way to obtain a proper measure of the GT decay strength. The

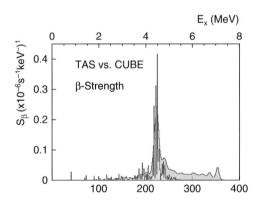

Fig. 18. Beta–strength as a function of excitation energy in the daughter nucleus following the β–decay of the 2^- ground state in ^{150}Ho measured with the CLUSTER CUBE (*sharp lines*) and the GSI-TAS (*continuous function*). See the text for details (See also Plate 8 in the Color Plate Section)

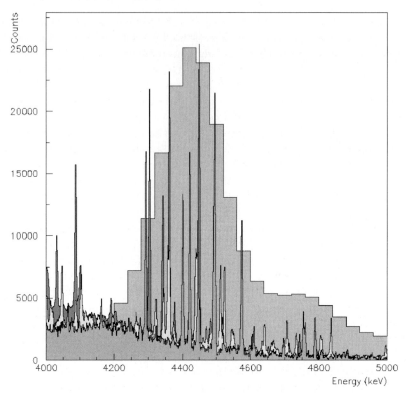

Fig. 19. Comparison of part of the GSI–TAS and CLUSTER CUBE spectra (various coincidence gates) for the decay of the 2^- ground state in ^{150}Ho (see text) (See also Plate 9 in the Color Plate Section)

high-resolution measurements are also essential if one wants the details of the daughter level scheme and the fine structure of the resonance.

4.4 Measurement of Nuclear Shapes in Beta Decay

The shape of the nucleus is one of the simplest of its macroscopic nuclear properties to visualise. In practice it turns out to be very difficult to measure. In general terms we now have a picture of nuclei at closed shells having spherical shapes and nuclei with even quite small numbers of valence nucleons being deformed. The nuclei with $A \sim 70$–80 and $N \sim Z$ are of particular interest in this context. These nuclei enjoy a particular symmetry since the neutrons and protons are filling the same orbits. This, together with a low single–particle level density, leads to rapid changes in deformation with the addition or subtraction of only a few nucleons. In terms of mean field models, these rapid changes occur because of the proximity in energy of large energy gaps for protons and neutrons at $Z, N = 34, 36$ on the oblate side and $Z, N = 38$

on the prolate side of the Nilsson diagram. Such models suggest [84, 85] the co–existence of states in these nuclei of quite different shape. There is experimental evidence to support this in the Se and Kr nuclei [86, 87] and it is also predicted for the lightest Sr nuclei.

As a result it is of considerable interest to map out the deformation of both the ground and excited states in these nuclei. In practice, this is not a simple task. There are a number of methods of measuring the ground state deformation in unstable nuclei based on the interaction of the electric quadrupole moment of the nucleus with an external electric field gradient [88–90]. However, these methods do not apply to nuclei with $J = 0$ or $^{1}/_{2}$. Apart from the nuclear re–orientation effect in Coulomb excitation, however, they do not give the sign of the quadrupole moment and thus cannot distinguish between oblate and prolate shapes.

In some cases, β–decay provides an alternative way of deducing whether the ground state of the parent nucleus is oblate or prolate. The basis of the method is an accurate measurement of the GT strength distribution, $B(GT)$, as a function of excitation energy in the daughter nucleus. The idea was first put forward by Hamamoto et al. [91] and was then pursued in more detail by Sarriguren et al. [92]. In essence they calculate the $B(GT)$ distributions for various nuclei in the region for the deformations minimising the ground state energy. In some cases, the calculated distributions within the β–decay window differ markedly with the shape of the ground state of the parent nucleus, especially for the light Kr and Sr isotopes.

A number of cases have been studied with the *Lucrecia* spectrometer described earlier. In this chapter we discuss the case of the even–even nucleus ^{76}Sr [31] where the ground state is amongst the most deformed known. This was based on the measurement [93] of the energy of the first excited 2^{+} state and Grodzin's formula [94], an empirical relationship between the deformation and the energy of the $2^{+} \rightarrow 0^{+}$ transition. This tells us nothing about the sign of the deformation.

CERN–ISOLDE is, at present, the ideal place for measuring the β–decay of ^{76}Sr, since it provides the most intense, mass–separated, low–energy beams of neutron–deficient Sr nuclei. The half life of ^{76}Sr is just 8.9 s. This is long enough for us to be able to implant the activity on the tape outside the shielding for *Lucrecia* and then transport it to the counting point. The tape system was moved every 15 s in order to avoid the build–up of the ^{76}Rb daughter activity, which has a half life of 36.8 s. The γ–ray spectrum was recorded in coincidence with positrons and X–rays using the ancillary detectors placed close to the implanted source in the central through hole in the NaI detector. The *upper* part of Fig. 20 shows the experimental total absorption spectrum of the β–decay of ^{76}Sr overlaid with the recalculated spectrum after the analysis. In the *lower panel* we see the $B(GT)$ distribution derived from this spectrum with the shading indicating the experimental uncertainty. These results were based on the singles spectra and did not use the recorded coincidences.

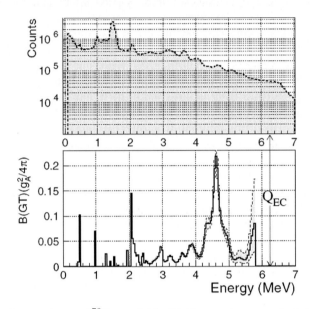

Fig. 20. Singles spectrum of ^{76}Sr overlaid with the spectrum recalculated after analysis (*upper panel*) and B(GT) distribution extracted from these data as a function of excitation energy in the daughter nucleus (*lower panel*) [31]. The shading indicates the experimental uncertainty

The reader should note a number of points. Firstly, the analysis of the TAGS spectrum was carried out as outlined in Sect. 3.1. Secondly, the proton separation energy is 3.5 MeV. As a result β–delayed proton emission has been observed at excitation energies from 4.8 to 5.8 MeV [95]. However, this contribution is only $\sim 2\%$ in B(GT), i.e. very small compared to decay via β–delayed γ–rays. Thirdly, the marked strength at 0.5, 1.0 and 2.1 MeV is to states already known [96] but the $B(GT)$ values reported earlier are in disagreement with the TAGS measurements as a result of the "Pandemonium" effect described earlier.

The theoretical derivation [92] of the $B(GT)$ distribution starts with the construction of the quasi–particle basis self–consistently from a deformed Hartree–Fock (HF) calculation with density-dependent Skyrme forces and pairing correlations in the BCS framework. From the minima in the total HF energy versus deformation plot they derive the possible ground state deformations. In the case of ^{76}Sr two minima are found; one is prolate with $\beta_2 = 0.41$, the other is oblate with $\beta_2 = -0.13$. Using these results the quasi–random–phase approximation (QRPA) equations are solved with a separable residual interaction derived from the same Skyrme force used in the HF calculation. To calculate the $B(GT)$ it is assumed that the states populated in the daughter nucleus have the same deformation as the parent state. Figure 21 shows the results of these calculations using the SK3 residual interaction. This plot shows the sum of the $B(GT)$ at any given energy bin up to that energy bin. The theoretical results are shown for both prolate and oblate states. It also shows the measured, accumulated $B(GT)$. The shading indicates the experimental uncertainty. The agreement of the experimental plot with the calculation for a prolate shape is very good over the energy range 0–5 to 6 MeV. In contrast there is no agreement with the results of the calculation based on an oblate shape. Thus our results confirm the large deformation, $\beta_2 \sim 0.4$, deduced from the in–beam studies and give the first definitive evidence that the deformation is of prolate character.

This result also validates this method of determining the ground state deformation. It was also applied to the case of ^{74}Kr [32], where earlier measurements [87] of the decay of the isomeric, first excited 0^+ state had indicated strong mixing of the oblate and prolate shapes. Again the TAGS measurements were made at CERN–ISOLDE. Figure 21 also shows the accumulated $B(GT)$ as a function of excitation energy for both theory and experiment for this case. This time it is clear that there is a mixture of prolate and oblate shapes in the ^{74}Kr ground state. This is confirmed by the Coulomb excitation of a beam of ^{74}Kr [97].

Summarising, it is clear that it is possible, in some cases, where there is sufficient difference between the calculated curves, to determine the shape of

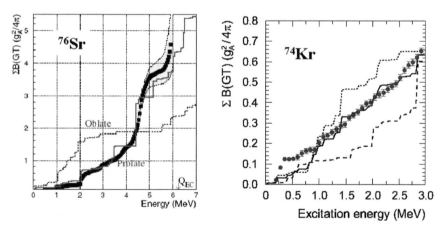

Fig. 21. Summed value of the measured $B(GT)$ as a function of the excitation energy in the daughter nucleus for the decay of ^{76}Sr compared with the theoretical distributions for oblate and prolate shapes (*left panel*) [31] and the corresponding data for ^{74}Kr [32] (*right panel*)

the nuclear ground state from measurements of the $B(GT)$ distribution in β–decay. As part of the same programme of measurements a number of other Kr and Sr decays were also studied. The results will appear in the literature in due course.

4.5 Reactor Decay Heat

Quite apart from its use as a tool for understanding nuclear structure a knowledge of β–decay is important in many practical applications. Here we will restrict our discussion to just one example.

There are more than 400 nuclear reactors operating world–wide, a number which will significantly increase in the near future given the effects of climate change and the desire to use electricity production methods which minimise the emission of "Greenhouse gases". During the operation of such reactors some 7–8% of the total power comes from the β–decay of the very large number of neutron–rich fission products of the fission process. When the reactor stops this reactor *decay heat*, as it is known, remains and dies away over a long period with a complex decay pattern which depends on the mix of decay products, which, in turn, depends on the initial composition of the fuel, the reactor configuration and the time for which it has been running. A knowledge of how the decay heat varies with time is important because (a) it is necessary for economic reasons to optimise the refuelling procedure, (b) one needs to maintain cooling because of the decay heating and (c) when the fuel is finally removed from the reactor it must be properly shielded. The form and extent of the cooling and also of the shielding needs to be specified on the basis of what are called *decay heat calculations*. Naturally, the decay heat varies as a function of time after shutdown and can, in principle, be calculated from known nuclear data. The calculations are based on the inventory of nuclei formed in reactor operation and the subsequent decays and a knowledge of their properties. They depend on libraries of nuclear cross–sections, fission yields and a detailed knowledge of radioactive decay schemes. A number of such databases are maintained by the U.S. [98], Japan [99] and Europe [100]. The question of importance to us here is how good the accumulated and recorded decay data are. It turns out that they suffer from two problems. The first we have already met, namely the Pandemonium effect. In many cases the studies have been carried out faithfully but rely on Ge detectors. As we explained in Sect. 3 the poor efficiency of such detectors means that many weak γ–rays, particularly of high energy, are not observed. As a result the mean γ–energies are underestimated and the mean β–energies overestimated. The second problem is that many of the isotopes produced in high yield in thermal fission belong to refractory elements. As a result it has been difficult or impossible to extract them from a standard ion source and ionise them. The solution to the first problem is the application of total absorption spectroscopy to the cases of interest and new techniques under development will gradually solve the second problem.

Not surprisingly a great deal of attention has been focussed on the efficacy of reactor decay heat calculations. Gradually the calculations have been refined and they can now reproduce integral measurements [101] reasonably well for a range of fissioning isotopes that are important in operating reactors. There are, however, differences in the approaches used in the main databases and there remains quite a large discrepancy in the calculations for the period 300–3,000 s after a fission pulse [101]. The authors of Ref. [101] dubbed this *the γ–ray discrepancy.* This discrepancy appears for ^{233}U, ^{235}U and ^{238}U as well as for the Pu isotopes.

In order to have a sufficiently reliable library of the relevant decay data it is not necessary to measure or re–measure all of the radioactive decays which contribute to the decay heat, some are more important than others because of the large variation in yield with mass in thermal fission. Recently, in [102], a careful analysis homed in on a more limited list of radioactive decays that should be measured with some care since they contribute strongly to the decay heat in the time period of interest. Amongst them are the isotopes of Tc.

Algora et al. [103] have used the total absorption technique to study the decay of ^{104}Tc. This is a species near the lower peak of the thermal fission mass curve and is produced in high abundance. The decay of this isotope and its neighbours is poorly known because Tc is a refractory element and thus difficult to extract from conventional ion sources. However, sources of such isotopes can be produced using an Ion–Guide Isotope Separator On–Line (IGISOL) [104]. In this method, the products of nuclear reactions (in the present case fission) from a thin target recoil into a gas, usually helium, where they are brought to rest. As a result of a complex series of collisions in the slowing down of the ion, including processes such as CE and collisions with impurity atoms, a significant fraction of the products end up in the 1^+ charge state. There is a continuous flow of helium gas and this carries the ions out of the chamber, through a nozzle, into a region where they are separated from neutral gas atoms prior to re–acceleration and mass separation. This process is fast, less than a ms overall, and chemically non–selective. It is an ideal way to produce and prepare sources of refractory elements such as Tc.

In the experiment carried out by Algora et al. [103] a 30 MeV proton beam from the Jyväskylä cyclotron was used to induce fission in a natural uranium target of 15 mg/cm^2 thickness. The beam current was typically about 4 μA. Following mass separation with the IGISOL system the beam of separated Tc was then moved to the centre of the total absorption spectrometer, the details of which are given in [103]. One difficulty in such experiments is to ensure that one has clean sources. This is a particular problem in TAS measurements since they are based on 4π scintillation detectors with modest intrinsic resolution (the best being NaI with typically $\Delta E/E \sim 8\%$ under experimental conditions) and we cannot distinguish between the activity of interest and their decay products or other contaminants using the detector. In this experiment it was possible to eliminate the isobaric contamination by means of selecting an appropriate collection/measuring cycle. Naturally this can only be done in cases with appropriate half lives for the parent and daughter nuclei. As a further check a second experiment was carried out with the separated beam from the IGISOL being injected into the Penning trap available at the Jyvaskyla IGISOL, which is called the JYFLTRAP [105]. The Penning trap acts as if it were a very HRS. These data have not yet been analysed for ^{104}Tc but this procedure is ideally suited to any experiment where one wants a very clean source.

The data from the first experiment but not yet the second, have been analysed using the methods [54, 55] outlined in Sect. 3.1. The details are given in [103]. Figure 22 shows the results for the feeding as a function of excitation energy. They are compared with what was known previously from experiments using Ge detectors [106]. One sees very clearly how these earlier measurements suffered from the Pandemonium effect. It is abundantly clear that a large amount of β–feeding is observed at high excitation in the daughter nucleus which was not previously seen in the high-resolution experiments. The mean γ– and β–energies derived from this experiment are 3263(65) and 915(35) keV, respectively, values which are quite different from those in the JEFF 3.1 library [100], where the corresponding values given are 1890(30) and 1595(75) keV. Not surprisingly this makes a large difference to the decay heat curves as we see in Fig. 23.

Without data from experiments with the TAGS technique the fit to the experimental decay heat is poor, particularly for the period of 10–3,000 s after the pulse. The agreement is better if one includes measurements with the TAGS technique on a series of nuclear species made by Greenwood et al. [52], however, this still leaves a considerable discrepancy for the period 200–3,000 s. Adding the new result on the single isotope ^{104}Tc removes a significant part

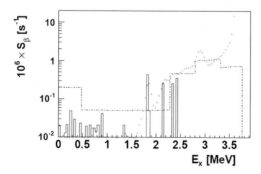

Fig. 22. Comparison of the deduced β–strength [101] for ^{104}Tc (*dots*) with the strength predicted by the gross theory of β–decay (*dot–dashed histogram*). The *discrete lines* represent the strength derived from β–decay studies with high-resolution detectors [100]

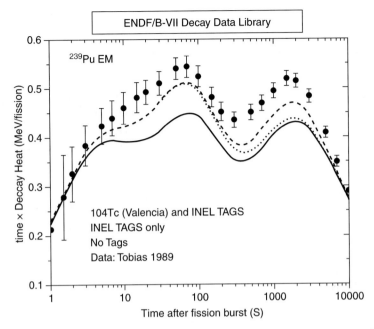

Fig. 23. Contribution of the electromagnetic radiation to the experimental decay heat as a function of time for a pulsed fission event and a ^{239}Pu target. (Figure courtesy of A. Sonzogni)

of this discrepancy and clearly makes further measurements of this kind a priority.

5 Future Measurements

One of the most important changes in studies of nuclear physics in recent years has been the steady improvement in our ability to accelerate radioactive nuclear species. It is this sea change which is the raison d'etre for the Euroschools on Exotic Nuclei. As a corollary this new–found ability has meant a considerable increase in our capacity to produce a wide range of short–lived nuclear species since the main aim is to re–accelerate them. This has allowed us and will continue to allow us to study the β–decay of more and more exotic nuclear species. As we have explained earlier, if we are to extract the fullest information from these studies we must use both the high resolution and total absorption techniques.

The examples in Sect. 4 show that experiments are already underway to exploit β–decay to the best of our present abilities. At the new international radioactive beam facilities FAIR, to be built at GSI, Darmstadt, and SPIRAL2, to be built at GANIL, France, plans are already underway [107, 108] to allow both types of study and also to allow studies of β–delayed neutrons as well. At these facilities and others which are under construction or on the drawing board, there will be a step change in beam intensities. Thus at FAIR the aim is to improve the intensities of the exotic beams delivered by three or four orders–of–magnitude. This will allow much better experiments on nuclei close to stability and also allow us to extend the range of nuclei that can be studied. It is not just a matter of an improvement in beam intensity. Considerable effort is being devoted to develop new methods of ionising chemical species or of other means of ionising radioactive species caught in a gas catcher [109] and in some cases transferred to a Penning trap to allow a pure beam/source to be produced (see Sect. 4.5).

At the same time considerable effort is being devoted to producing a range of equipment tailor–made to measure sources of short–lived species. It would be a foolish man or woman who forecasts where this will lead. We will certainly greatly extend our knowledge of β–decay far from stability but what phenomena we will uncover and what applications will be derived from our measurements and techniques we will leave to our youthful readers to discover for themselves.

References

1. S. Hofmann, On beyond Uranium in *Science Spectra*, ed. by V. Moses (Taylor and Francis, London 2002);
2. Hofmann, S.: Superheavy elements. In: Al-Khalili, J.S., Roeckl, E. (eds.) Lecture Notes Physics, vol. 764, Springer, Heidelberg (2009)
3. Gelletly, W., Eberth, J.: Gamma–ray arrays: past, present and future. Lect. Notes Phys. **700**, 79–117 (2006)
4. M. Jung et al.: Phys. Rev. Lett. **69**, 2164 (1992)
5. F. Bosch et al.: Phys. Rev. Lett. **77**, 5190 (1996)
6. Bosch, F.: Measurement of mass and beta lifetime of stored exotic nuclei. Lect. Notes Phys. **651**, 137–168 (2004)
7. Blank, B.: One– and Two–proton radioactivity. In: Al-Khalili, J.S., Roeckl, E. (eds.) Lecture Notes Physics, vol. 764, Springer, Heidelberg (2009)
8. M.V. Stoitsov et al.: Phys. Rev. C **68**, 054312 (2003)
9. Severijns, N.: Weak interaction studies by precision experiments in nuclear beta decay. Lect. Notes Phys. **651**, 339–377 (2004)
10. E.J. Konopinski, M.E. Rose, Alpha–, Beta– and Gamma–Ray Spectroscopy, Vol. 2, ed. by K.Siegbahn, (North Holland, Amsterdam, New York, Oxford 1968) pp. 1327–1364
11. R.J. Blin–Stoyle, Fundamental Interactions and the Nucleus (North Holland/Elsevier, New York, 1973)
12. W. Pauli, Rapports du Septieme Conseil de Physique Solvay, Brussels, 1933
13. E. Fermi, Z. Phys. **88**, 161 (1934)
14. R.B. Firestone et al.: Table of Isotopes, 8th edition (John Wiley and Sons, New York 1995)
15. http://www.nndc.bnl.gov/ensdf/dec_form.jsp

16. G. Gamow, E. Teller, Phys. Rev. **49**, 895 (1936)
17. K.S. Krane, Introductory Nuclear Physics (John Wiley and Sons, New York 1988)
18. K. Heyde, Basic Ideas and Concepts in Nuclear Physics, Fundamental and Applied Nuclear Physics Series (Institute of Physics Publishing 1994)
19. B. Singh et al.: Nuclear Data Sheets **84**, 487(1998)
20. C.S. Wu et al.: Phys. Rev. **105**, 1413 (1957)
21. J.H. Christenson et al.: Phys. Rev. Lett. **13**, 138 (1964)
22. R.J. Blin–Stoyle, in *Isospin in Nuclear Physics*, ed. by D.H. Wilkinson (North Holland, Amsterdam, New York, Oxford 1969)
23. Lenzi, S.M., Bentley, M.A.: Test of isospin symmetry along the N = Z line. In: Al-Khalili, J.S., Roeckl, E. (eds.) Lecture Notes Physics, vol. 764, Springer, Heidelberg (2009)
24. Z. Hu et al.: Phys. Rev. C **60**, 024315 (1999)
25. Z. Hu et al.: Phys. Rev. C, **62**, 064315 (2000)
26. M. Gierlik et al.: Nucl. Phys. A **724**, 313 (2003)
27. P. Kleinheinz et al.: Phys. Rev. Lett. **55**, 2664 (1985)
28. A. Algora et al.: Phys. Rev. C **68**, 034301 (2003)
29. B. Rubio, Frontiers of Collective Motions, ed. by H. Sagawa. H. Iwasaki (World Scientific, Singapore 2002)
30. E. Nacher et al.: in *GSI Sci. Rep. 2003*, GSI Rep. 2003–1 (2003), p. 8 (http://www-aix.gsi.de/annrep2002/Files/8.pdf)
31. E. Nacher et al.: Phys. Rev. Lett. **92**, 232501 (2004)
32. E. Poirier et al.: Phys. Rev.C **69**, 034307 (2004)
33. J.C. Hardy et al.: Nucl. Phys. A **509**, 429 (1990)
34. I.S. Towner, J.C.Hardy, Proc. of 5th Int. WEIN Symposium: Physics beyond the Standard Model, Santa Fe, 1998, ed. by P. Herczeg, C.M. Hoffmann, H.V. Klapdor–Kleingrothaus (World Scientific, Singapore, 1999) p. 338
35. J.C. Hardy, I.S. Towner, Phys. Rev. Lett. **94**, 092502 (2005)
36. K. Ikeda, S. Fujii, J.I. Fujita, Phys. Lett. **3**, 271 (1963)
37. K. Ikeda, Prog. Theor. Phys. **31**, 434 (1964)
38. A. Arima, Phys. Lett. B **122**, 126 (1983)
39. A. Arima et al.: Adv.in Nucl. Phys. **18**, 1 (1987)
40. C.D. Goodman et al.: Phys.Rev. Lett. **44**, 1755 (1980)
41. C. Gaarde et al.: Nucl. Phys. A **369**, 238 (1981)
42. H. Abele et al.: Phys. Rev. Lett. **88**, 211801 (2002)
43. A.M. Lane, J.M. Soper, Nucl. Phys. **35**, 676 (1955)
44. G. Audi et al.: Nucl. Phys. A **624**, 1 (1997)
45. N.B. Gove, N.J. Martin, At. Data and Nucl. Data Tables **10**, 205 (1971)
46. J.C. Hardy et al.: Phys. Lett. B **71**, 307 (1977)
47. B. Rubio et al.: J. Phys. G, Nucl. Part. Phys. **31**, S1477 (2005)
48. B. Rubio, W. Gelletly, Romanian Reports on Physics **59**, 635 (2007)
49. C.L. Duke et al.: Nucl. Phys. A **151**, 609 (1970)
50. K.H. Johansen et al.: Nucl. Phys. A **203**, 481 (1973)
51. G. Alkhazov et al.: Phys. Lett. B **157**, 35 (1985)
52. R.C. Greenwood et al.: Nucl. Instr. and Meth. in Phys. Res. A **314**, 514 (1992)
53. M. Karny et al.: Nucl. Instr. and Meth. in Phys. Res. B **126**, 411 (1997)
54. J.L. Tain, D. Cano–Ott, Nucl. Instr. and Meth. in Phys. Res. A **571**, 719 (2007)
55. J.L.Tain, D. Cano–Ott, Nucl. Instr. and Meth. in Phys. Res. A **571**, 728 (2007)

56. D. Cano–Ott, Ph. D.Thesis, University of Valencia (2002)
57. A. Pérez–Cerdán, B. Rubio, priv.communication
58. D. Cano–Ott et al.: Nucl. Instr. and Meth. in Phys. Res. A **430**, 333 (1999)
59. D. Cano–Ott et al.: Nucl. Instr. and Meth. in Phys. Res. A **430**, 488 (1999)
60. Van Duppen, P.: Isotope separation on line and post acceleration, Lect. Notes Phys. **700**, 377 (2006)
61. J.A. Nolen et al.: in Heavy Ion Acceleration Technology, ed. By K.W.Shepherd, AIP Conference Proceedings No. **473**, 477 (1999) (AIP, New York)
62. U. Koster et al.: Nucl. Instr. and Meth. in Phys. Res. B **204**, 347 (2003)
63. I. Dillman et al.: Phys.Rev.Lett. **91**, 162503(2003)
64. K. Takahashi et al.: At .Nucl. Data Tables **12**, 101 (1973)
65. M. Hannawald et al.: Nucl. Phys. A **688**, 578c (2001)
66. B. Pfeiffer et al.: Nucl. Phys. A **693**, 282 (2001)
67. B.A. Brown, A. Etchegoyan, W.D.M. Rae, OXBASH – the Oxford–Buenos Aires–MSU Shell Model Code, Internal report No. MSUCL–524, Michigan State University Cyclotron Laboratory, unpublished
68. E.M. Burbidge et al.: Rev.Mod.Phys., **29**, 547 (1957)
69. Langanke, K., Thielemann, F.-K., Wiescher, M.: Nuclear astrophysics and nuclei far from stability, Lect. Notes Phys. **651**, 383–453 (2004)
70. P. Dendooven, Nucl. Instr. and Meth. in Phys. Res. B **126**, 182 (1997)
71. Morrisey, D.J., Sherrill, B.M.: In–flight separation of projectile fragments, Lect. Notes Phys. **651**, 113–133 (2004)
72. H. Geissel et al.: Nucl. Instr. and Meth. in Phys. Res. A **282**, 247 (1989)
73. D.J. Morrisey et al.: Nucl. Instr. and Meth. in Phys. Res. B **204**, 90 (2003)
74. A.C. Mueller, R. Anne, Nucl. Instr. and Meth. in Phys. Res. B **56/57**, 559 (1991)
75. T. Kubo et al.: Nucl. Instr. and Meth. in Phys. Res. B **204**, 97 (2003)
76. P. Hosmer et al.. Phys. Rev. Lett. **94**, 112501 (2005)
77. M. Bernas et al.. Z. Phys. A **336**, 41 (1990)
78. R. Schneider et al.: Nucl. Phys. A **558c**, 191 (1995)191
79. See e.g. K. Langanke, G. Martinez–Pinedo, Rev. Mod. Phys. **75**, 819 (2003)
80. S.N. Liddick et al.: Phys. Rev. C **73**, 044322 (2006)
81. W.F. Mueller et al.: Nucl. Instr. and Meth. in Phys. Res. A **466**, 492 (2001)
82. R.D. Evans, The Atomic Nucleus (McGraw–Hill, New York, Toronto, London 1955)
83. J. Eberth et al.: Nucl. Instr. and Meth. in Phys. Res. A **369**, 135 (1996)
84. W. Nazarewicz et al.: Nucl. Phys. A **435**, 397 (1985)
85. P. Bonche et al.: Nucl. Phys. A **443**, 39 (1985)
86. J.H. Hamilton et al.: Phys. Rev. Lett. **32**, 239 (1974)
87. C. Chandler et al.: Phys. Rev. C **56**, R2924 (1997)
88. R. Neugart, G. Neyens, Nuclear Moments, Lec. Notes Phys. **700**, 135–189 (2006)
89. E. Davni et al.: Phys. Rev. Lett. **50**, 1652 (1983)
90. F. Hardeman et al.: Phys. Rev. C **43**, 130 (1991)
91. I. Hamamoto et al.: Z. Phys.A **353**, 145 (1995)
92. P. Sarriguren et al.: Nucl. Phys. A **691**, 631 (2001)
93. C.J. Lister et al.: Phys. Rev.C **42**, R1191 (1990)
94. L. Grodzins, Phys. Lett. **2**, 88 (1966)

95. Ch. Miehe et al.: in *New Facet of Spin Giant Resonances in Nuclei*, ed. by H. Sakai (1997) p. 140
96. Ph. Dessagne et al.: Eur. Phys. J. A **20**, 405 (2004)
97. W. Korten et al.: Nucl. Phys. **746**, 90c (2004)
98. http://www.nndc.bnl.gov/nndc/endf/
99. http://wwwndc.tokai-sc.jaea.go.jp/ftpnd/evlret.html
100. http://www.nea.fr/html/dbdata/projects/nds_jef.htm
101. T. Yoshida et al.: J.Nucl.Sci.Tech. **36**, 135 (1999)
102. http://www-nds.iaea.org/beta_decay/Presentations2/index.html
103. A. Algora et al.: Eur. Phys. J.- Special Topics 150, 383 (2007)
104. J. Äystö, Nucl. Phys. A **693**, 477 (2001)
105. V. Kolhinen et al.: Nucl. Instr. and Meth. in Phys. Res. A **528**, 776 (2004)
106. J. Blachot, Nucl. Data Sheets **64**, 1 (1991)
107. http://www.gsi.de/fair/
108. http://www.ganil.fr/research/developments/spiral2/files/WB_SP2_Final.pdf
109. W. Trimble et al.: Nucl. Phys. A **746**, 415 (2004)

One- and Two-Proton Radioactivity

B. Blank

Centre d'Etudes Nucléaires de Bordeaux-Gradignan, Université Bordeaux I –
CNRS/IN2P3, Chemin du Solarium, F-33175 Gradignan Cedex, France

Abstract For many years one-proton radioactivity studies have been a powerful
tool to investigate nuclear structure close to the proton drip line. In many cases,
they are the only means to give access to structural information about unbound
quantum states. In particular, these investigations can be performed with rather
low-production rates. Selected examples from the wealth of experimental information
and their theoretical understanding will be presented to demonstrate the potential of
these investigations. Recently, two-proton radioactivity was discovered and opens a
new window to study the properties of atomic nuclei even further away from nuclear
stability. The potential of this new nuclear decay mode will be outlined as well.
Finally, we present some information on other exotic decay modes at the proton
drip line.

1 Introduction

The study of the structure of the atomic nucleus started soon after the discovery of radioactivity and of the atomic nucleus itself. The first experiments
dealt with the investigation of radioactive nuclear decay. The use of nuclear
reactions was initiated only when Irène and Frédéric Joliot-Curie produced
the first human-made radioactivity by bombarding an aluminium foil with
α particles, which produced the phosphorus isotope ^{30}P and a neutron. ^{30}P
finally decays into ^{30}S.

Still today the investigation of nuclear decay modes as well as the use
of nuclear reactions are the most powerful tools to study nuclear structure.
However, investigations with nuclear reactions with a particular isotope can
only be performed reasonably well with a significant number of this isotope
produced per time interval. A typical number is 10–100 counts per second
needed to get a meaningful result. Nuclear decay studies can be performed
with as little as one count per day. It is evident that the quality of the information gained increases with the statistics obtained in a particular experiment.
However, a rule of thumb is that decay measurements can be performed with
much lower rates than reaction studies.

Blank, B.: *One- and Two-Proton Radioactivity.* Lect. Notes Phys. **764**, 153–201 (2009)
DOI 10.1007/978-3-540-85839-3_5   © Springer-Verlag Berlin Heidelberg 2009

This is the main reason why investigations close to the limits of nuclear stability are preferentially performed with nuclear decay. The nuclei of interest are generally very weakly produced and thus the experiments do not yield sufficient statistics for a meaningful reaction-type experiment.

After the observation and study of the "classical" radioactivities, i.e. α, β, and γ decay as well as fission, theoreticians, notably Ya. B. Zel'dovich [158] and V. Goldanskii [60], proposed at the beginning of the 1960s the occurrence of new types of nuclear decay modes, which were predicted to happen once the nuclear forces are no longer able to bind all nucleons in nuclei with a strong excess of either protons or neutrons. In particular at the proton drip line, the limit of nuclear binding for proton-rich nuclei, one-proton (1p) and two-proton (2p) radioactivity were expected to occur. For nuclei with an odd number of protons Z, 1p radioactivity, the emission of a proton from a nuclear ground state with a certain half-life, was predicted, whereas even-Z nuclei were predicted to decay by 2p radioactivity, once the proton drip line is reached. According to the definition of Goldanskii, 1p emission should not be an open decay channel for 2p radioactive nuclei, because only in this case the two protons are emitted simultaneously. Such a situation was expected to appear for medium-mass nuclei where the Coulomb barrier is strong enough to create long-lived states sufficiently narrow so that the parent and the different possible daughter states do not overlap. Then due to nuclear pairing, the 1p daughter nucleus has a higher mass excess than the parent nucleus and the 1p channel is not open. It is interesting to note that ^{45}Fe, the nucleus for which this radioactivity was discovered (see below) was already in the list of nuclei of possible candidates proposed by Goldanskii [59, 61]. The occurrence of 2p radioactivity was also extensively discussed by Jaenecke [82].

One-proton radioactivity was discovered at the beginning of the 1980's [78, 90] and has since developed into a powerful nuclear structure tool. In many cases, 1p radioactivity studies are today the only means to access structural information of the most proton-rich nuclei. As will be shown below, these experiments allow a test of the nuclear mass surface, they give access to the sequence of single-particle levels – a fundamental input for the nuclear shell model (see, e.g. [63]), the most successful model describing the atomic nucleus – they allow a de-composition of the nuclear wave function, they give access to nuclear deformation, and they are a tool to studying the tunnelling through the Coulomb and centrifugal barrier, in particular in cases of changing nuclear deformation. Basically all these experimental observables are only accessible in these investigations.

Two-proton radioactivity was discovered more than 20 years later [58, 124]. These studies are today still in its infancy. However, with future developments on the experimental as well as on the theoretical side they have an enormous potential and could address questions similar to those investigated by 1p radioactivity for even more exotic, i.e. more proton-rich nuclei. In addition, due to the presence of two protons in the decay channel, pairing studies might give new insights into the forces governing the atomic nucleus.

In the following, we will describe the experiments which led to the discovery of one- and two-proton radioactivity. For 1p radioactivity, we turn our attention then to recent studies which allowed accumulating a wealth of nuclear structure data and which are today a powerful tool to investigate the most exotic nuclei. As 2p radioactivity was discovered more recently, most of the 2p data available to date will be presented and put into a general context in particular as compared to nuclear models.

2 One-Proton Radioactivity

Proton emission from atomic nuclei was observed already in the early years of nuclear physics, even before it was clear that the atomic nucleus is constituted of protons and neutrons [102]. They were called at this time "H rays". Clearer indications of proton emission were obtained at the beginning of the 1960s, when proton emission was observed after β decay of ^{25}Si [8]. However, in these cases the half-life of the nuclear quantum state which emits the proton is very short (10^{-15} s and shorter). Therefore this state itself cannot be observed directly.

What one generally measures in studies of 'direct' (in contrast to β-delayed) 1p decay is the total half-life of the parent state, its partial half-life for proton emission which requires the knowledge of the branching ratios for competing disintegration modes, and the energy (E_p) of the proton emitted. If the radioactive atoms of interest are available as thin source, E_p can be measured by means of a suitable detector viewing the source. The Q value for 1p decay (Q_p) can then be deduced from E_p by applying a correction for the (undetected) energy of the recoiling daughter atom. In case of deep implantation of the radioactive atoms into a detector the electronic signal of 1p decay contains both the energy (loss) of the proton and the energy of the related recoil.

Direct 1p emission from a long-lived state was first observed in studies of ^{53}Co, where an isomer with a half-life of 247 ms was identified to decay by proton emission to the ground state of ^{52}Fe. Despite its high-decay energy, this decay has such a long half-life, because (i) the proton has to "carry away" a large amount of angular momentum and (ii) there is only a small overlap between the nuclear wave functions of the initial and final state. In fact, the decay occurs between an $I^\pi = (19/2^-)$ high-spin state and the $I^\pi = 0^+$ ground state of ^{52}Fe. The penetration through this high angular-momentum barrier slows the decay significantly down. As a comparison let us mention that the half-life would be about 10^{-18} s, if there were no angular momentum to carry away and if there were a perfect match of the wave functions of the initial and final states.

2.1 Discovery of One-Proton Radioactivity

One-proton radioactivity was actively searched for since its prediction in the beginning of the 1960s by Zel'dovich [158] and Goldanskii [60]. Following the above-mentioned discovery of ^{53m}Co, early experimental results were presented by Karnaukhov and co-workers at the Leysin conference [86]. These authors bombarded different targets with beams of ^{32}S and ^{35}Cl. For the ^{96}Ru target, counts possibly corresponding to proton emission with an energy of 0.83 MeV were observed and tentatively attributed to the decay of light praseodymium (or lanthanum) isotopes. ^{121}Pr was discussed as a possible emitter. However, today it is not clear what the real observation was as from new experiments [129] ^{121}Pr can be excluded as the possibly observed proton emitter.

A clear and well-defined signal of proton radioactivity was finally observed in experiments at GSI in the beginning of the 1980s. In experiments at the velocity filter SHIP (Fig. 1), ^{151}Lu could be identified by Hofmann et al. to emit protons with an energy of about 1230 keV and a half-life of 80 ms [78]. Shortly after, in an experiment at the GSI on-line separator (Fig. 1) Klepper et al. observed a second proton radioactive nucleus, ^{147}Tm [90]. Both nuclei were produced in fusion–evaporation reactions, where a heavy-ion beam impinges on a suitable target with energies around the Coulomb barrier, which is the energy where, in a simple picture, the two nuclei have a sufficiently high velocity to overcome the mutual repulsion due to their positive charge and the surfaces of the two nuclei touch each other. In the first experiment [78], the fusion–evaporation residues were separated in-flight from the beam impinging on the target by means of their reduced velocity. In the second of these pioneering experiments [90], the fusion products were stopped, ionised, re-accelerated and mass analysed. In both cases, the decay was observed in silicon detectors.

Faestermann et al. [46] at the Munich TANDEM could identify proton radioactivity from ^{109}I and ^{113}Cs a few years later. In the early 1990s, extensive

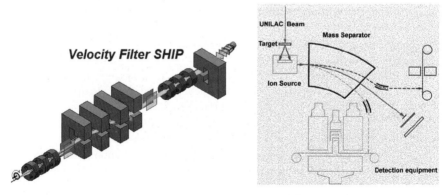

Fig. 1. Schematic view of the GSI velocity filter SHIP (*left panel*) at which the first case of proton radioactivity could be successfully observed, and of the GSI on-line mass separator (*right panel*) where shortly after the second 1p emitter could be identified

studies were performed at Daresbury Laboratory and many more ground-state proton emitters could be observed (see, e.g. [97, 98, 119, 120, 136, 137]). This wealth of experimental data could be acquired due to the combination of high-resolution separators and efficient detection setups which included for the first time the use of silicon strip detectors which allowed a rather sensitive correlation of implantation events and subsequent decays. In fact, this combination enables an efficient separation of the fusion products of interest and a correlation in space and time of the fusion products implanted in these silicon strip detectors with their radioactive decay, which has to occur "shortly" after their arrival in the same position as the implantation.

After the shut-down of the accelerator at Daresbury Laboratory, these studies were continued at Argonne National Laboratory and at Oak Ridge National Laboratory. In both laboratories, powerful separators – the Fragment Mass Analyser (FMA) in Argonne and the Recoil Mass Spectrometer in Oak Ridge were installed and combined with detection setups including double-sided silicon strip detectors (DSSSD), which allowed achieving high sensitivities for proton-radioactivity studies (see, e.g. [37, 134]). More recently successful experiments have also been performed at Legnaro National Laboratory [143] and at the Accelerator Laboratory of the University of Jyväskylä [88]. All isotopes above $Z=50$, for which 1p radioactivity from their ground state or from a long-lived isomer has been observed, are represented in the nuclear chart in Fig. 10. A summary of the experimental results is also given in Table 1.

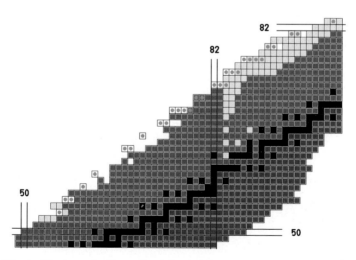

Fig. 2. Chart of isotopes in the region of ground-state proton emitters. The isotopes for which proton emission from the ground state or from long-lived isomers was observed are indicated by *full circles*. From $Z=53$ up to $Z=83$, proton radioactivity was observed for all odd-Z elements but one (See also Plate 10 in the Color Plate Section)

Table 1. Summary of experimental information on long-lived ground- and isomeric-state proton emitters. Given are the proton energies (E_p), the half-life of the proton emitter ($T_{1/2}^{total}$), the proton-emission branching ratio (B.R.), the orbital and/or the spin of the state emitting the proton (for deformed emitters the Nilsson-model quantum numbers are given) and the angular momentum of proton emission as well as the references for the work. For ^{146}Tm, conflicting assignments of the proton groups are given in the literature. The most likely assignment was chosen. However, the ground state could also be a 1^+ state

Emitter	E_p (keV)	$T_{1/2}^{total}$ (ms)	B.R.	Parent-state characteristics	Angular momentum	Reference
^{53}Co^m	1569(19)	245(8)	≈0.014	$(19/2^-)$	$(l = 9)$	[24, 25, 81]
^{54}Ni^m	1260(50)	≈220×10^{-6}		10^+	$l = 5$	[45]
^{94}Ag^m	790(30)	390(40)	0.019(5)	(21^+)		[113]
	1010(30)	390(40)	0.022(4)	(21^+)		
^{109}I	811(5)	0.0935(5)	1.	d5/2, 5/2$^+$	$l = 2$	[46, 53, 76] [73, 103, 137]
^{112}Cs	807(7)	0.500(100)	1.	d5/2		[120]
^{113}Cs	958(3)	0.0167(7)	1.	d5/2, 5/2$^+$	$l = 2$	[53, 74, 76] [9, 120]
^{117}La	797(11)	22.4(19)	0.939(7)	d5/2, 3/2$^+$[411] or h11/2, 3/2$^-$[541]		[101, 144]
^{117}La^m	933(6)	10(5)	0.974(13)	g9/2, 9/2$^+$[404]		[144]
^{121}Pr	882(10)	10^{+6}_{-3}	1.	g7/2, 3/2$^+$[422] or h11/2, 3/2$^-$[541]		[129]
^{130}Eu	1020(15)	$0.90^{+0.49}_{-0.29}$	1.	d5/2, 3/2$^+$[411], 1$^+$		[36]
^{131}Eu	940(9)	$18.8^{+1.8}_{-1.7}$	0.879(13)	d5/2, 3/2$^+$[411]		[38, 143]
	811(7)	23^{+10}_{-6}		or g7/2, 5/2$^+$[413]		
^{135}Tb	1179(7)	$0.94^{+0.33}_{-0.22}$	1.	h11/2, 7/2$^-$[523]		[155]
^{140}Ho	1086(7)	6(3)	1.	h11/2, 7/2$^-$[523]		[134]
^{141}Ho	1169(8)	4.1(1)	0.992(3)	h11/2, 7/2$^-$[523]		[11, 38, 134]

Table 1. (continued)

Emitter	E_p (keV)	$T_{1/2}^{total}$ (ms)	B.R.	Parent-state characteristics	Angular momentum	Reference
$^{141}\mathrm{Ho}^m$	970		0.007(2)			[11]
$^{144}\mathrm{Tm}$	1234(8)	$6.6^{+0.9}_{-0.7}$	1.	d3/2, 1/2$^+$[441]		[134, 142]
$^{144}\mathrm{Tm}$	1700(16)	$0.0019^{+0.0012}_{-0.0005}$	0.71(27)	h11/2, 5$^-$ or 10$^+$	$l = 5$	[71]
	1430(25)		0.29(11)			
$^{145}\mathrm{Tm}$	1728(7)	0.00313(29)	0.904(15)	h11/2, 11/2$^-$	$l = 5$	[9, 87]
	1400(20)		0.096(15)	h11/2, 11/2$^-$	$l = 3$	
$^{146}\mathrm{Tm}$	1189(4)	77.6(23)	0.66(3)	h11/2, 5$^-$	$l = 5$	[10, 55, 98, 129]
	1015(8)		0.19(3)	h11/2, 5$^-$	$l = 3$	
	937(7)		0.15(3)	h11/2, 5$^-$	$l = 0$	
$^{146}\mathrm{Tm}^m$	1119(4)	199(4)	0.98(2)	h11/2, 10$^+$	$l = 5$	[10, 55, 98, 129]
	889(7)		0.018(3)	h11/2, 10$^+$	$l = 3$	
$^{147}\mathrm{Tm}$	1053(5)	567(23)	0.15(5)	h11/2, 11/2$^-$	$l = 5$	[76, 90, 98] [94, 137, 152]
$^{147}\mathrm{Tm}^m$	1113(3)	0.360(36)	1.	d3/2, 3/2$^+$	$l = 2$	[76, 137]
$^{150}\mathrm{Lu}$	1261(3)	44.8(30)	0.70(4)	h11/2, > 5$^-$	$l = 5$	[54, 76, 128, 137]
$^{150}\mathrm{Lu}^m$	1284(5)	$0.043^{+0.007}_{-0.005}$	1.	d3/2	$l = 2$	[54, 55, 128]
$^{151}\mathrm{Lu}$	1231(3)	80.6(2)	0.634(9)	h11/2, 11/2$^-$	$l = 5$	[12, 78, 137]
$^{151}\mathrm{Lu}^m$	1310(10)	0.016(1)	1.	d3/2, 3/2$^+$	$l = 2$	[12]
$^{155}\mathrm{Ta}^{(m)}$	1444(15)	$2.9^{+1.5}_{-1.1}$	1.	h11/2, 11/2$^-$	$l = 5$	[118]
$^{156}\mathrm{Ta}$	1009(5)	145^{+24}_{-22}	0.59(7)	d3/2	$l = 2$	[119, 121]
$^{156}\mathrm{Ta}^m$	1106(7)	358(45)	0.042(9)	h11/2	$l = 5$	[98, 121]
$^{157}\mathrm{Ta}$	927(7)	10.1(4)	0.034(12)	s1/2, 1/2$^+$	$l = 0$	[80]
$^{159}\mathrm{Re}^m$	1805(20)	0.0202(37)	1.	h11/2, 11/2$^-$	$l = 5$	[83, 118]

Table 1. (continued)

Emitter	E_p (keV)	$T_{1/2}^{total}$ (ms)	B.R.	Parent-state characteristics	Angular momentum	Reference
^{160}Re	1262(5)	$0.79^{+0.12}_{-0.10}$	0.91(5)	d3/2	$l = 2$	[119, 121]
^{161}Re	1192(6)	0.440(2)	1.	s1/2, 1/2$^+$	$l = 0$	[80, 93]
^{161}Re^m	1315(7)	14.8(3)	0.066(9)	h11/2, 11/2$^-$	$l = 5$	[77, 80, 93]
^{164}Ir$^{(m)}$	1817(9)	$0.113^{+0.062}_{-0.030}$	1.	h11/2, (9$^+$)	$l = 5$	[89]
^{165}Ir^m	1707(7)	0.30(6)	0.87(4)	h11/2 11/2$^-$	$l = 5$	[35]
^{166}Ir	1145(8)	10.5(22)	0.069(29)	d3/2	$l = 2$	[35]
^{166}Ir^m	1316(8)	15.1(9)	0.018(6)	h11/2	$l = 5$	[35]
^{167}Ir	1064(6)	35.2(20)	0.32(4)	s1/2, 1/2$^+$	$l = 0$	[35]
^{167}Ir^m	1238(7)	30.0(60)	0.004(1)	h11/2, 11/2$^-$	$l = 5$	[35]
^{170}Au	1463(12)	$0.286^{+0.050}_{-0.040}$	0.89(10)	d3/2, (2$^-$)	$l = 2$	[88]
^{170}Au^m	1743(6)	$0.617^{+0.050}_{-0.040}$	0.59(6)	h11/2, (9$^-$)	$l = 5$	[88]
^{171}Au	1439(10)	$0.0245^{+0.0047}_{-0.0031}$	1.	s1/2, 1/2$^+$	$l = 0$	[3, 88, 126]
^{171}Au^m	1693(4)	1.02(1)	0.46(4)	h11/2, 11/2$^-$	$l = 5$	[3, 35, 88]
^{176}Tl	1258(18)	$5.2^{+3.0}_{-1.4}$	1.	s1/2, (3$^-$,4$^-$,5$^-$)	$l = 0$	[88]
^{177}Tl	1156(20)	18(5)	0.27(13)	s1/2, 1/2$^+$	$l = 0$	[126]
^{177}Tl^m	1956(8)	$0.206^{+0.036}_{-0.029}$	0.52(7)	h11/2, 11/2$^-$	$l = 5$	[88, 126]
^{185}Bi^m	1598(16)	0.049(7)	0.85(6)	s1/2, 1/2$^+$	$l = 0$	[37, 127]

Before we will address some selected topics of 1p radioactivity, we will turn our attention to a basic model used to analyse results from proton radioactivity and to the production of proton emitters and the detection of proton radioactivity with modern techniques.

2.2 Barrier-Penetration Half-Life Calculations and the Interpretation of Proton Radioactivity

The protons which are emitted by proton radioactive nuclei are quasi bound. This means that, although the Q_p value is positive, i.e. from a purely energetics point of view they are unbound, they are bound "for some time", the time it takes to penetrate the Coulomb and centrifugal barrier. This is schematically shown in Fig. 3. The barrier penetration is a quantum mechanical tunnelling process and its half-life increases with the barrier height and decreases with increasing Q_p value.

The half-life for tunnelling can be determined with more or less sophisticated models. In a rather simple model, the half-life is a function of the integral of the one-dimensional barrier above the energy available for the decay and the frequency with which the proton knocks on the barrier, which is calculated from the time it takes for the proton to travel from one side of the nucleus to the other:

$$1/\tau = \frac{v/2}{1.17 \times (A_1 + A_2)^{1/3}} \times \exp\left(-\frac{2}{\hbar c} \times \sqrt{2A_1 A_2/(A_1 + A_2) \times m_0 \times S_{\text{int}}}\right)$$

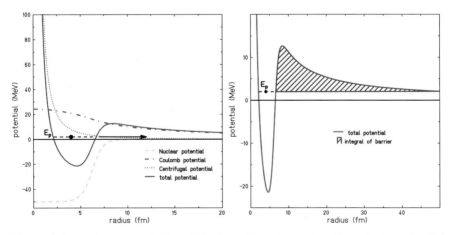

Fig. 3. Schematic representation of the tunnelling of a proton through the potential barrier of a nucleus. E_p, the energy available for the penetration of the barrier, is linked to the Q_p value via a recoil correction. *Left-hand side*: The different contributions to the potential and their sum are shown. *Right-hand side*: The total potential is shown and the area under the potential which the proton has to tunnel through is indicated

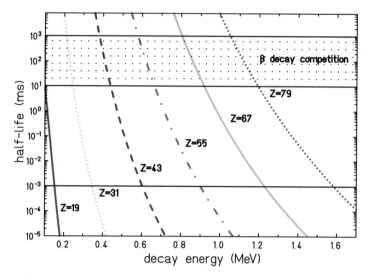

Fig. 4. Half-life for proton emission as a function of the decay energy determined with spherical barrier-penetration calculations with angular momentum $\ell=0$. The *lower horizontal line* gives the experimental detection limit due to technical limitations whereas the *hatched area* is the region of typical β-decay half-lives

Here, v is the velocity of the proton and $m_0 = 931.5\,\mathrm{MeV/c}$. S_{int} is the integral over the barrier as shown schematically in Fig. 3. The potential is the sum of the nuclear potential, the Coulomb potential, and the centrifugal potential.

This simple text book model can be used to determine the barrier penetration times for different Q_p values, the mass difference between the parent and daughter system, in different nuclei. A similar, although somewhat more advanced calculation is shown in Fig. 4. As can be seen, for a given nuclear charge Z, the barrier penetration times are rather steep functions of the available decay energy. With increasing nuclear charge, these curves become less steep which is due to the increase of the barrier height and width. The *lower line* in the figure gives the lower limit of experimentally accessible half-lives. For long barrier penetration half-lives, proton radioactivity is in competition with β^+ decay. Therefore, the hatched area in Fig. 4 gives typical β-decay half-lives in the vicinity of the proton drip line, which extend from a few milliseconds to many seconds.

From this figure it becomes clear that 1p radioactivity has to be searched for in the half-life region between a few micro-seconds and a few hundred milliseconds. As the barrier penetration half-life depends very sensitively on the available decay energy, Q values around $800\,\mathrm{keV}$ are observed for the lightest known proton emitters in the $Z=50$ region, whereas energies as high as $2\,\mathrm{MeV}$ were observed for the heaviest proton emitters in the lead region. The fact that today no ground-state proton emitter lighter than about $A=100$ has been observed comes simply from the fact that in this low-mass region Nature did not provide nuclei with masses such that the difference between the parent and daughter nucleus mass is in the right region.

As can be seen from Fig. 4, the Q_p value window which allows to observe 1p radioactivity is much smaller for light nuclei than for heavier ones. The addition of angular momentum changes the picture only slightly. For a given nuclear state, the Q_p value window which can give rise to proton radioactivity widens with increasing nuclear charge. For light nuclei, this window is smaller than 50 keV, whereas for the heaviest nuclei a window width of close to 500 keV is reached. From this figure, it becomes evident why the probability to find ground-state proton radioactivity in heavier nuclei is much higher than in light nuclei, a fact which agrees with the experimental observation of proton radioactivity for all odd-Z elements but two for elements above tin ($Z=50$), but for none below (see Fig. 10).

2.3 Production of Proton Emitters and Detection of Proton Radioactivity

Proton-rich nuclei at the proton drip line can be produced by two conceptually different methods: fragmentation/spallation as well as fusion–evaporation reactions. In the former, a stable nucleus heavier than the final product searched for is fragmented by the impact of a high-energy ($E \approx 1000$ MeV) proton (spallation) or a energetic heavy ion with energies from a few tens MeV to a few GeV per nucleon is fragmented in a suitable target (fragmentation) [108]. In the latter case, the nuclei of interest are produced by means of a fusion reaction of a medium-mass stable-nuclei beam with an energy around the Coulomb barrier impinging on a suitable target. In any case, the nuclei of interest have to be separated from the bulk of other products.

As fragmentation reactions allow easily to reach the proton drip line only for lighter nuclei (up to $Z=50$) and fusion–evaporation reactions do a much better job for heavier nuclei, basically only fusion–evaporation reactions have been used in proton-radioactivity experiments. In all recent studies, the proton radioactive nuclei were produced at the entrance of an in-flight separator, which allows separating first of all the primary beam from the reaction products and in addition to select reaction products with a suitable mass-over-charge ratio A/Q. Although such a separator reduces dramatically the number of nuclei one has to deal with, the nuclei of interest are nonetheless accompanied by a large number and a large variety of other nuclei.

Therefore, a powerful detection setup is required which allows to observe and to some extent identify these nuclei and which enables one to detect their decay. This task became much easier once DSSSDs (Fig. 5) were introduced in this research [136]. These detectors allow locating each event in space and time and correlating thus the implantation of a proton emitter with its decay. For this purpose, the implantation and decay have to take place in the same X and Y strip of the detector, which comes down to treating the DSSSD as $x \times y$ independent detectors, where x is the number of strips in X direction

Fig. 5. Photograph of a DSSSD as used at the FMA of Argonne National Laboratory

and y, the number of orthogonal strips in Y direction. The experimenter has, however, to make sure that the implantation rate in each such sub-detector is sufficiently low, so that on average the decay of the species implanted can take place before a new implantation occurs. This means that the implantation rate f should be smaller than or comparable to the inverse of the half-life of the species studied.

The DSSSDs available today can have pitches of strips as low as $100\,\mu$m or less. Therefore, rather highly pixelated detectors can be reached. The only disadvantages of such high subdivisions are inter-strip dead zones, where part of the energy signal from implantation or decay may be lost, and the need for a large number of electronics channels to readout all strips of the detector.

2.4 Sequence of Single-Particle Orbitals

One of the basic ingredients to the nuclear shell model is the sequence and the energy of single-particle levels. The sequence of these levels determines which orbitals are filled for which nucleus. The levels are well known close to the valley of nuclear stability, but their energy is also known to change as a function of the proton-to-neutron ratio of the nuclei. Therefore, far away from stability, inversions of these single-particle levels occur frequently and in order to correctly model nuclei, their sequence and their energy have to be measured.

Proton radioactivity is a powerful tool to determine the sequence of single-particle levels. This is done by analysing the proton-emission decay energy and its half-life. As briefly mentioned above, the barrier penetration half-life is related to the barrier height. In the case of 1p emission, this barrier is constituted of the Coulomb barrier due to the charge of the protons and the nucleus and the angular momentum barrier, which depends on the angular momentum of the orbital from which the proton is emitted. A comparison of the experimental partial half-life for a particular proton emission, which is the total half-life of the nucleus divided by the branching ratio for this particular proton-emission branch, with the half-life calculated within any well-suited theoretical model allows to determine the angular momentum of the emitting orbital.

Figure 6 presents such a comparison. As the barrier penetration calculations contain only the decay dynamics but no nuclear-structure effects which still slow down the emission thus yielding longer half-lives, the theoretical value has to be shorter than the experimental datum. Knowing which single-particle levels are close to the Fermi surface in the region of the nuclei studied, one can in many cases clearly decide which single-particle level must be the emitting level.

Once the single-particle configuration is chosen, one can use the comparison between theoretical prediction and experimental result to determine the "experimental" spectroscopic factor:

$$S^{\mathrm{exp}} = \frac{t^{\mathrm{calc}}_{1/2,p}}{t^{\mathrm{exp}}_{1/2,p}} \tag{1}$$

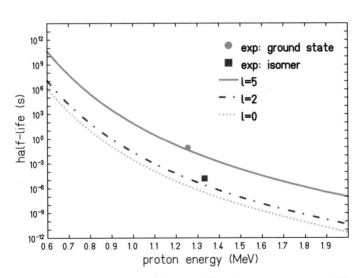

Fig. 6. Comparison of experimental data and barrier penetration half-lives calculated with a simple spherical model [75] which uses a nuclear, a Coulomb and a centrifugal potential to relate decay energy, half-life, and angular momentum of the emitted proton in the case of proton radioactivity of ^{151}Lu. The fact that ^{151}Lu is most likely modestly oblate deformed [47, 138] does not change the conclusion significantly in the present case. The ground state is found to have an $h_{11/2}$ structure, whereas the emitted proton in the isomer is in an $d_{3/2}$ state

This spectroscopic factor, which includes nuclear structure effects like the radial wave function mismatch, may be compared to theoretical predictions, e.g. from the nuclear shell model. There the spectroscopic factor is defined as

$$S^{\text{theo}} = | < \Psi_i(Z+1, A+1)|a^{j+}|\Psi_f(Z, A) > |^2, \tag{2}$$

where Ψ_i and Ψ_f are the wave functions of the initial and final states, respectively. a^{j+} is the creation operator for a proton in orbital j. For a perfect overlap of the initial and the final wave function, the spectroscopic factor is unity.

Davids et al. [35] calculated this theoretical spectroscopic factor in a low-seniority shell model (LSSM) for the spherical region between $Z = 64$ and 82 and found that the spectroscopic factor depends only on the number of pairs of proton holes p below the $Z = 82$ closed shell in the proton radioactivity daughter, i.e. $S^{\text{theo}} = p/9$. In Fig. 7, we compare experimentally determined spectroscopic factors with those determined from the LSSM. Evidently, the theoretical prediction is in nice agreement with the experimental values.

Such a relatively simple model works only in this region, where the $h_{11/2}, d_{3/2}$, and $s_{1/2}$ orbitals are almost degenerate and the nuclei are spherical. The fact that some of the experimental points do not lie on the theoretical curve may be linked, e.g. to deformation effects which are not included in this simple model. This deficiency, mainly observed for the $d_{3/2}$ orbital, can be overcome by including in more sophisticated models the coupling of the motion of the proton to vibrations of the daughter nucleus. This type of models were proposed by Davids and Esbensen [34] or by Hagino [72].

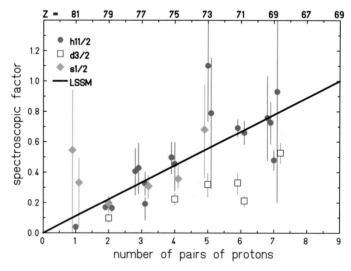

Fig. 7. Experimental spectroscopic factors for proton emitters from elements between $Z = 69$ and 81 are compared to the spectroscopic factor determined within the LSSM

A comparison between model calculations with and without this coupling clearly shows the necessity of this coupling and nice agreement is obtained for any angular momentum for odd-mass proton emitters, which decay to close to spherical daughter nuclei. In odd–odd nuclei, however, the treatment of the unpaired neutron as a pure spectator is another approximation which has to be overcome, if one has a "closer look" into these nuclei. In these cases, coupling of the unpaired neutron to the even–even core is most likely also necessary to achieve better results.

In the case of strongly deformed nuclei in the region between $Z = 63$ and 67, but also for nuclei with modest deformation ($Z = 53$–55), these approaches are no longer valid and deformed models have to be used [7, 33, 48, 84, 100]. In addition, Coriolis coupling, which is a kinematical coupling of the degrees of freedom of the proton and the core and which mixes different Nilsson configurations, becomes important (see, e.g. [7, 44, 49, 154]). However, in order to achieve reasonable results, the pairing residual interaction, which reduces the Coriolis effect, has to be taken into account [49]. These models usually achieve nice agreement with experimental data, when they assume deformations rather close to the predictions from the Möller et al. model [107].

All these models assume the same deformation for the parent and the daughter state. However, e.g. in the case of proton emission from $^{185}\text{Bi}^m$, the deformation is expected to change from a moderately oblate $1/2^+$ intruder state to a spherical 0^+ state in ^{184}Pb. This deformation change is beyond the scope of most theoretical models. The time-dependent Schrödinger equation approach naturally includes this change of deformation [147–149]. However, none of the studies performed up to now with this model really took advantage from this fact and tried to describe proton emission from $^{185}\text{Bi}^m$.

2.5 "Fine Structure" in Proton Emission

Most proton emitters have decay branches from the proton emitting state to the ground state of the daughter nucleus. This is true, independent of the fact whether it is an isomer or a ground state which decays. In these decays, the proton often has to "carry away" a large amount of angular momentum and a decay to an excited state might reduce this angular momentum. However, the decay to the ground state takes profit from the large decay Q value which speeds up the decay.

The decay to an excited state in the daughter nucleus is favoured, if the excitation energy of this final state is low and the decay Q value stays high. This is the case in strongly deformed nuclei. Indeed, first evidence for this so-called "fine structure" of proton radioactivity was obtained from the decay of ^{131}Eu [143]. Two proton lines with an energy difference of 121(3) keV were observed with the same half-life. From the same half-life, from 2^+ energy systematics from neighbouring nuclei, and from the observation of peaks for γ decay and internal conversion with correct branching ratios, the authors concluded on the first observation of "fine structure" in proton radioactivity.

The observation of fine structure helps to clarify the nuclear structure of the decaying nucleus. From the observation of only the ground-state decay branch it was not possible to determine, whether in the decay of ^{131}Eu the proton was emitted from the $3/2^+$[411] or the $5/2^+$[413] Nilsson orbital [38]. A comparison of the experimental branching ratio for emission to the ground and the first excited state to theoretical predictions clearly designates the first configuration to be the one which emits the proton, in agreement with predictions from Möller et al. [107]. Detailed calculations also allowed the different contributions from the $d_{3/2}$, $d_{5/2}$, $g_{7/2}$, and $g_{9/2}$ orbitals to the decay to be determined. From the excitation energy of the 2^+ state, the authors concluded on a quadrupole deformation of $\beta_2 = 0.34$ of ^{130}Sm. This is in agreement with the expectation of strong prolate deformation in this region [107].

This first observation of fine structure triggered more work on this subject. Thus, before its experimental observation, fine structure was predicted for ^{141}Ho [91, 99]. From a deformation of $\beta_2 = 0.29$ [107], the first paper [99] predicted an excitation energy of 140 keV for the first excited 2^+ state of the daughter nucleus and, by means of their model, a branching ratio of about 5%. A similar work, although with a different model [91], deduced a branching ratio for fine structure of 6%, while using a deformation of $\beta_2 = 0.29$ and an excitation energy of 160 keV.

Its experimental observation followed shortly after in experiments at Oak Ridge [11, 133]. From the branching ratio for fine structure observed and the excitation energy of the first excited 2^+ state in the daughter nucleus of 202 keV, a deformation of $\beta_2 = 0.24$ was inferred.

Other nuclei for which fine structure was observed are ^{144,145,146}Tm. The composition of the ^{145}Tm ground-state wave function was deduced and a deformation of $\beta_2 = 0.18$ was inferred. Although the theoretical models used to describe the decay [34, 72] do not yield exactly the same wave function composition, the general picture of both models is similar and indicates that the fine structure is due to a small admixture to the wave function from the $f_{7/2}$ proton orbital coupled to the 2^+ state of the daughter, whereas the main contribution to the proton decay comes from the $h_{11/2}$ orbital coupled to the 0^+ daughter ground state.

The situation for the two odd–odd proton emitters is much less clear. As mentioned above, the treatment of the unpaired neutron as a spectator or as a nucleon participating in the decay may significantly change the picture. In the case of ^{146}Tm, as much as five different proton groups were identified and attributed to proton emission with fine structure [10, 11, 55, 129]. However, the detailed assignments and therefore also the details of the nuclear structure information deduced are different and the situation is yet unclear.

In the case of ^{144}Tm [71], the experimental information is rather scarce and the interpretation was strongly influenced by the results obtained on ^{146}Tm. The dominant decay should come from the proton – even–even core configuration $h_{11/2} \otimes 0^+$, whereas the weaker decay to the excited state in ^{143}Er is due to the proton even–even core configuration $f_{7/2} \otimes 0^+$. This simplified picture assumes the unpaired neutron to be a spectator.

In conclusion, one can say that fine structure is a valuable tool to study details of the wave function of the initial and final state. However, in particular in the case of odd–odd nuclei, high-statistics and high-resolution data are needed to disentangle the different decaying states and make correct assignments for the experimental observation. These experimental requirements are not always fulfilled and hamper thus the interpretation.

2.6 Proton Radioactivity as a Spectroscopic Tool

As mentioned in the preceding chapters, proton radioactivity can be used to test the wave function of the emitting state, the barrier penetration, or some aspects of the daughter nucleus. However, in recent years, proton radioactivity has also been used as a tag to identify nuclei and to study other characteristics of the nuclei. This tagging of an implantation event in the focal plane of a separator allows correlating it with its production at the target station of the recoil separator and thus filter out a few events of interest out of the overwhelming amount of less exotic, i.e. less interesting events.

In these applications of proton radioactivity, the proton decay branch is used to identify an evaporation residue by a delayed coincidence of proton emission and implantation signal. As in the experiments described before, implantation and decay have to take place in the same pixel of a DSSSD. Then the implantation event can be correlated with γ rays observed at the production target during the formation of the evaporation residue. As this formation process is induced by a fusion–evaporation reaction (fragmentation reactions can be used as well, but for the moment only fusion–evaporation was used in this type of studies; see, e.g. [145] for similar experiments not related to proton radioactivity), the reaction products are formed in high-spin states which de-excite through γ-ray cascades, most often as an yrast cascade. These γ-ray cascades are beneath the most interesting messengers of nuclear structure and allow to determine, e.g. the deformation of the emitting nucleus and its moment of inertia.

The first example of recoil decay tagging (RDT) with proton radioactivity was proposed by Paul et al. [122], who studied excited states of the proton emitter ^{109}I. Several γ lines from the $h_{11/2}$ band could be observed for the first time and compared with the γ-ray decay scheme of neighbouring odd-mass iodine isotopes. From a drop of the $15/2^-$ state in ^{109}I, an onset of deformation for this extremely proton-rich nucleus was concluded.

After this first use of proton radioactivity to tag events, new experiments were performed for ^{109}I [157], ^{113}Cs [70], ^{117}La [96], ^{131}Eu [142], ^{141}Ho [142], ^{145}Tm [139], ^{146}Tm [129, 140], ^{147}Tm [140, 141], ^{151}Lu [96, 156], ^{161}Re [83], and ^{167}Ir [135]. The RDT study of ^{145}Tm allowed to observe for the first time proton-γ coincidences at the focal plane of the FMA at Argonne [139]. In this experiment, protons from the ground-state to ground-state decay as

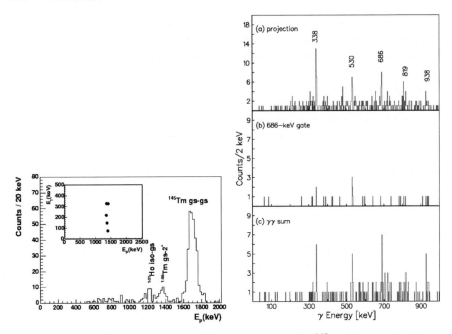

Fig. 8. *Left-hand side*: Proton spectrum as observed for ^{145}Tm. The decay of the ^{145}Tm ground state to the proton daughter ground and first excited state is observed. The *inset* shows the proton–γ-ray coincidence spectrum from the focal plane of the FMA. *Right-hand side*: Gamma-ray spectrum from the de-excitation of high-spin states of ^{145}Tm. These γ rays are observed at the target position of the FMA and tagged with protons from proton radioactivity observed at the FMA focal plane in delayed coincidences (from [139])

well as the decay to the first excited 2^+ state were used as tags. Figure 8 shows the observed proton spectrum from the focal plane of the FMA and the γ rays correlated to the proton emitters. The inset shows the proton–γ-ray correlations.

In general, these investigations allowed to observe positive and negative parity bands and to determine the moments of inertia and thus the deformation of these proton emitters. The basic limitation to these studies is the production cross section necessary to achieve sufficient statistics. With the γ-ray arrays available today like, e.g. GAMMASPHERE, production cross sections of about 100 nb are necessary to observe the most important γ rays within a reasonable experiment time.

2.7 Search for New Proton Emitters

The proton-emission studies presented in the preceding sections allowed investigating nuclear structure close and beyond the proton drip line. As can be seen from Table 1, 28 ground-state proton emitters and 19 proton-emitting isomers are known experimentally. The experimental together with the theoretical investigations allow improving our understanding of the evolution of single-particle levels in the vicinity of the proton drip line, of configuration mixing in this region, of the delineation of regions of deformation, of high-spin states and the moments of inertia of proton drip-line nuclei and of many other aspects.

These investigations were possible, because, unlike, e.g. α emission, 1p emission is a relatively simple process involving only two "particle" in the exit channel, the proton and the heavy recoil. Therefore, only two-body kinematics is necessary and no pre-formation like in α decay is needed. Another aspect is certainly also the fact that many proton emitters are now known experimentally and systematic trends could be explored. Therefore, the question has to be asked whether new proton emitters will still enrich our understanding of nuclear structure at the proton drip-line.

The possibility to discover and study a new proton emitter depends mainly on two aspects: (i) its production cross section and (ii) its life time. The production cross section depends, e. g. on the projectile–target combination and the number of particles to be evaporated to produce the nucleus of interest. As a rule of thumb, the further away from stability a nucleus, the more difficult it is to produce, because its production usually involves more evaporated neutrons.

A plot showing the dependence of the production cross sections on the number of evaporated particles is shown in Fig. 11. From this figure, one can

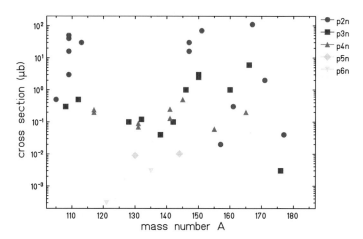

Fig. 9. Production cross sections for fusion–evaporation reactions used to produce ground-state proton emitters. The cross sections vary from $10\,\mu$b for the least exotic proton emitters produced via $p2n$ reactions to $1\,$nb for the most exotic proton emitters produced by means of $p6n$ reactions. Each additional neutron to be emitted "costs" about an order of magnitude in cross section (See also Plate 11 in the Color Plate Section)

see that ground-state proton emitters which were produced by means of a $p2n$ reaction, meaning that the fusion compound nucleus has emitted one proton and two neutrons to form the final proton emitter, have a production cross section of about 10 µb. For each additional neutron to be evaporated, the cross sections drop by about an order of magnitude to reach a cross section of about 1 nb for $p6n$ reactions. This fact has to be kept in mind in the search for new proton emitters and might pose serious problems for their detection.

This clearly shows that the choice of the projectile–target combination is crucial for the experimental success. However, this choice is today limited to stable-isotope projectiles and targets. Although radioactive beams are now available in many laboratories across the world, their intensity is still orders of magnitude too small to reach, with the help of more favourable production cross section, the production rates obtained with stable-isotope projectile beams. One may hope that new radioactive beam installations will produce the radioactive projectiles needed in amounts which can compete with stable beams.

The second factor, the life time of a proton emitter, is first of all dictated by the Q_p value. Figure 12 shows this quantity for nuclei close to the proton drip-line between proton number $Z = 51$ and 83. The data are taken from Table 1 or from the atomic mass evaluation [2].

From an experimental point of view, it is particularly puzzling that for $Z = 61$ no proton emitter could be observed (see Fig. 10 and Table 1). This, of course, might just be an "accident" of Nature which did not provide a

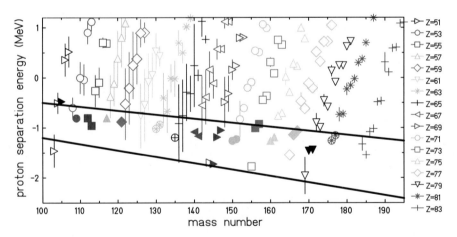

Fig. 10. Ground-state proton-separation energies for proton-rich nuclei. The *open symbols* show values taken from the 2003 mass evaluation [2], whereas the *full symbols* show the known ground-state proton emitters. The *full lines* show calculations for barrier penetration half-lives of 100 ms (ℓ=0) and 1 µs (ℓ=5) with a spherical WKB model, a semiclassical approximation of quantum mechanics (See also Plate 12 in the Color Plate Section)

praseodymium isotope with a mass and binding energy to have a half-life in the accessible range between about $1\,\mu s$ and $100\,ms$. $^{134,135}Pm$ might be possible proton emitters, but up to now experiments to find them were unsuccessful [139].

There are also other regions, for which it is astonishing that no proton emitters could be identified. The case of light nuclei below $Z = 50$ has been discussed already above. Here again the most likely explanation is that Nature did not provide nuclei in the range of proton separation energies, which would have half-lives in the accessible range. The only possibility seems to be to develop new detection setups, which should allow for the identification of much shorter half-lives. This is no longer compatible with the detection of proton emission at the focal plane of recoil separators. Instead one might imagine observing the daughter nucleus at the exit of such a separator and to detect the proton just behind a production target at the entrance of this separator. However, in this case the proton must be observed in a rather hostile environment in the vicinity of the target. Double-reaction experiments as performed, e.g. in the study of 2p radioactivity [92, 159] (see below) might be a way out.

In the region between $Z = 50$ and 82, more proton emitters might also be expected. The *lines* indicated in Fig. 12 show the limits for half-lives between $100\,ms$ and $1\,\mu s$. These *lines* have been calculated with a spherical WKB model assuming an $\ell=0$ emission for the $100\,ms$ line and an $\ell=5$ emission for the $1\,\mu s$ line. This figure evidences that new proton ground-state emitters might be expected for ^{116}La, ^{120}Pr, ^{134}Tb, ^{155}Ta, ^{164}Ir, $^{174,175}Tl$, $^{184,185}Bi$. In addition, proton emission from isomers might be expected, too. However, such an extrapolation based just upon the proton separation energies and their extrapolation is most likely too simple. As has been shown, proton emission depends very sensitively on the angular momentum and on nuclear structure as, e.g. the overlap between the parent and daughter wave functions. Nonetheless, this figure can serve as a guide for a more detailed theoretical study to calculate possible proton-emission Q_p values and half-lives.

More proton emitters might also be expected above lead. However, these nuclei are more and more difficult to access. In addition, α decay becomes increasingly important and might thus prevent from observing other proton emitters.

2.8 Conclusions on One-Proton Radioactivity

The study of proton emitters as well as their use as a tag for other investigations have allowed acquiring a wealth of new experimental data, often only accessible by means of proton emitters, and to deepen our understanding of nuclear structure close to the proton drip-line. In particular in the region where the $s_{1/2}$, $d_{3/2}$, and $h_{11/2}$ orbitals are quasi-degenerate, these investigations allowed to determine the sequence of these single-particle orbitals which is an important input, e.g. for shell-model calculations. The experimental data also triggered new developments of nuclear-structure models or at least their application to this field. These models have not been described in the present lecture and are beyond its scope. A recent review can be found in [39].

Proton emitters have been used more recently as a tag to study high-spin states above the proton-emitting ground state. This method is certainly very promising and will find more applications with the increase of the performance of γ-ray spectrometers like AGATA or GRETINA and the increase in production rates expected at new radioactive beam installations.

At these installations, one may also hope to find new proton emitters, in particular by using proton-rich radioactive beams with high intensity to take profit from higher production cross sections when less neutrons have to be evaporated. In parallel, new detection methods will have to be developed to access proton emitters with half-lives below the 1 μs range. Digital electronics with flash ADCs are probably one way to go.

3 Two-Proton Radioactivity

Similar to the situation in 1p decay studies (see introductory text of Sect. 2), the general aim of investigations of the 2p disintegration is to gain information on the total and partial half-life of the parent state and on the energy of the two protons. The energy of the two protons and of the recoiling daughter atom determine the Q value for 2p decay (Q_{2p}). A particularly interesting experimental task of 2p experiments is to measure the correlations between the two protons and to thus get insight into the disintegration mechanism.

As mentioned in Sect. 1, 2p radioactivity was first proposed at the beginning of the 1960s [60, 158]. Goldanskii [59, 61] was the first to give a list of possible candidates. Later Jaenecke [82] provided another list of possible ground-state 2p emitters. However, these lists could not take profit from well-known mass surfaces, in particular close to the drip lines. Much more refined proposals were made in the 1990s by Brown [19], Ormand [115, 117], and Cole [29–31]. These allowed to single out nuclei like ^{39}Ti, ^{42}Cr, ^{45}Fe, ^{49}Ni, ^{48}Ni, and ^{54}Zn as promising candidates. The uncertainities associated with these predictions and in particular the differences beneath them did not allow determining which of the candidates would be the best nucleus to discover 2p radioactivity.

To illustrate this fact, Fig. 11 shows the tunnelling half-life as calculated with a simple di-proton model (see, e.g. [19]) as a function of Q_{2p} for the case of ^{42}Cr. The figure gives the relation between half-life and Q_{2p} value, the latter quantity being taken from the model predictions [19, 29, 117] or from other mass models described in the literature. It is evident from this figure that these predictions scatter over a wide range of Q_{2p} values et thus of barrier-penetration half-lives so that only experiment could help to find the best candidates.

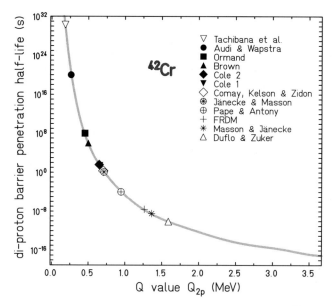

Fig. 11. Barrier penetration half-life for a di-proton emission of ^{42}Cr as a function of the available decay energy. The *different symbols* correspond to different Q-value predictions from the literature. The half-lives are calculated from Coulomb wave functions using the Wigner single-particle width (see, e.g. [19])

As the lightest of the nuclei mentioned above were the first to become accessible experimentally, they were also the first to be studied. It was Détraz and co-workers who investigated the decay characteristics of ^{39}Ti and found that it decays by β decay [40]. The next nucleus to be investigated was ^{42}Cr [57]. Its decay was found to be dominated by β decay with a half-life

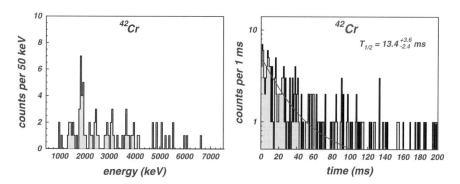

Fig. 12. *Left-hand side*: Charged-particle decay-energy spectrum obtained after ^{42}Cr implantation. The 1.9 MeV peak is interpreted as a β-delayed decay. *Right-hand side*: Half-life spectrum for the decay of ^{42}Cr obtained by correlating implantation and subsequent decay. (From [57])

of $13.4^{+3.6}_{-2.4}$ ms. The decay energy and time spectra obtained during this experiment for ^{42}Cr are shown in Fig. 12. The prominent peak at 1.9 MeV was excluded to be of 2p nature, as such a high energy would lead to an extremely short barrier penetration time for the two protons (see Fig. 11), in contradiction with the ^{42}Cr half-life.

Finally, also ^{49}Ni was found to decay mainly by β decay [57]. These early experiments could not exclude that these nuclei decay with very small branches by 2p radioactivity. In fact, 2p radioactivity is in competition with β^+ decay and the dominance of one or the other depends mainly on the Q values for 2p and β^+ decay. The values determine the partial half-lives for the different decay modes and the branching ratio is directly related to these partial half-lives. The shorter a partial half-life, the larger the branching ratio for this decay branch. In order to observe both decay branches, 2p emission and β^+ decay, the two partial half-lives have to have similar values.

Two-proton radioactivity was finally discovered in experiments at the LISE3 separator of GANIL [58] and at the FRS of GSI [124]. These experiments are now described in some detail.

3.1 Discovery of Two-proton Radioactivity

Exotic nuclei can be produced by different means. Today the most powerful methods are the ISOL (isotope separation on-line) method [153] and the projectile fragmentation method [108]. Very schematically, the ISOL method is better suited for precision measurements, whereas the projectile fragmentation method allows the more exotic nuclei to be studied. This latter method allowed accessing the nuclei of interest and the most proton- and neutron-rich nuclei we know today were produced by this method (see, e.g. [15, 79, 95]).

The investigation of ^{45}Fe was possible due to the availability of high-intensity primary beams at energies of a few tens of MeV/nucleon or higher, energies necessary to induce projectile fragmentation. In both experiments, at GANIL and at GSI, ^{45}Fe was produced by fragmentation of a primary ^{58}Ni beam impinging on the target of a fragment separator. The fragments of interest, i.e. proton-drip line nuclei around mass number A=45, were selected by the LISE3 separator [109] at GANIL or the FRS [51] at GSI and finally implanted in a silicon detector telescope at the focal plane of these separators. Figure 13 shows schematic layouts of these two separators. Both use a first magnetic selection by means of the magnetic rigidity $B\rho \approx A \times v/Q$, where v is the velocity of the fragments and Q their charge. Then the fragments are slowed down, as a function of their energy and their charge, by a specially shaped degrader. All fragments have similar velocities after the target, close to the velocity of the 'unreacted' beam. After the degrader this is no longer true and the fragments have velocities which are functions of their mass and their charge. Therefore, a second selection according to their magnetic rigidity will allow to significantly reducing again the number of nuclei transmitted. At relativistic energies as at the FRS, these two selections are sufficient to reduce

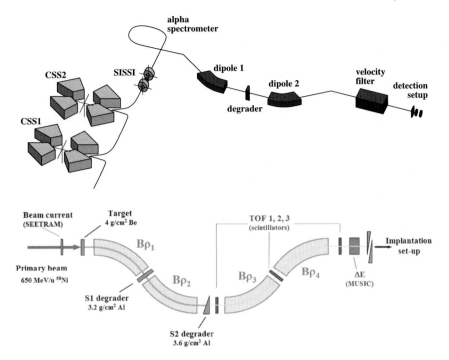

Fig. 13. SISSI/ALPHA/LISE3 complex of GANIL (*upper panel*) for the production of exotic nuclei as used in the experiments on 2p radioactivity [109] and FRS (*lower panel*) which was used in the 2p radioactivity studies at GSI [51]

the number of fragments arriving at the focal plane and thus at the detection setup. At the LISE3 separator, an additional velocity filter selecting nuclei with a similar velocity by crossed electric and magnetic fields finally reduces the number of nuclei so that the detectors can handle the counting rates.

The fragment separators allow decreasing the number of fragments which arrive at the focal plane to a value which can be handled by both the data acquisitions used and the detector system. These counting rates are typically a few hundred per second at maximum. At this rate, nuclei are implanted in the detection setup. To study their decay, these implantations have to be correlated with the subsequent decay of the nuclei. To allow high- implantation rates without losing the correlation, highly sub-divided setups are necessary. This can be achieved either by a large number of individual detectors or by a high granularity of one detector. This last condition is fulfilled by highly pixelated DSSSDs. These detectors consist of orthogonal strips on either side of the detector. The crossing of a front and a back strip forms a pixel which can be treated as an independent detector. Therefore, a 16 x 16 strip detector as used in the GANIL experiment has a total of 256 pixels. The GSI experiment used a telescope of seven large-surface silicon detectors in which the

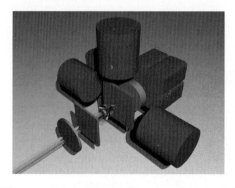
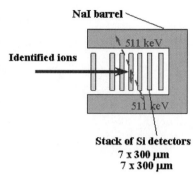

Fig. 14. *Left-hand side*: Setup used at the SISSI/ALPHA/LISE3 complex of GANIL for experiments on 2p radioactivity [58]. The silicon detectors allow one to determine the parameters of implantation events such as energy loss, residual energy, time-of-flight, and position as well as the characteristics of the decays such as charged-particle sum energy and half-life. The DSSSD was the implantation device where the nuclei of interest stopped. The subsequent decay had to occur in the same x–y pixel of the DSSSD. *Right-hand side*: Setup used at the FRS in the studies of 2p radioactivity [124]. The nuclei of interest stopped according to their range in one of the silicon detectors which registered their energy loss. The subsequent decay had to take place in the same detector and the charged-particle energy and the half-life could be measured

fragments were implanted as a function of their range. In order to keep track of implantation and decay events, the implantation and decay rate in a single detector element should not much exceed the inverse of the half-life of the nuclei of interest.

The setups used at GANIL and at GSI are shown in Fig. 14. The silicon detectors register the energy loss of an ion for an implantation event and the energy of the charged particles emitted in the subsequent decays. In the GANIL experiment [58], the adjacent detectors allowed to search for a β^+ particle which would be emitted in the concurrent β^+ decay. In the GSI setup [124], the NaI barrel allows to detect γ radiation from either the annihilation of positrons from β decay or from other γ radiation. Both setups allowed therefore distinguishing with a high degree of confidence between ground-state 2p emission (no β or γ radiation) or β^+ decay (β and γ radiation).

The spectra accumulated with these detectors are shown and explained in the following. The fragments arriving at the focal plane of the separators are usually identified by their energy-loss, their time-of-flight, and their magnetic rigidity. These quantities allow preparing plots which unambiguously identify and separate all fragments. Such a plot is shown for the experiment performed at GANIL (see Fig. 15). Although only two parameters, namely energy-loss and time-of-flight, are shown, eight other independent parameters were used to purify the spectrum and obtain this almost background-free spectrum. A similar spectrum was also obtained in the GSI experiment.

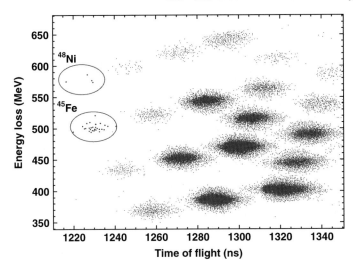

Fig. 15. Fragment identification spectrum from the GANIL experiment [58]. The energy loss in the first silicon detector of the setup (see Fig. 14) is plotted as a function of the time-of-flight from the production target in the SISSI device to the detection setup. Individual nuclei correspond to the different clusters of events. Nuclei absent in the plot are known to be particle unstable with half-lives much shorter than the flight time through the separator

Each of the implantation events shown in Fig. 15 can then be correlated to subsequent decays in the same detector element. The signal heights registered for the decay events yield the charged-particle energy spectrum, whereas the time elapsed between implantation and decay yields the decay-time spectrum. Both decay-energy and decay-time spectra are shown in Fig. 16. These spectra contain the decay characteristics from the GANIL and the GSI experiments which lead to the discovery of 2p radioactivity [58, 124] as well as from an experiment performed in GANIL in 2005 [43]. They yield an energy of 1.151(15) MeV. The half-life average of all experiments is $T_{1/2} = 1.75^{+0.49}_{-0.28}$ ms and the 2p branching ratio is BR $= 59(7)\%$. This gives a partial 2p half-life of $T_{1/2}$ / BR $= T_{1/2}^{2p} = 3.0^{+0.9}_{-0.6}$ ms.

A first indication that the charged-particle peak at 1.151 MeV is indeed of 2p radioactivity nature comes from the fact that this energy fits nicely the prediction from recent Q-value calculations. Brown [19] gives a value of 1.154(94) MeV, whereas Ormand finds a Q value of 1.279(181) MeV [115]. Both values are in nice agreement with our experimental value. Cole's calculations result in a value of 1.218(49) MeV, which is in reasonable agreement with our experimental result [29].

A second experimental observable was the absence of any additional radiation such as β or γ rays. As explained above, both the GANIL as well as the

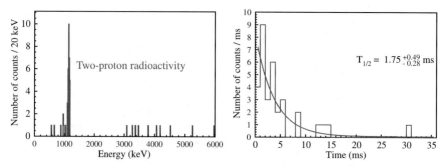

Fig. 16. Decay energy of ^{45}Fe obtained from three experiments [43, 58, 124], taking into account only decay events occurring less than 15 ms after the associated implantation event (*left panel*) and decay-time spectrum for the events having a decay energy around the 1.151 MeV peak (*right panel*). The half-life resulting from the latter data is $T_{1/2} = 1.75^{+0.49}_{-0.28}$ ms

GSI setup were equipped with detectors capable to register such radiation. Although the detection efficiency was far from 100%, the fact that none of the events with a decay energy around 1.151 MeV was in coincidence with such radiation leaves only a probability of about 1% to miss all radiation if it is present (see Fig. 17). As an example we explain the situation from the GANIL experiment.

The β detection efficiency was determined by means of neighbouring nuclei which emit β-delayed protons to be 30%. This means that there is a 70% probability to miss a β particle, if present, in one decay event. To miss both β particles in two decay events, the probability is 0.7×0.7, i.e. 49%. It is easy to determine that the probability to miss all β particles in the 12 decays of the first GANIL experiment which contribute to the 1.151 MeV peak is 1.4%. A similar reasoning for the GSI experiment yielded a comparable result [124].

If the peak observed at 1.151 MeV is due to, e.g., a β-delayed proton emission, the silicon detector which registers the decay detects not only the proton, but also the β particle which leaves part of its energy in the detector. As shown in Fig. 18, this leads to a broadening of the observed peak. The energy deposited by the proton is constant, but the energy loss of the β particle depends on the emission point in the detector and in particular on the emission direction. A comparison of the 1.151 MeV peak with β-delayed proton peaks in neighbouring nuclei showed that these latter peaks are up to 50% larger. The narrow width of the peak in Fig. 16 is therefore another indication for the 2p radioactivity nature of the decay of ^{45}Fe.

These different pieces of evidence leave only little room for another explanation of the events observed than 2p radioactivity. Maybe the most convincing piece is the observation of the daughter decay. For this purpose, one analyses the second radioactive decay after implantation of an isotope

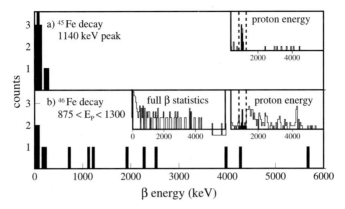

Fig. 17. (a) Spectrum as registered by the Si(Li) detector behind the implantation device from the GANIL experiment [58] for the decay of ^{45}Fe gated by the 2p peak. The spectrum shows that beyond the noise of the detector, no signal from β particles is observed. The *inset* shows the decay-energy spectrum from the implantation detector with the peak corresponding to 2p radioactivity which was used to gate the β-particle spectrum. (b) Same spectrum as in (a) but for ^{46}Fe. The main spectrum shows the events due to β particles in coincidence with the region indicated in the *right inset* which represents the charged-particle spectrum from the decay of ^{46}Fe with the gate region indicated. The *left inset* shows the full spectrum of the Si(Li) detector for all events from ^{46}Fe decays

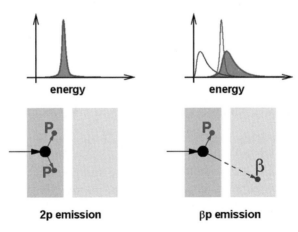

Fig. 18. *Left-hand side*: Sketch of a 2p emission without associated β particle; the total decay energy being registered by the implantation silicon detector yields a narrow peak. *Right-hand side*: Sketch of a β-delayed proton emission; the proton is absorbed in the silicon detector but the β particle adds more or less energy to the signal depending on its emission direction and thus broadens the peak

of ^{45}Fe with a gate on the 1.151 MeV peak from the first decay. The analysis sequence is the following: (i) one identifies a ^{45}Fe implantation by means of energy-loss, residual-energy, and time-of-flight measurements. (ii) one observes, correlated in space, i.e. in the same detector element, and time with this implantation, the first decay of the implanted isotope. (iii) by gating on the 1.151 MeV peak one searches for the second decay again subsequent in time, but in the same detector element. This second decay is then the decay of the daughter nucleus produced in the first decay with an energy release of 1.151 MeV.

The events between 3 and 5 MeV in the decay-energy spectrum of ^{45}Fe (see Fig. 16) were from the beginning interpreted as being due to the decay of ^{43}Cr, the 2p daughter of ^{45}Fe. This was only indicative, as the charged-particle spectrum from the decay of ^{43}Cr is not very distinct. It shows a large distribution of events between 2 and 5 MeV without any really pronounced peak. What yielded more convincing arguments was the daughter-decay half-life. When taking into account all decay events after ^{45}Fe implantation and performing a parent–daughter fit for the half-life, i.e. including the decay of the parent nucleus (^{45}Fe) and the grow-in and decay of the daughter (^{43}Cr), one can determine the daughter-decay half-life. This half-life was determined in the first experiment [58] to be $T_{1/2} = 16.7(70)$ ms, which may be compared to the half-life of ^{43}Cr of 21.1(4) ms [42].

This argument could be refined with increased statistics. Figure 19 compares the second-decay half-life after ^{45}Fe implantation and in coincidence with the 1.151 MeV peak with the half-lives [42] of all possible daughters of ^{45}Fe. It is evident from this figure that only the half-life of ^{43}Cr, the 2p daughter of ^{45}Fe, is in agreement with the observed daughter-decay half-life.

The argument of the daughter-decay half-life together with the other pieces of evidence presented before leaves no other room for the interpretation of the experimental observation than ground-state 2p radioactivity. This establishes 2p radioactivity as a new nuclear decay mode.

3.2 Direct Observation of Two Protons in the Decay of ^{45}Fe

The experiments presented up to now did not allow to observe directly the two protons emitted in the decay of ^{45}Fe. Nonetheless, they yielded irrefutable evidence of this new decay mode. However, to study in more detail the decay of nuclei by 2p emission and in particular to investigate the decay mechanism, it is necessary to observe the two protons and to determine the energy sharing between them and the proton–proton angle.

For this purpose, the principal deficiency of silicon detectors had to be overcome. These detectors do not allow the protons to escape from the detector. Therefore, only their sum energy could be determined. To directly observe the two protons, gas detectors were built which function as time projection chambers (TPC). In these devices, the isotope to be studied is stopped in the

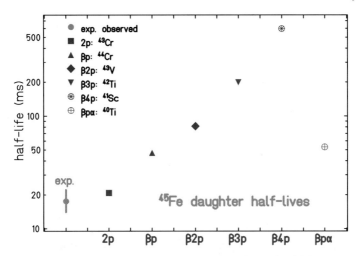

Fig. 19. The experimentally determined daughter-decay half-life conditioned by the 1.151 MeV peak (see Fig. 16) is compared to the half-lives [42] of all possible daughter decays for ^{45}Fe. Only the half-life of ^{43}Cr, the 2p daughter of ^{45}Fe, is in agreement with the experimental value

gas volume where it emits the two protons. The charges generated by these charged particles are projected onto a two-dimensional (2D) detector which produces a 2D picture of the decay. The third dimension is obtained by means of the arrival time of the charges on this 2D detector.

Such a device (the Bordeaux TPC [16]) was recently commissioned at the LISE3 separator and the decay of ^{45}Fe was studied [56]. One of the ^{45}Fe decay events registered with this chamber is shown in Fig. 20. The *top* of the figure shows the implantation event and allows to determine the stopping position of the ^{45}Fe nucleus. The *lower part* presents the subsequent decay event, where the two-hump structure can only be explained by the emission of two protons.

A similar device was developed at Warsaw University. It combines the principles of a TPC with techniques of digital photography. In this optical time projection chamber (OTPC), primary ionisation charges created by ions and particles stopped within the detector's active volume drift in a uniform electric field towards the amplification stage where they induce emission of light. This light is recorded by a CCD camera and by a photomultiplier (PM). The camera image provides a 2D projection of the particles' tracks while the digitised PM signal delivers information on the position along the drift direction. Combination of these two allows a full reconstruction of a decay event in three dimensions. More details on the operation of this detector are given in [104].

The OTPC was used to study the decay of ^{45}Fe in an experiment performed at the NSCL of Michigan State University, East Lansing, USA. The ions of ^{45}Fe produced by the fragmentation reaction of a ^{58}Ni beam at 160 A×MeV

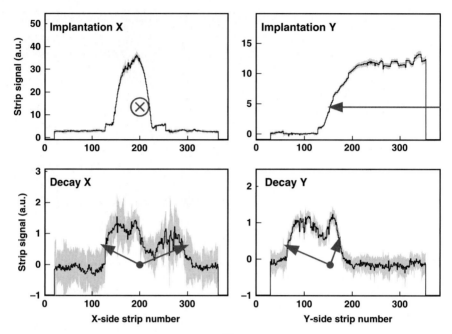

Fig. 20. A single event of 2p decay of ^{45}Fe from a recent experiment performed with the Bordeaux TPC at the LISE3 facility of GANIL. The *top row* shows the ^{45}Fe implantation event, where the ion enters the chamber parallel to the X direction and stops in the center of the chamber. The decay (*bottom*) takes place at the point where the ion is stopped. The double-hump structure is clear evidence for the emission of two protons in the decay of ^{45}Fe (from [16])

on a nickel target were selected using the A1900 separator, identified in-flight and stopped inside the OTPC. The events of 2p radioactivity of ^{45}Fe were clearly identified. An example is shown in Fig. 13.

The β-decay channels accompanied by emission of protons were also recorded. Figure 13 shows also a CCD image of an event interpreted as a β-delayed three-proton emission. This represents the first observation of such an exotic decay channel [106]. The reconstruction of 75 events of 2p decay of ^{45}Fe allowed to determine the angular and energy correlations between emitted protons for the first time [105]. They were found to be in good agreement with the predictions of the three-body model of Grigorenko and co-workers [68] (see Fig. 22). These findings open a new possibility to examine the structure of nuclei at the neutron-deficient limit of nuclear existence.

3.3 Other Two-Proton Emitters

The decay of ^{45}Fe was the first to evidence 2p radioactivity. However, from theoretical calculations [19, 22, 29, 115, 116] it was evident that other nuclei,

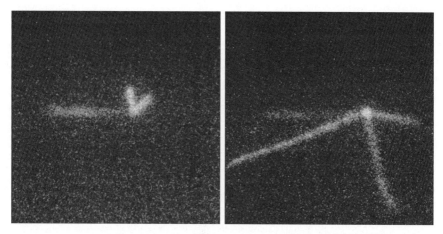

Fig. 21. *Left-hand side*: Example of a CCD image of a two-proton decay event of ^{45}Fe taken in a 25 ms exposure. A track of a ^{45}Fe ion entering the chamber from the left is seen. The *two bright short tracks* are protons of about 0.6 MeV. They were emitted 0.62 ms after implantation of the ion (from [105]). *Right-hand side*: Example of a CCD image of a β-decay event of ^{45}Fe. A weak track of a ^{45}Fe ion entering from the left is seen. The *three long bright tracks* are consistent with high-energy protons escaping the active volume of the detector. The decay occurred 3.33 ms after the implantation. The event is interpreted as a β-delayed three-proton decay of ^{45}Fe (from [123]) (See also Plate 13 in the Color Plate Section)

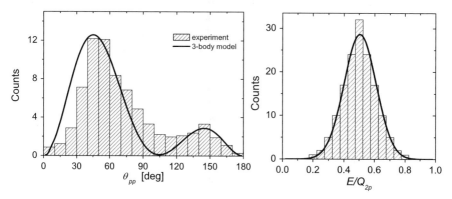

Fig. 22. *Left-hand side*: Proton–proton correlations in the 2p decay of ^{45}Fe. The experimental distribution of the opening angle between the two protons (*left*) and the energy distribution of the emitted protons in units of the total decay energy Q_{2p} (*right*) are shown as histograms. The predictions of the three-body model of Grigorenko and co-workers [68] are shown by the *solid lines* (from [105])

in particular ^{48}Ni and ^{54}Zn, could also show the new decay mode. The decay of these nuclei was investigated in experiments at the LISE3 separator of GANIL.

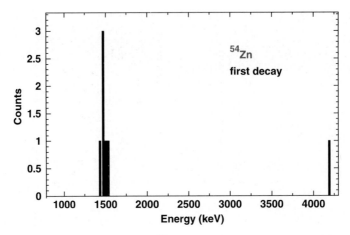

Fig. 23. Decay-energy spectrum from ^{54}Zn. The spectrum is generated from the first decay events after implantations of unambiguously identified ^{54}Zn. Implantation and decay are correlated in space and time by means of a DSSSD as the implantation device. The peak at 1.48(2) MeV is due to 2p radioactivity

Due to its high-production rates, clear evidence for 2p radioactivity could be obtained for ^{54}Zn. The decay-energy spectrum for ^{54}Zn is shown in Fig. 23. The different arguments used to establish that the identified decays are indeed due to 2p radioactivity are the same as just developed for ^{45}Fe: (i) the decay energy which matches modern theoretical predictions [22, 29, 116], (ii) the narrow width of the 2p peak, (iii) the absence of β radiation, and iv) the daughter decay characteristics.

For ^{48}Ni, the situation is much more complicated. The production rate of this nucleus is so low that only about one nucleus can be produced per day. Nonetheless, the decay of this nucleus could be studied to some extent in a recent GANIL experiment [43]. Four implantations could be observed and correlated with subsequent decays in the same DSSSD pixel. The decay-energy and decay-time spectra obtained are shown in Fig. 14. The event with a decay energy of 1.35(2) MeV has no β particle in coincidence, whereas all other events have a β particle detected in one of the silicon detectors adjacent to the implantation detector. Therefore, this event may be due to 2p radioactivity. It is followed very shortly by a second decay, which is in agreement with expectations for a ^{46}Fe decay [42], the 2p daughter of ^{48}Ni [43].

Although the observation of a single possible 2p radioactivity event is far from being sufficient for establishing this decay mode of ^{48}Ni, this event has nonetheless all characteristics of a 2p event. However, it is evident that this calls for confirmation. To achieve higher-statistics data, one needs either much longer beam times, e.g. at GANIL or new facilities like the RIBF installation recently commissioned at RIKEN, Japan have to be used.

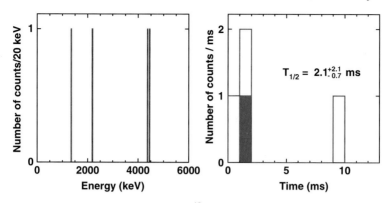

Fig. 24. *Left-hand side*: Decay energy of ^{48}Ni obtained from a recent GANIL experiment [43]. The event with the lowest energy is most likely due to 2p radioactivity of ^{48}Ni. The other events were observed in coincidence with β particles in adjacent detectors. *Right-hand side*: The decay-time spectrum for the four decay events of ^{48}Ni is plotted. The event shown as a full histogram is the low-energy event. The half-life is $T_{1/2} = 2.1^{+2.1}_{-0.7}$ ms (See also Plate 14 in the Color Plate Section)

3.4 Comparison with Theoretical Models

The first model applied to 2p radioactivity is the so-called di-proton model (see, e.g. [19, 115]). In this model, the two protons are supposed to form a structure-less "particle" with mass $A = 2$ and charge $Z = 2$. No considerations are made about possible binding or resonance energies. The di-proton just has no internal structure. Therefore, the total energy available for the decay Q_{2p}, i.e. the mass difference between the parent state (e.g. ^{45}Fe) and the daughter state (^{43}Cr+2p), is available as kinetic energy for the decay. A di-proton with energy Q_{2p} tunnels through the Coulomb barrier and separates outside the nucleus into two independent protons. This simple di-proton picture was also used to calculate the curve relating the decay energy and the tunnelling half-life in the case of ^{42}Cr in Fig. 12.

It is evident that neglecting any interaction between the two protons which are known to form a resonance is not a good approximation. Therefore, this model was extended to include the proton–proton interaction [5, 6]. In this extended R-matrix model, the decay is sub-divided into two sequential decays, a first decay where the parent nucleus emits two protons which form a resonance state. In the second step, this resonance disintegrates by emitting the two protons. Nuclear-structure effects like the imperfect overlap of the parent and the daughter wave functions are included via spectroscopic factors calculated by means of the nuclear shell model.

This model was recently applied to 2p radioactivity for ^{45}Fe [21, 43], ^{48}Ni [43], and ^{54}Zn [13]. As can be seen from Fig. 25, reasonable agreement is obtained for all three nuclei.

The extended R-matrix model contains to a large extent the nuclear structure needed to describe 2p radioactivity. However, the dynamics of the emission process is completely neglected. A model which treats this dynamics part in a much more thorough way is the three-body model developed by Grigorenko and co-workers [64, 65]. Nuclear structure is included only at the rather basic level of single-particle orbitals. This model describes 2p radioactivity from a resonance consisting of an inert core with mass $A-2$, where A is the mass number of the 2p emitter, and two protons. These three particles are quasi-bound due to realistic interactions. Therefore, this model allows calculating the energy sharing between the three particles and the relative emission angles in the 2p-emitter center-of-mass frame (see Fig. 22). However, it is not quite clear how the observed correlations can be linked to the nucleon–nucleon correlation inside the nucleus.

This model was used to relate the decay energy to the decay time for different single-particle levels [64, 65] in the cases of ^{45}Fe, ^{48}Ni, and ^{54}Zn. The results are shown as the lines in Fig. 25. The experimental results for

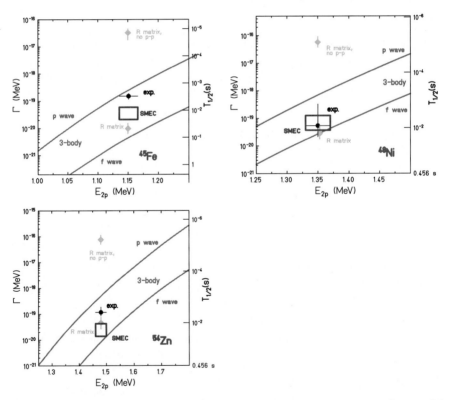

Fig. 25. Experimental information on ground-state 2p emission compared to model predictions from the extended R-matrix model of Brown and Barker [21], the three-body model of Grigorenko et al. [64, 66, 67], and the SMEC of Rotureau et al. [131, 132] for ^{45}Fe, ^{48}Ni, and ^{54}Zn

all nuclei lay between the theoretical predictions from this model for f and p single-particle levels.

The most recent model developed for 2p radioactivity is the shell model embedded in the continuum (SMEC). This model was originally developed for one particle in the continuum to couple bound states and continuum states. The model was then extended to two particles in the continuum to describe 2p radioactivity [131, 132]. It has three subspaces, one for the bound states and two for the coupling to the continuum states. The results from this model depend strongly on the Q value for one-proton emission which is not known experimentally.

The results from the SMEC are also plotted in Fig. 25. The sizes of the rectangles in the figure correspond to the uncertainties from this model which are due to uncertainties from the experimental Q_{2p} and Q_p values and the different effective interactions used.

The comparison of the experimental data with the model predictions as shown in Fig. 25 demonstrates that these models are able to describe this new nuclear decay mode reasonably well. Part of the discrepancies comes from a lack of theoretical input in the models. As mentioned above, the extended R-matrix model does not contain the dynamics of the emission process. The same is also true for the SMEC. The three-body model which treats best the emission dynamics has large deficiencies concerning nuclear structure. Future developments certainly have to try to treat both aspects, nuclear structure and emission dynamics, at the same level.

3.5 Conclusions on Two-Proton Radioactivity

The experimental results presented in the preceding paragraphs have demonstrated that 2p radioactivity is established as a new nuclear decay mode (see [17] for a recent review of the field). Although the experimental information is still scarce, this decay mode was shown to be the main decay branch for ^{45}Fe and ^{54}Zn. In the case of ^{48}Ni, more decays of this nucleus have to be studied to definitively demonstrate that 2p radioactivity is also present as a decay branch.

The theoretical models succeed to describe this new phenomenon astonishingly well. However, none of the models contain all necessary inputs to completely describe 2p radioactivity. New developments are needed to go beyond the present descriptions of 2p radioactivity.

To understand 2p radioactivity and to develop it into a powerful nuclear structure tool, more 2p emitters have also to be studied. Possible candidates are ^{59}Ge, ^{63}Se or ^{67}Kr which can be reached at new facilities like the Japanese RIBF at RIKEN or the FAIR facility of GSI in Germany.

4 Other Exotic Decay Channels

Besides 1p and 2p radioactivity, proton-rich nuclei may also decay by other decay channels. These decay channels are, e.g. β-delayed single- or multi-proton emission, β-delayed α emission, or emission of protons from excited states populated by other means. The exhaustive treatment of these channels is beyond the scope of the present lecture notes, however, we will nonetheless discuss a few selected examples shortly.

4.1 Beta-Delayed Decay Modes

Beta-delayed one-proton (β1p) emission is today almost a standard decay mode. It was observed in close to hundred different nuclei (see [14] for a recent review). This decay mode is extensively used to determine nuclear masses [42], parameters of fundamental interactions [1], isospin mixing [50], nuclear spins [150], the Gamow–Teller strength distribution in nuclear β decay [151], parameters of nuclear astrophysics [23], and much more.

In principle, β-delayed two-proton (β2p) emission can be used just for the same research areas. However, these β2p emitters are still further away from the valley of stability than the β1p emitters and therefore more difficult to produce. Only nine β2p emitters are known experimentally. However, more than ten are expected to be observable up to mass number $A=100$ [14]. The main interest of these β2p emitters is most likely the possibility to study proton–proton correlations. Although all decays investigated up to now turned out to be purely sequential decays, i.e. one proton is emitted after the other, a small correlated branch is expected from theoretical considerations [20]. If observable, these correlations may give valuable insights in, e.g. pairing correlations in the atomic nucleus.

As mentioned earlier (see Fig. 13 (*right*)), the most exotic of these β-delayed decay modes is the β-delayed three-proton decay recently observed in the decay of ^{45}Fe [106]. However, beyond the interest of its observation itself, the nuclear structure information which can be gained from this decay is probably limited. As it is most likely an emission from the isobaric analogue state in ^{45}Mn, it would be, however, interesting to see whether nuclear structure models will be able to predict or reproduce the branching ratios for one-, two-, three-, and maybe even four-proton emission from this state.

Beta-delayed α emission is an important decay mode only for a handful of light nuclei [14]. It plays a role in determining certain astrophysical parameters (see, e.g. [26, 52]). Its importance for nuclear structure comes from the fact that light nuclei often break up into several α particles [41]. These break-up channels are the inverse of astrophysical reactions like the triple-α capture reaction to form ^{12}C which is at the origin of life.

4.2 Two-Proton Emission from Short-lived States

The term 2p radioactivity was coined by Goldanskii [60] in order to describe a phenomenon where 2p emission is possible, but the 1p emission channel is energetically not accessible. In addition, he requested, somewhat arbitrarily, a half-life of more than 10^{-12} s for the 2p emitting state.

Beyond 2p emission as described above, i.e. 2p radioactivity and β2p emission, other 2p emission modes exist and have been observed. Thus 2p emission with a half-life as short as nuclear reaction times ($t \approx 10^{-21}$ s) has been observed from the ground states of ^{6}Be [18], ^{12}O [92], and ^{19}Mg [110]. However, all experimental information is in agreement with a three-body decay limited just by phase space, which most likely means that the decay mainly proceeds through the tails of broad intermediate states in the 1p daughter nucleus.

Two-proton emission from nuclear states populated in inelastic reactions has been observed in the decay of levels in ^{14}O [4], ^{17}Ne [27, 28, 159], ^{18}Ne [62], and ^{19}Na [114]. All these observations are again in agreement with an uncorrelated emission pattern, except for ^{17}Ne where a proton–proton angular correlation could be found for higher-lying excited states [159].

The interpretation of this decay pattern is still controversial. Other work suggested that some of the excited states of ^{17}Ne may have a pronounced 2p halo structure and a large overlap with the ^{15}O ground state leading to much larger spectroscopic factors for a direct 2p decay than for a sequential decay [69, 85]. However, the measured energies of the two emitted protons from ^{17}Ne show large differences which cannot be reconciled with a di-proton scenario where roughly equal energies are expected for the two protons. Another explanation of the 2p emission pattern in ^{17}Ne could be a large deformation in higher-lying states. Strong anisotropy of the Coulomb barrier could be a source of dynamical correlations between emitted protons even if this 2p decay is a sequence of two successive 1p emissions. Obviously, for a consistent interpretation of the ^{15}O + 2p data, higher statistics is mandatory.

4.3 Two-Proton Emission from ^{94}Ag

The decay of ^{94}Ag^m is particular in several respects. It possesses two long-lived β-decaying isomers and has the highest spin state ever observed to β decay. One of these isomeric levels decays by several different decay branches: β-delayed γ decay [32, 125], β-delayed proton emission [111], direct proton decay [113], and direct 2p emission [112].

In particular, the 2p emission branch is peculiar. The experimental setup allowed the measurement of the individual proton energies and the relative proton–proton angle. Although the statistics of the 2p experiment is rather low [112], the authors identified proton–proton correlations. They interpreted these data as due to simultaneous 2p emission from a strongly deformed ellipsoidal nucleus. In this case, the strongly asymmetric Coulomb barrier favours the emission of two protons within narrow cones around the poles of the ellipsoid, either from the same pole or on opposite sides.

It is not clear to which extent the observed proton–proton correlations in such a scenario are related to proton–proton correlations inside the nucleus. As only microscopic theoretical studies of the ^{94}Ag^m decay could disentangle an internal structure of decaying states from dynamical effects of the anisotropic Coulomb barrier, providing thus more details about the 2p decay pattern in the nucleus, and these studies are impossible using present-day computers, a detailed interpretation of the observed data is not possible today.

From the experimental side, further experiments of the ^{94}Ag^m decay cannot be made without new technical developments to produce ^{94}Ag^m as the on-line separator of GSI unfortunately has been dismantled a few years ago. Therefore, a detailed understanding of the decay properties of this isomeric state based on new high-statistics experimental data has to await the production of ^{94}Ag^m at facilities other than the GSI on-line separator [130].

5 Conclusions

Nuclear structure studies in the vicinity of the proton drip line have yielded a wealth of information on the organisation of protons and neutrons in the atomic nucleus. For many elements, they have allowed to reach the limits of nuclear stability and therefore to test our theoretical understanding of nuclear structure of the most proton-rich species which exist.

One-proton radioactivity has developed, since its discovery at the beginning of the 1980s, into an indispensable tool of nuclear structure investigations beyond the proton drip line. To a large extent, the experimental information and the theoretical understanding which went along with it, e.g. the sequence of single-particle states, deformation effects, or the composition of the nuclear wave function, can only be obtained by means of these studies. Therefore, this decay mode will most likely keep its importance in the next decade, as higher-intensity primary beams but possibly also the use of radioactive beams will still enhance the quality of the data achievable and allow thus an even more detailed investigation of nuclear structure close to or beyond the proton drip line.

Two-proton radioactivity, recently discovered, suffers from the rather low production rates of the 2p emitters. It is needless to say that future facilities will bring an increase by at least one, maybe several orders of magnitude of these rates. This will enable us to refine the studies performed with 2p radioactivity to the same degree as one-proton radioactivity today. Therefore, basically all the investigations performed with one-proton radioactivity could be also tried with these 2p emitters. In addition, the 2p emission process may ultimately allow also to investigate proton–proton correlations inside the atomic nucleus.

The β-delayed decay modes have impressively demonstrated their usefulness in many experiments. They are a rather universal tool applied in investigations concerning the nuclear mass surface, astrophysics, fundamental-interaction studies or other topics of nuclear structure. Their potential will continue to increase with increased production rates for the most exotic of these emitters.

For all these reasons, the study of very proton-rich nuclei in the vicinity of the limits of nuclear stability is an extremely powerful tool of nuclear structure physics and of related research fields. The efficient use of present-day facilities and the construction of new installations for nuclear research can only increase its usefulness.

References

1. E.G. Adelberger, C. Ortiz, A. García, H.E. Swanson, M. Beck, O. Tengblad, M.J.G. Borge, I. Martel-Bravo, H. Bichsel, Phys. Rev. Lett. **83**, 1299 (1999)
2. G. Audi, O. Bersillon, J. Blachot, A.H. Wapstra, Nucl. Phys. **A729**, 3 (2003)
3. T. Bäck, B. Cederwall, K. Lagergren, R. Wyss, A. Johnson, D. Karlgren, P. Greenlees, D. Jenkins, P. Jones, D. Joss, R. Julin, S. Juutinen, A. Keenan, H. Kettunen, P. Kuusiniemi, M. Leino, A.P. Leppänen, M. Muikku, P. Nieminen, J. Pakarinen, P. Rahkila, J. Uusitalo, Eur. Phys. J. A **16**, 489 (2003)
4. C. Bain, P. Woods, R. Coszach, T. Davinson, P. Decrock, M. Gaelens, W. Galster, M. Huyse, R. Irvine, P. Leleux, E. Lienard, M. Loiselet, C. Michotte, R. Neal, A. Ninane, G. Ryckewaert, A. Shotter, G. Vancraeynest, J. Vervier, J. Wauters, Phys. Lett. **B373**, 35 (1996)
5. F.C. Barker, Phys. Rev. C **59**, 535 (1999)
6. F.C. Barker, Phys. Rev. C **63**, 047303 (2001)
7. B. Barmore, A.T. Kruppa, W. Nazarewicz, T. Vertse, Phys. Rev. C **62**, 054315 (2000)
8. R. Barton, R. McPherson, R.E. Bell, W.R. Frisken, W.T. Link, R.B. Moore, Can. J. Phys. **41**, 2007 (1963)
9. J.C. Batchelder, C.R. Bingham, K. Rykaczewski, K.S. Toth, T. Davinson, J.A. McKenzie, P.J. Woods, T.N. Ginter, C.J. Gross, J.W. McConnell, E.F. Zganjar, J.H. Hamilton, W.B. Walters, C. Baktash, J. Greene, J.F. Mas, W.T. Milner, S.D. Paul, D. Shapira, X.J. Xu, C.H. Yu, Phys. Rev. C **57**(3), R1042 (1998)
10. J.C. Batchelder, M. Tantawy, C.R. Bingham, M. Danchev, D.J. Fong, T.N. Ginter, C.J. Gross, R. Grzywacz, K. Hagino, J.H. Hamilton, M. Karny, W. Krolas, C. Mazzocchi, A. Piechaczek, A.V. Ramayya, K.P. Rykaczewski, A. Stolz, J.A. Winger, C.H. Yu, E.F. Zganjar, Eur. Phys. J. A **25**, 149 (2005)
11. C. Bingham, M. Tantawy, J. Batchelder, M. Danchev, T. Ginter, C. Gross, D. Fong, R. Grzywacz, K. Hagino, J. Hamilton, Nucl. Instrum. Meth. **241**, 185 (2005)
12. C.R. Bingham, J.C. Batchelder, K. Rykaczewski, K.S. Toth, C.H. Yu, T.N. Ginter, C.J. Gross, R. Grzywacz, M. Karny, S.H. Kim, B.D. MacDonald, J. Mas, J.W. McConnell, P.B. Semmes, J. Szerypo, W. Weintraub, E.F. Zganjar, Phys. Rev. C **59**, R2984 (1999)

13. B. Blank, A. Bey, G. Canchel, C. Dossat, A. Fleury, J. Giovinazzo, I. Matea, N. Adimi, F.D. Oliveira, I. Stefan, G. Georgiev, S. Grévy, J. Thomas, C. Borcea, D. Cortina, M. Caamano, M. Stanoiu, F. Aksouh, B.A. Brown, F. Barker, W.A. Richter, Phys. Rev. Lett. **94**, 232501 (2005)

14. B. Blank, M.J.G. Borge, Prog. Nucl. Part. Phys. **4560**, 403 (2008)

15. B. Blank, M. Chartier, S. Czajkowski, J. Giovinazzo, M.S. Pravikoff, J.C. Thomas, G. de France, F. de Oliveira Santos, M. Lewitowicz, C. Borcea, R. Grzywacz, Z. Janas, M. Pfützner, Phys. Rev. Lett. **84**, 1116 (2000)

16. B. Blank, L. Audirac, G. Canchel, F. Delalee, C.E. Demonchy, J. Giovinazzo, L. Hay, P. Hellmuth, J. Huikari, S. Leblanc, S. List, C. Marchand, I. Matea, J.-L. Pedroza, J. Pibernat, A. Rebii, L. Serani, F. de Oliveira Santos, S. Grévy, L. Perrot, C. Stodel, J.C. Thomas, C. Borcea, C. Dossat, R. de Oliveira, Nucl. Instrum. Meth. B, accepted for publication (2008)

17. B. Blank, M. Płoszajczak, Rep. Prog. Phys. **71**, 046301 (2008)

18. O.V. Bochkarev, A.A. Korsheninnikov, E.A. Kuz'min, I.G. Mukha, L.V. Chulkov, G.B. Yan'kov, Sov. J. Nucl. Phys. **49**, 941 (1989)

19. B.A. Brown, Phys. Rev. C **43**, R1513 (1991)

20. B.A. Brown, Phys. Rev. C **44**, 924 (1991)

21. B.A. Brown, F.C. Barker, Phys. Rev. C **67**, 041,304 (2003)

22. B.A. Brown, R.R.C. Clement, H. Schatz, A. Volya, W.A. Richter, Phys. Rev. C **65**, 045802 (2002)

23. L. Buchmann, E. Gete, J.C. Chow, J.D. King, D.F. Measday, Phys. Rev. C **63**, 034303 (2001)

24. J. Cerny, J. Esterl, R.A. Gough, R.G. Sextro, Phys. Lett. **B33**, 284 (1970)

25. J. Cerny, R.A. Gough, R.G. Sextro, J. Esterl, Nucl. Phys. A **188**, 666 (1972)

26. J.C. Chow, J.D. King, N.P.T. Bateman, R.N. Boyd, L. Buchmann, J.M. D'Auria, T. Davinson, M. Dombsky, E. Gete, U. Giesen, C. Iliadis, K.P. Jackson, A.C. Morton, J. Powell, A. Shotter, Phys. Rev. C **66**, 064316 (2002)

27. M. Chromik, B. Brown, M. Fauerbach, T. Glasmacher, R. Ibbotson, H. Scheit, M. Thoennessen, P. Thirolf, Phys. Rev. C **55**, 1676 (1997)

28. M. Chromik, P. Thirolf, M. Thoennessen, B. Brown, T. Davinson, D. Gassmann, P. Heckman, J. Prisciandro, P. Reiter, E. Tryggestad, P. Woods, Phys. Rev. C **66**, 024313 (2002)

29. B.J. Cole, Phys. Rev. C **54**, 1240 (1996)

30. B.J. Cole, Phys. Rev. C **56**, 1866 (1997)

31. B.J. Cole, Phys. Rev. C **59**, 726 (1999)

32. M.L. Commara, K. Schmidt, H. Grawe, J. Doring, R. Borcea, S. Galanopoulos, M. Gorska, S. Harissopulos, M. Hellstrom, Z. Janas, R. Kirchner, C. Mazzocchi, A.N. Ostrowski, C. Plettner, G. Rainovski, E. Roeckl, Nucl. Phys. A **708**, 167 (2002)

33. C.N. Davids, H. Esbensen, Phys. Rev. C **61**, 054302 (2000)

34. C.N. Davids, H. Esbensen, Phys. Rev. C **64**, 034317 (2001)

35. C.N. Davids, P.J. Woods, J.C. Batchelder, C.R. Bingham, D.J. Blumenthal, L.T. Brown, B.C. Busse, L.F. Conticchio, T. Davinson, S.J. Freeman, D.J. Henderson, R.J. Irvine, R.D. Page, H.T. Penttilä, D. Seweryniak, K.S. Toth, W.B. Walters, B.E. Zimmerman, Phys. Rev. C **55**, 2255 (1997)

36. C.N. Davids, P.J. Woods, H. Mahmud, T. Davinson, A. Heinz, J.J. Ressler, K. Schmidt, D. Seweryniak, J. Shergur, A.A. Sonzogni, W.B. Walters, Phys. Rev. C **69**, 011302 (2004)

37. C.N. Davids, P.J. Woods, H.T. Penttilä, J.C. Batchelder, C.R. Bingham, D.J. Blumenthal, L.T. Brown, B.C. Busse, L.F. Conticchio, T. Davinson, D.J. Henderson, R.J. Irvine, D. Seweryniak, K.S. Toth, W.B. Walters, B.E. Zimmerman, Phys. Rev. Lett. **76**, 592 (1996)
38. C.N. Davids, P.J. Woods, D. Seweryniak, A.A. Sonzogni, J.C. Batchelder, C.R. Bingham, T. Davinson, D.J. Henderson, R.J. Irvine, G.L. Poli, J. Uusitalo, W.B. Walters, Phys. Rev. Lett. **80**, 1849 (1998)
39. D. Delion, R. Liotta, R. Wyss, Phys. Rep. **424**, 113 (2006)
40. C. Détraz, R. Anne, P. Bricault, D. Guillemaud-Mueller, M. Lewitowicz, A.C. Mueller, Y.H. Zhang, V. Borrel, J.C. Jacmart, F. Pougheon, A. Richard, D. Bazin, J.P. Dufour, A. Fleury, F. Hubert, M.S. Pravikoff, Nucl. Phys. **A519**, 529 (1990)
41. C. Diget, F. Barker, M. Borge, J. Cederkäll, V. Fedosseev, L. Fraile, B. Fulton, H. Fynbo, H. Jeppesen, B. Jonson, U. Köster, M. Meister, T. Nilsson, G. Nyman, Y. Prezado, K. Riisager, S. Rinta-Antila, O. Tengblad, M. Turrion, K. Wilhelmsen, J. Äystö, Nucl. Phys. **A760**, 3 (2005)
42. C. Dossat, N. Adimi, F. Aksouh, F. Becker, A. Bey, B. Blank, C. Borcea, R. Borcea, A. Boston, M. Caamano, G. Canchel, M. Chartier, D. Cortina, S. Czajkowski, G. de France, F. de Oliveira Santos, A. Fleury, G. Georgiev, J. Giovinazzo, S. Grévy, R. Grzywacz, M. Hellström, M. Honma, Z. Janas, D. Karamanis, J. Kurcewicz, M. Lewitowicz, M.L. Jiménez, C. Mazzocchi, I. Matea, V. Maslov, P. Mayet, C. Moore, M. Pfützner, M. Pravikoff, M. Stanoiu, I. Stefan, J. Thomas, Nucl. Phys. A **792**, 18 (2007)
43. C. Dossat, A. Bey, B. Blank, G. Canchel, A. Fleury, J. Giovinazzo, I. Matea, N. Adimi, F.D. Oliveira, I. Stefan, G. Georgiev, S. Grévy, J. Thomas, C. Borcea, D. Cortina, M. Caamano, M. Stanoiu, F. Aksouh, B.A. Brown, F. Barker, W.A. Richter, Phys. Rev. C **72**, 054315 (2005)
44. H. Esbensen, C.N. Davids, Phys. Rev. C **63**, 014315 (2000)
45. D. Rudolph, R. Hoischen, M. Hellström, S. Pietri, Zs. Podolyák, P. H. Regan, A. B. Garnsworthy, S. J. Steer, F. Becker, P. Bednarczyk, L. Cáceres, P. Doornenbal, J. Gerl, M. Górska, J. Grebosz, I. Kojouharov, N. Kurz, W. Prokopowicz, H. Schaffner, H. J. Wollersheim, L.-L. Andersson, L. Atanasova, D. L. Balabanski, M. A. Bentley, A. Blazhev, C. Brandau, J. R. Brown, C. Fahlander, E. K. Johansson, A. Jungclaus, S. M. Lenzi, Phys. Rev. C 78, 021301 (2008)
46. T. Faestermann, A. Gillitzer, K. Hartel, P. Kienle, E. Nolte, Phys. Lett. B **137**, 23 (1984)
47. L.S. Ferreira, E. Maglione, Phys. Rev. C **61**, 021304 (2000)
48. L.S. Ferreira, E. Maglione, Phys. Rev. Lett. **86**, 1721 (2001)
49. G. Fiorin, E. Maglione, L.S. Ferreira, Phys. Rev. C **67**, 054302 (2003)
50. M. Battacharya, D. Melconian, A. Komives, S. Triambak, A. García, E. G. Adelberger, B. A. Brown, M. W. Cooper, T. Glasmacher, V. Guimaraes, P. F. Mantica, A. M. Oros-Peusquens, J. I. Prisciandaro, M. Steiner, H. E. Swanson, S. L. Tabor, M. Wiedeking, Phys. Rev. C 77, 065503 (2008)
51. H. Geissel, P. Armbruster, B. Franczak, B. Langenbeck, O. Klepper, F. Nickel, E. Roeckl, D. Schardt, K.H. Schmidt, D. Schüll, K. Sümmerer, G. Münzenberg, J.P. Dufour, M.S. Pravikoff, H.G. Clerc, E. Hanelt, T. Schwab, H. Wollnik, B. Sherrill, Nucl. Instrum. Meth. **B70**, 286 (1992)

52. E. Gete, L. Buchmann, R.E. Azuma, D. Anthony, N. Bateman, J.C. Chow, J.M. D'Auria, M. Dombsky, U. Giesen, C. Iliadis, K.P. Jackson, J.D. King, D.F. Measday, A.C. Morton, Phys. Rev. C **61**, 064310 (2000)
53. A. Gillitzer, T. Faestermann, K. Hartel, P. Kienle, E. Nolte, Z. Phys. A **326**, 107 (1987)
54. T.N. Ginter, J.C. Batchelder, C.R. Bingham, C.J. Gross, R. Grzywacz, J.H. Hamilton, Z. Janas, M. Karny, S.H. Kim, J.F. Mas, J.W. McConnell, A. Piechaczek, A.V. Ramayya, K. Rykaczewski, P.B. Semmes, J. Szerypo, K.S. Toth, R. Wadsworth, C.H. Yu, E.F. Zganjar, Phys. Rev. C **61**, 014308 (1999)
55. T.N. Ginter, J.C. Batchelder, C.R. Bingham, C.J. Gross, R. Grzywacz, J.H. Hamilton, Z. Janas, M. Karny, A. Piechaczek, A.V. Ramayya, K.P. Rykaczewski, W.B. Walters, E.F. Zganjar, Phys. Rev. C **68**, 034330 (2003)
56. J. Giovinazzo, B. Blank, C. Borcea, G. Canchel, C. Demonchy, F. de Oliveira Santos, C. Dossat, S. Grévy, L. Hay, J. Huikari, S. Leblanc, I. Matea, J.L. Pedroza, J. Pibernat, L. Serani, C. Stodel, J.C. Thomas, Phys. Rev. Lett. **99**, 102501 (2007)
57. J. Giovinazzo, B. Blank, C. Borcea, M. Chartier, S. Czajkowski, G. de France, R. Grzywacz, Z. Janas, M. Lewitowicz, F. de Oliveira Santos, M. Pfützner, M. Pravikoff, J.C. Thomas, Eur. Phys. J. **A10**, 73 (2001)
58. J. Giovinazzo, B. Blank, M. Chartier, S. Czajkowski, A. Fleury, M.L. Jimenez, M. Pravikoff, J.C. Thomas, F. de Oliveira Santos, M. Lewitowicz, V. Maslov, M. Stanioiu, R. Grzywacz, M. Pfützner, C. Borcea, B. Brown, Phys. Rev. Lett. **89**, 102501 (2002)
59. V.I. Goldanskii, Nuovo Cimento **25, suppl. 2**, 123 (1962)
60. V.I. Goldansky, Nucl. Phys. **19**, 482 (1960)
61. V.I. Goldansky, Nucl. Phys. **27**, 648 (1961)
62. J. Gomez del Campo, A. Galindo-Uribarri, J. Beene, C. Gross, J. Liang, M. Halbert, D. Stracener, D. Shapira, R. Varner, E. Chavez-Lomeli, M. Ortiz, Phys. Rev. Lett. **86**, 43 (2001)
63. H. Grawe, Lect. Notes Phys. **651**, 33 (2004)
64. L. Grigorenko, R. Johnson, I. Mukha, I. Thompson, M. Zhukov, Phys. Rev. Lett. **85**, 22 (2000)
65. L. Grigorenko, R. Johnson, I. Mukha, I. Thompson, M. Zhukov, Phys. Rev. C **64**, 054002 (2001)
66. L. Grigorenko, I. Mukha, M. Zhukov, Nucl. Phys. **A714**, 425 (2003)
67. L. Grigorenko, M. Zhukov, Phys. Rev. C **68**, 054005 (2003)
68. L. Grigorenko, M. Zhukov, Phys. Rev. C **68**, 054005 (2003)
69. L.V. Grigorenko, Y.L. Parfenova, M.V. Zhukov, Phys. Rev. C **71**, 051604R (2005)
70. C.J. Gross, K.P. Rykaczewski, D. Shapira, J.A. Winger, J.C. Batchelder, C.R. Bingham, R.K. Grzywacz, P.A. Hausladen, W. Krolas, C. Mazzocchi, A. Piechaczek, E.F. Zganjar, Eur. Phys. J. A **25**, 115 (2005)
71. R. Grzywacz, M. Karny, K.P. Rykaczewski, J.C. Batchelder, C.R. Bingham, D. Fong, C.J. Gross, W. Krolas, C. Mazzocchi, A. Piechaczek, M.N. Tantawy, J.A. Winger, E.F. Zganjar, Eur. Phys. J. A **25**, 145 (2005)
72. K. Hagino, Phys. Rev. C **64**, 041304 (2001)
73. F. Heine, T. Faestermann, A. Gillitzer, J. Homolka, M. Kopf, W. Wagner, Z. Phys. A **340**, 225 (1991)

74. F. Heine, T. Faestermann, A. Gillitzer, H. Körner, Proc. Int. Conf. on Nuclei far from Stability, p.331 (1992)
75. S. Hofmann, Nuclear Decay Modes, IOP Publishing, p.143 (1996)
76. S. Hofmann, J. Agarwal, P. Armbruster, F. Hessberger, P. Larsson, G. Münzenberg, K. Poppensieker, W. Reisdorf, J. Schneider, H. Schött, Proc. 7th Int. Conf. on Atomic Masses and Fundamental Constants AMCO, p.184 (1984)
77. S. Hofmann, W. Faust, G. Münzenberg, W. Reisdorf, P. Armbruster, K. Güttner, H. Ewald, Z. Phys. A **291**, 53 (1979)
78. S. Hofmann, W. Reisdorf, G. Münzenberg, F. Hessberger, J. Schneider, P. Armbruster, Z. Phys. **A305**, 111 (1982)
79. P.T. Hosmer, H. Schatz, A. Aprahamian, O. Arndt, R.R.C. Clement, A. Estrade, K.L. Kratz, S.N. Liddick, P.F. Mantica, W.F. Mueller, F. Montes, A.C. Morton, M. Ouellette, E. Pellegrini, B. Pfeiffer, P. Reeder, P. Santi, M. Steiner, A. Stolz, B.E. Tomlin, W.B. Walters, A. Wöhr, Phys. Rev. Lett. **94**, 112501 (2005)
80. R.J. Irvine, C.N. Davids, P.J. Woods, D.J. Blumenthal, L.T. Brown, L.F. Conticchio, T. Davinson, D.J. Henderson, J.A. Mackenzie, H.T. Penttilä, D. Seweryniak, W.B. Walters, Phys. Rev. C **55**, R1621 (1997)
81. K.P. Jackson, C.U. Cardinal, H.C. Evans, N.A. Jelley, J. Cerny, Phys. Lett. **B33**, 281 (1970)
82. J. Jänecke, Nucl. Phys. **61**, 326 (1965)
83. D. Joss, I. Darby, R. Page, J. Uusitalo, S. Eeckhaudt, T. Grahn, P. Greenlees, P. Jones, R. Julin, S. Juutinen, Phys. Lett. B **641**, 34 (2006)
84. S. Kadmensky, V. Bugrov, Phys. Atomic Nuclei **59**, 399 (1996)
85. R. Kanungo, M. Chiba, S. Adhikari, D. Fang, N. Iwasa, K. Kimura, K. Maeda, S. Nishimura, Y. Ogawa, T. Ohnishi, A. Ozawa, C. Samanta, T. Suda, T. Suzuki, Q. Wang, C. Wu, Y. Yamaguchi, K. Yamada, A. Yoshida, T. Zheng, I. Tanihata, Phys. Lett. **B571**, 21 (2003)
86. V.A. Karnaukhov, D. Bogdanov, L. Petrov, Proc. Leysin conference on Nuclei far from stability, p.457 (1970)
87. M. Karny, R.K. Grzywacz, J.C. Batchelder, C.R. Bingham, C.J. Gross, K. Hagino, J.H. Hamilton, Z. Janas, W.D. Kulp, J.W. McConnell, M. Momayezi, A. Piechaczek, K.P. Rykaczewski, P.A. Semmes, M.N. Tantawy, J.A. Winger, C.H. Yu, E.F. Zganjar, Phys. Rev. Lett. **90**, 012502 (2003)
88. H. Kettunen, T. Enqvist, T. Grahn, P.T. Greenlees, P. Jones, R. Julin, S. Juutinen, A. Keenan, P. Kuusiniemi, M. Leino, A.P. Leppanen, P. Nieminen, J. Pakarinen, P. Rahkila, J. Uusitalo, Phys. Rev. C **69**, 054323 (2004)
89. H. Kettunen, P. Greenlees, K. Helariutta, P. Jones, R. Julin, S. Juutinen, P. Kuusiniemi, M. Leino, M. Muikku, P. Nieminen, J. Uusitalo, Acta Phys. Pol. B **32**, 989 (2001)
90. O. Klepper, T. Batsch, S. Hofmann, R. Kirchner, W. Kurewicz, W. Reisdorf, E. Roeckl, D. Schardt, G. Nyman, Z. Phys. **A305**, 125 (1982)
91. A.T. Kruppa, B. Barmore, W. Nazarewicz, T. Vertse, Phys. Rev. Lett. **84**, 4549 (2000)
92. R.A. Kryger, A. Azhari, M. Hellström, J.H. Kelley, T. Kubo, R. Pfaff, E. Ramakrishnan, B.M. Sherrill, M. Thoennessen, S. Yokoyama, R.J. Charity, J. Dempsey, A. Kirov, N. Robertson, D.G. Sarantites, L.G. Sobotka, J.A. Winger, Phys. Rev. Lett. **74**, 860 (1995)

93. K. Lagergren, D.T. Joss, R. Wyss, B. Cederwall, C.J. Barton, S. Eeckhaudt, T. Grahn, P.T. Greenlees, B. Hadinia, P.M. Jones, R. Julin, S. Juutinen, D. Karlgren, H. Kettunen, M. Leino, A.P. Leppanen, P. Nieminen, M. Nyman, R.D. Page, J. Pakarinen, E.S. Paul, P. Rahkila, C. Scholey, J. Simpson, J. Uusitalo, D.R. Wiseman, Phys. Rev. C **74**, 024316 (2006)

94. P. Larsson, T. Batsch, R. Kirchner, O. Klepper, W. Kurcewicz, E. Roeckl, D. Schardt, W. Feix, G. Nyman, P. Tidemand-Petersson, Z. Phys. A **314**, 9 (1983)

95. M. Lewitowicz, R. Anne, G. Auger, D. Bazin, C. Borcea, V. Borrel, J. Corre, T. Dörfler, A. Fomichov, R. Grzywacz, D. Guillemaud-Mueller, R. Hue, M. Huyse, Z. Janas, H. Keller, S. Lukyanov, A. Mueller, Y. Penionzhkevich, M. Pfützner, F. Pougheon, K. Rykaczewski, M. Saint-Laurent, K. Schmidt, W. Schmidt-Ott, O. Sorlin, J. Szerypo, O. Tarasov, J. Wauters, J. Zylicz, Phys. Lett. **B332**, 20 (1994)

96. Z. Liu, private communication

97. K. Livingston, P.J. Woods, T. Davinson, N.J. Davis, S. Hofmann, A.N. James, R.D. Page, P.J. Sellin, A.C. Shotter, Phys. Rev. C **48**, R2151 (1993)

98. K. Livingston, P.J. Woods, T. Davinson, N.J. Davis, S. Hofmann, A.N. James, R.D. Page, P.J. Sellin, A.C. Shotter, Phys. Lett. B **312**, 46 (1993)

99. E. Maglione, L.S. Ferreira, Phys. Rev. C **61**, 047307 (2000)

100. E. Maglione, L.S. Ferreira, R.J. Liotta, Phys. Rev. Lett. **81**, 538 (1998)

101. H. Mahmud, C.N. Davids, P.J. Woods, T. Davinson, A. Heinz, G.L. Poli, J.J. Ressler, K. Schmidt, D. Seweryniak, M.B. Smith, A.A. Sonzogni, J. Uusitalo, W.B. Walters, Phys. Rev. C **64**, 031303 (2001)

102. E. Mardsen, W. Lantsberry, Phil. Mag. **30**, 240 (1915)

103. C. Mazzocchi, R. Grzywacz, S. Liddick, K. Rykaczewski, H. Schatz, J.C. Batchelder, C.R. Bingham, C.J. Gross, J.H. Hamilton, J.K. Hwang, S. Ilyushkin, A. Korgul, W. Krolas, K. Li, R.D. Page, D. Simpson, J.A. Winger, Phys. Rev. Lett. **98**, 212501 (2007)

104. K. Miernik, W. Dominik, H. Czyrkowski, R. Dabrowski, A. Fomitchev, M. Golovkov, Z. Janas, W. Kusmierz, M. Pfützner, A. Rodin, S. Stepantsov, R. Slepniev, G.M. Ter-Akopian, R. Wolski, Nucl. Instr. Meth. A581, 194 (2007)

105. K. Miernik, W. Dominik, Z. Janas, M. Pfützner, L. Grigorenko, C. R. Bingham, H. Czyrkowski, M. Ćwiok, I. G. Darby, R. Dabrowski, T. Ginter, R. Grzywacz, M. Karny, A. Korgul, W. Kuśmierz, S. N. Liddick, M. Rajabali, K. Rykaczewski, A. Stolz, Phys. Rev. Lett. 99, 192501 (2007)

106. K. Miernik, W. Dominik, Z. Janas, M. Pfützner, C. R. Bingham, H. Czyrkowski, M. Ćwiok, I. G. Darby, R. Dabrowski, T. Ginter, R. Grzywacz, M. Karny, A. Korgul, W. Kuśmierz, S. N. Liddick, M. Rajabali, K. Rykaczewski, A. Stolz, Phys. Rev. C 76, 041304 (2007)

107. P. Möller, J.R. Nix, K.L. Kratz, At. Data Nucl. Data Tables **66**, 131 (1997)

108. D. Morrissey, B. Sherrill, Lect. Notes Phys. **651**, 113 (2004)

109. A.C. Mueller, R. Anne, Nucl. Instrum. Meth. **B56**, 559 (1991)

110. I. Mukha, private communication

111. I. Mukha, L. Batist, E. Roeckl, H. Grawe, J. Döring, A. Blazhev, C.R. Hoffman, Z. Janas, R. Kirchner, M.L. Commara, S. Dean, C. Mazzocchi, C. Plettner, S.L. Tabor, M. Wiedeking, Phys. Rev. C **70**, 044311 (2004)

112. I. Mukha, E. Roeckl, L. Batist, A. Blazhev, J. Doring, H. Grawe, L. Grigorenko, M. Huyse, Z. Janas, R. Kirchner, M.L. Commara, C. Mazzocchi, S.L. Tabor, P.V. Duppen, Nature **439**, 298 (2006)

113. I. Mukha, E. Roeckl, J. Doring, L. Batist, A. Blazhev, H. Grawe, C.R. Hoffman, M. Huyse, Z. Janas, R. Kirchner, M.L. Commara, C. Mazzocchi, C. Plettner, S.L. Tabor, P.V. Duppen, M. Wiedeking, Phys. Rev. Lett. **95**, 022501 (2005)

114. F. de Oliveira Santos, P. Himpe, M. Lewitowicz, I. Stefan, N. Smirnova, N.L. Achouri, J.C. Angelique, C. Angulo, L. Axelsson, D. Baiborodin, F. Becker, M. Belleguic, E. Berthoumieux, B. Blank, C. Borcea, A. Cassimi, J.M. Daugas, G. de France, F. Dembinski, C.E. Demonchy, Z. Dlouhy, P. Dolegieviez, C.D.G. Georgiev, L. Giot, S. Grévy, D. Guillemaud-Mueller, V. Lapoux, E. Lienard, M.L. Jimenez, K. Markenroth, I. Matea, W. Mittig, F. Negoita, G. Neyens, N. Orr, F. Pougheon, P. Roussel-Chomaz, M.G. Saint-Laurent, F. Sarazin, H. Savajols, M. Sawicka, O. Sorlin, M. Stanoiu, C. Stodel, G. Thiamova, D. Verney, A. Villari, Eur. Phys. J. **A24**, 237 (2005)

115. W.E. Ormand, Phys. Rev. C **53**, 214 (1996)

116. W.E. Ormand, Phys. Rev. C **55**, 2407 (1997)

117. W.E. Ormand, B.A. Brown, Phys. Rev. C **52**, 2455 (1995)

118. R.D. Page, L. bianco, I.G. Darby, J. Uusitalo, D.T. Joss, T. Grahn, R.D. Herzberg, J. Pakarinen, J. Thomson, S. Eeckhaudt, P.T. Greenlees, P.M. Jones, R. Julin, S. Juutinen, S. Ketelhut, M. Leino, A.P. Leppanene, M. Nyman, P. Rahkila, J. Saren, C. Scholey, A. Steer, M.B.G. Hornillos, J.S. Al-Khalili, A.J. Cannon, P.D. Stevenson, S. Eturk, B. Gall, B. Hadinia, M. Venhart, J. Simpson, Phys. Rev. C **75**, 061302(R) (2007)

119. R.D. Page, P.J. Woods, R.A. Cunningham, T. Davinson, N.J. Davis, S. Hofmann, A.N. James, K. Livingston, P.J. Sellin, A.C. Shotter, Phys. Rev. Lett. **68**, 1287 (1992)

120. R.D. Page, P.J. Woods, R.A. Cunningham, T. Davinson, N.J. Davis, A.N. James, K. Livingston, P.J. Sellin, A.C. Shotter, Phys. Rev. Lett. **72**, 1798 (1994)

121. R.D. Page, P.J. Woods, R.A. Cunningham, T. Davinson, N.J. Davis, A.N. James, K. Livingston, P.J. Sellin, A.C. Shotter, Phys. Rev. C **53**, 660 (1996)

122. E.S. Paul, P.J. Woods, T. Davinson, R.D. Page, P.J. Sellin, C.W. Beausang, R.M. Clark, R.A. Cunningham, S.A. Forbes, D.B. Fossan, A. Gizon, J. Gizon, K. Hauschild, I.M. Hibbert, A.N. James, D.R. LaFosse, I. Lazarus, H. Schnare, J. Simpson, R. Wadsworth, M.P. Waring, Phys. Rev. C **51**, 78 (1995)

123. M. Pfützner, private communication

124. M. Pfützner, E. Badura, C. Bingham, B. Blank, M. Chartier, H. Geissel, J. Giovinazzo, L. Grigorenko, R. Grzywacz, M. Hellström, Z. Janas, J. Kurcewicz, A. Lalleman, C. Mazzocchi, I. Mukha, G. Münzenberg, C. Plettner, E. Roeckl, K. Rykaczewski, K. Schmidt, R. Simon, M. Stanoiu, J.C. Thomas, Eur. Phys. J. **A14**, 279 (2002)

125. C. Plettner, H. Grawe, I. Mukha, J. Doring, F. Nowacki, L. Batist, A. Blazhev, C.R. Hoffman, Z. Janas, R. Kirchner, M.L. Commara, C. Mazzocchi, E. Roeckl, R. Schwengner, S.L. Tabor, M. Wiedeking, On the Nucl. Phys. A **733**, 20 (2004)

126. G.L. Poli, C.N. Davids, P.J. Woods, D. Seweryniak, J.C. Batchelder, L.T. Brown, C.R. Bingham, M.P. Carpenter, L.F. Conticchio, T. Davinson, J. DeBoer, S. Hamada, D.J. Henderson, R.J. Irvine, R.V.F. Janssens, H.J. Maier, L. Müller, F. Soramel, K.S. Toth, W.B. Walters, J. Wauters, Phys. Rev. C **59**, R2979 (1999)

127. G.L. Poli, C.N. Davids, P.J. Woods, D. Seweryniak, M.P. Carpenter, J.A. Cizewski, T. Davinson, A. Heinz, R.V.F. Janssens, C.J. Lister, J.J. Ressler, A.A. Sonzogni, J. Uusitalo, W.B. Walters, Phys. Rev. C **63**, 044304 (2001)
128. A.P. Robinson, C.N. Davids, G. Mukherjee, D. Seweryniak, S. Sinha, P. Wilt, P.J. Woods, Phys. Rev. C **68**, 054301 (2003)
129. A.P. Robinson, P.J. Woods, D. Seweryniak, C.N. Davids, M.P. Carpenter, A.A. Hecht, D. Peterson, S. Sinha, W.B. Walters, S. Zhu, Phys. Rev. Lett. **95**, 032502 (2005)
130. E. Roeckl, Int. J. Mod. Phys. E **15**, 368 (2006)
131. J. Rotureau, J. Okolowicz, M. Ploszajczak, Phys. Rev. Lett. **95**, 042503 (2005)
132. J. Rotureau, J. Okolowicz, M. Ploszajczak, Nucl. Phys. **A767**, 13 (2006)
133. K. Rykaczewski, Eur. Phys. J. A **15**, 81 (2002)
134. K. Rykaczewski, J.C. Batchelder, C.R. Bingham, T. Davinson, T.N. Ginter, C.J. Gross, R. Grzywacz, M. Karny, B.D. MacDonald, J.F. Mas, J.W. McConnell, A. Piechaczek, R.C. Slinger, K.S. Toth, W.B. Walters, P.J. Woods, E.F. Zganjar, B. Barmore, L.G. Ixaru, A.T. Kruppa, W. Nazarewicz, M. Rizea, T. Vertse, Phys. Rev. C **60**, 011301 (1999)
135. C. Scholey, M. Sandzelius, S. Eeckhaudt, T. Grahn, P. Greenlees, P. Jones, R. Julin, S. Juutinen, M. Leino, A.P. Leppanen, P. Nieminen, M. Nyman, J. Perkowski, J. Pakarinen, P. Rahkila, P. Rahkila, J. Uusitalo, K.V. de Vel, B. Cederwall, B. Hadinia, K. Lagergren, D. Joss, D. Appelbe, C. Barton, J. Simpson, D. Warner, I. Darby, R. Page, E. Paul, D. Wiseman, J. Phys. G **31**, S1719 (2005)
136. P.J. Sellin, P.J. Woods, D. Branford, T. Davinson, N.J. Davis, D.G. Ireland, K. Livingston, R.D. Page, A.C. Shotter, S. Hofmann, R.A.H.A.N. James, M.A.C. Hotchkis, M.A. Freer, S.L. Thomas, Nucl. Instrum. Meth. **311**, 217 (1992)
137. P.J. Sellin, P.J. Woods, T. Davinson, N.J. Davis, K. Livingston, R.D. Page, A.C. Shotter, S. Hofmann, A.N. James, Phys. Rev. C **47**, 1933 (1993)
138. P. Semmes, Nucl. Phys. A **682**, 239c (2001)
139. D. Seweryniak, private communication
140. D. Seweryniak, C.N. Davids, A. Robinson, P.J. Woods, B. Blank, M.P. Carpenter, T. Davinson, S.J. Freeman, N. Hammond, N. Hoteling, R.V.F. Janssens, T.L. Khoo, Z. Liu, G. Mukherjee, J. Shergur, S. Sinha, A.A. Sonzogni, W.B. Walters, A. Woehr, J. Phys. G **31**, S1503 (2005)
141. D. Seweryniak, C.N. Davids, W.B. Walters, P.J. Woods, I. Ahmad, H. Amro, D.J. Blumenthal, L.T. Brown, M.P. Carpenter, T. Davinson, S.M. Fischer, D.J. Henderson, R.V.F. Janssens, T.L. Khoo, I. Hibbert, R.J. Irvine, C.J. Lister, J.A. Mackenzie, D. Nisius, C. Parry, R. Wadsworth, Phys. Rev. C **55**, R2137 (1997)
142. D. Seweryniak, P.J. Woods, J.J. Ressler, C.N. Davids, A. Heinz, A.A. Sonzogni, J. Uusitalo, W.B. Walters, J.A. Caggiano, M.P. Carpenter, J.A. Cizewski, T. Davinson, K.Y. Ding, N. Fotiades, U. Garg, R.V.F. Janssens, T.L. Khoo, F.G. Kondev, T. Lauritsen, C.J. Lister, P. Reiter, J. Shergur, I. Wiedenhöver, Phys. Rev. Lett. **86**, 1458 (2001)
143. A.A. Sonzogni, C.N. Davids, P.J. Woods, D. Seweryniak, M.P. Carpenter, J.J. Ressler, J. Schwartz, J. Uusitalo, W.B. Walters, Phys. Rev. Lett. **83**, 1116 (1999)
144. F. Soramel, A. Guglielmetti, L. Stroe, L. Müller, R. Bonetti, G.L. Poli, F. Malerba, E. Bianchi, A. Andrighetto, J.Y. Guo, Z.C. Li, E. Maglione, F.

Scarlassara, C. Signorini, Z.H. Liu, M. Ruan, M. Ivascu, C. Broude, P. Bednarczyk, L.S. Ferreira, Phys. Rev. C **63**, 031304 (2001)

145. M. Stanoiu, F. Azaiez, F. Becker, M. Belleguic, C. Borcea, C. Bourgeois, B.A. Brown, Z. Dlouhy, Z. Dombradi, Z. Fulop, H. Grawe, S. Grevy, F. Ibrahim, A. Kerek, A. Krasznahorkay, M. Lewitowicz, S. Lukyanov, H. van der Marel, P. Mayet, J. Mrazek, S. Mandal, D. Guillemaud-Mueller, F. Negoita, Y.E. Penionzhkevich, Z. Podolyak, P. Roussel-Chomaz, M.G.S. Laurent, H. Savajols, O. Sorlin, G. Sletten, D. Sohler, J. Timar, C. Timis, A. Yamamoto, Eur. Phys. J. A **20**, 95 (2004)

146. P. Talou, N. Carjan, C. Negrevergne, D. Strottman, Phys. Rev. C **62**, 014609 (2000)

147. P. Talou, N. Carjan, D. Strottman, Phys. Rev. C **58**, 3280 (1998)

148. P. Talou, N. Carjan, D. Strottman, Nucl. Phys. A **647**, 21 (1999)

149. P. Talou, D. Strottman, N. Carjan, Phys. Rev. C **60**, 054318 (1999)

150. J. Thaysen, L. Axelsson, J. Äystö, M. Borge, L. Fraile, H. Fynbo, A. Honkanen, P. Hornshoj, Y. Jading, A. Jokinen, B. Jonson, I. Martel, I. Mukha, T. Nilsson, G. Nyman, M. Oinonen, K. Riisager, T. Siiskonen, M. Smedberg, O. Tengblad, F. Wenander, Phys. Lett. **B467**, 194 (1999)

151. J. Thomas, L. Achouri, J. Äystö, R. Beraud, B. Blank, G. Canchel, S. Czajkowski, P. Dendooven, A. Ensallem, J. Giovinazzo, N. Guillet, J. Honkanen, A. Jokinen, A. Laird, M. Lewitowicz, C. Longour, F. de Oliveira Santos, K. Peräjärvi, M. Stanoiu, Eur. Phys. J. **A21**, 419 (2004)

152. K.S. Toth, D.C. Sousa, P.A. Wilmarth, J.M. Nitschke, K.S. Vierinen, Phys. Rev. C **47**, 1804 (1993)

153. P. Van Duppen, Lect. Notes Phys. **700**, 37 (2006)

154. A. Volya, C. Davids, Eur. Phys. J. A **25**, 161 (2005)

155. P.J. Woods, P. Munro, D. Seweryniak, C.N. Davids, T. Davinson, A. Heinz, H. Mahmud, F. Sarazin, J. Shergur, W.B. Walters, A. Woehr, Phys. Rev. C **69**, 051302 (2004)

156. C.H. Yu, J.C. Batchelder, C.R. Bingham, R. Grzywacz, K. Rykaczewski, K.S. Toth, Y. Akovali, C. Baktash, A. Galindo-Uribarri, T.N. Ginter, C.J. Gross, M. Karny, S.H. Kim, B.D. MacDonald, S.D. Paul, D.C. Radford, J. Szerypo, W. Weintraub, Phys. Rev. C **58**, R3042 (1998)

157. C.H. Yu, A. Galindo-Uribarri, S.D. Paul, M.P. Carpenter, C.N. Davids, R.V.F. Janssens, C.J. Lister, D. Seweryniak, J. Uusitalo, B.D. MacDonald, Phys. Rev. C **59**, R1834 (1999)

158. Y.B. Zel'dovich, Sov. Phys. JETP **11**, 812 (1960)

159. T. Zerguerras, B. Blank, Y. Blumenfeld, T. Suomijärvi, M. Thoennessen, D. Beaumel, B. Brown, M. Chartier, M. Fallot, J. Giovinnazzo, C. Jouanne, V. Lapoux, I. Lhenry-Yvon, W. Mittig, P. Roussel-Chomaz, H. Savajols, J.A. Scarpaci, A. Shrivastava, Eur. Phys. J. **A20**, 389 (2004)

Superheavy Elements

S. Hofmann[1,2]

[1] Gesellschaft für Schwerionenforschung (GSI), D-64291 Darmstadt, Germany
[2] Institut für Kernphysik, Goethe-Universität, D-60438 Frankfurt, Germany
S.hofmann@gsi.de

Abstract The nuclear shell model predicts that the next doubly magic shell closure beyond ^{208}Pb is at a proton number Z=114, 120, or 126 and at a neutron number N=172 or 184. The outstanding aim of experimental investigations is the exploration of this region of spherical 'SuperHeavy Elements' (SHEs). Experimental methods have been developed which allowed for the identification of new elements at production rates of one atom per month. Using cold fusion reactions which are based on lead and bismuth targets, relatively neutron-deficient isotopes of the elements from 107 to 113 were synthesized at GSI in Darmstadt, Germany, and/or at RIKEN in Wako, Japan. In hot fusion reactions of ^{48}Ca projectiles with actinide targets more neutron-rich isotopes of the elements from 112 to 116 and even 118 were produced at the Flerov Laboratory of Nuclear Reactions (FLNR) at the Joint Institute for Nuclear Research (JINR) in Dubna, Russia. Recently, part of these data which represent the first identification of nuclei located on the predicted island of SHEs were confirmed in two independent experiments. The decay data reveal that for the heaviest elements, the dominant decay mode is α emission rather than fission. Decay properties as well as reaction cross-sections are compared with results of theoretical studies. Finally, plans are presented for the further development of the experimental set-up and the application of new techniques. At a higher sensitivity, the detailed exploration of the region of spherical SHEs will be in the center of interest of future experimental work. New data will certainly challenge theoretical studies on the mechanism of the synthesis, on the nuclear decay properties, and on the chemical behavior of these heaviest atoms at the limit of stability.

1 Introduction

Searching for new chemical elements is an attempt to answer questions of partly fundamental character: How many elements may exist? How long is their lifetime? Which properties determine their stability? How can they be synthesized? What are their chemical properties? How are the electrons arranged in the strong electric field of the nucleus?

Searching for new elements beyond uranium by the process of neutron capture and succeeding β^- decay, Hahn and Straßmann [1] discovered a novel

Hofmann, S.: *Superheavy Elements*. Lect. Notes Phys. **764**, 203–252 (2009)
DOI 10.1007/978-3-540-85839-3_6   © Springer-Verlag Berlin Heidelberg 2009

disintegration mode which they assigned to the possibility that a heavy nucleus might 'divide itself into two nuclei'. The correct interpretation of the observed phenomena was given by Meitner and Frisch [2], and the term 'fission' was coined for this process. By applying the charged liquid-drop model of the nucleus [3, 4], nuclear fission was explained, and it was shown that this disintegration mode will most likely limit the number of chemical elements. At that time, the maximum number of elements was expected to be about 100. Such an estimate results from the balance of two fundamental nuclear parameters, the strength of the attractive nuclear force which binds neutrons and protons together and creates a surface tension, and the repulsive electric force.

The properties of nuclei are not smooth uniform functions of the proton and neutron numbers, but show non-uniformities as evidenced by variations in the measured atomic masses. Like the electrons in an atom, also the nucleons in a nucleus – described by quantum mechanical laws – form closed shells called 'magic' numbers. At the magic proton or neutron numbers 2, 8, 20, 28, 50, and 82, the nuclei have an increased binding energy relative to the average trend. For neutrons, $N=126$ is also identified as a magic number. However, the highest stability is observed in the case of the 'doubly magic' nuclei with a closed shell for both protons and neutrons. Amongst other special properties, the doubly magic nuclei are spherical and resist deformation.

The magic numbers were successfully explained by the nuclear shell model [5–7], and an extrapolation into unknown regions was thus undertaken. The numbers 126 for the protons and 184 for the neutrons were predicted to be the next shell closures. Instead of 126 for the protons also 114 or 120 were calculated as closed shells. The term superheavy elements (SHEs) was coined for these elements.

The prediction of magic numbers, although not unambiguous, was less problematic than the calculation of the stability of those doubly closed shell nuclei against fission. As a consequence, predicted half-lives based on various calculations differed by many orders of magnitude [8–14]. Some of the half-lives approached the age of the universe, and attempts have been made to discover SHEs occurring in nature [15, 16]. Although the corresponding discoveries were announced from time to time, none of them could be substantiated after more detailed inspection.

There was also great uncertainty on the production yields for SHEs. Closely related to the fission probability of SHEs in the ground state, the survival of the compound nuclei formed after complete fusion was difficult to predict. Even the best choice of the reaction mechanism, fusion or transfer of nucleons, was critically debated. However, as soon as relevant experiments could be performed, it turned out that the most successful methods for the laboratory synthesis of heavy elements are fusion–evaporation reaction using heavy-element targets, recoil-separation techniques, and the identification of the nuclei by generic ties to known daughter decays after implantation into position-sensitive detectors [17–19].

The newly developed detection methods considerably extended the range of half-lives that can be reached by experiment. The lower half-life limit of about 1 μs is determined by the flight time through the separator whereas the upper limit of about one day is given by the rate of implanted reaction products and background considerations. The detectors are sensitive for measuring all radioactive decays based on particle emission like proton radioactivity, α and β decay, and spontaneous fission (SF).

A further extension of the measuring possibilities was achieved with γ-ray, X-ray, or particle detectors mounted around the target. If these detectors are operated in delayed coincidence with signals from the implantation of reaction products and their radioactive decay in the focal plane of the separator, the sensitivity of 'in-beam' spectroscopy is significantly improved. This so-called recoil-decay-tagging (RDT) method was first applied in a study of the heavy ion radiative capture mechanism using the reaction ^{90}Zr + ^{90}Zr → ^{180}Hg [20]. Meanwhile the method has become a standard tool in nuclear in-beam spectroscopy.

In the following sections a detailed description is given for the set-ups of the physics experiments used for the investigation of SHEs. Moreover, chemistry methods are described which were used in recent discoveries of new isotopes in the heavy element region or in confirmation experiments. Experiments are presented, in which cold and hot fusion reactions were used for the synthesis of SHEs. These experiments resulted in the identification of elements 107–112 at the Gesellschaft für Schwerionenforschung (GSI), in the production of a new isotope of element 113 at the RIKEN laboratory in Wako, Japan, and in the recent synthesis of neutron-rich isotopes of element 112 and the identification of the new elements from 113 to 116 and 118 at the Flerov Laboratory of Nuclear Reactions (FLNR) at the Joint Institute for Nuclear Research (JINR) in Dubna, Russia. We also report on synthesis of new isotopes and confirmation experiments performed at Lawrence Berkeley National Laboratory (LBNL) in Berkeley, USA. Recent results are reported from spectroscopy experiments in-beam as well as in the focal plane of separators. Of great interest for such studies are isomeric states which are located in the regions of deformed heavy nuclei near $Z=100$, $N=152$ and $Z=108$, $N=162$. The subsequent sections contain a theoretical description of properties of nuclei in the region of SHEs and phenomena which influence the yield for the synthesis of SHEs. Empirical descriptions of hot and cold fusion reactions are outlined. Finally, a summary and an outlook are given.

2 Experimental Techniques

2.1 Ion Source and Accelerator

First attempts to synthesize transuranium elements were based on the idea to produce nuclei by neutron capture, which β^- decay into an isotope of the next heavier, so far unknown element. Up to fermium, this method made it possible to climb up the periodic table element by element. While from neptunium to californium, some isotopes can be produced in amounts of kilograms or at least grams in high neutron flux reactors, the two heaviest species, ^{254}Es and ^{257}Fm, are available only in quantities of micrograms and picograms, respectively. At fermium, however, the method ends due to the lack of β^- decay and too short α and fission half-lives of the heavier elements. Sufficiently thick enough targets cannot be manufactured from these elements.

The region beyond fermium is best accessible using heavy-ion fusion reactions, the bombardment of heavy-element targets with heavy ions from an accelerator. The cross-section is less than in the case of neutron capture, and values are considerably below the geometrical size of the nuclei. Moreover, only thin targets with thicknesses of the order of $0.5\,\mathrm{mg/cm^2}$ can be used. This limitation arises from the energy loss of the ion beam in the target, which results (using thicker targets) in an energy distribution that is too wide for both the production of fusion products and their in-flight separation. On the other hand, the use of thin targets in combination with well defined beam energies from accelerators results in unique information about the reaction mechanism. The data are obtained by measuring excitation functions, i.e., the yield as a function of the beam energy.

Various combinations of projectiles and targets are in principle possible for the synthesis of heavy elements: Actinide targets irradiated by light projectiles of elements in the range from neon to nickel, targets of lead and bismuth irradiated by projectiles from calcium to krypton, and symmetric combinations like tin plus tin up to samarium plus samarium. Also inverse reactions using, e.g., lead or uranium isotopes as projectile are possible and may have technical advantages in specific cases.

Historically, the first accelerators used for the production of heavy elements were the cyclotrons in Berkeley and later in Dubna. They were only able to accelerate light ions up to about neon with sufficient intensity and up to an energy high enough for fusion reactions. Later on, larger and more powerful cyclotrons were built in Dubna for the investigation of reactions using projectiles near calcium. These were the U300 and U400 cyclotrons, which have 300 and 400 cm diameter. In Berkeley a linear accelerator HILAC (Heavy Ion Linear ACcelerator), later upgraded to the SuperHILAC, was built. The shutdown of this accelerator in 1992 led to a revival of heavy element experiments at the 88-inch Cyclotron at Berkeley. Aiming at the acceleration of ions as heavy as uranium, the UNILAC (UNIversal Linear ACcelerator) was constructed in Darmstadt, Germany, during the years 1969–1974 [21].

In order to compensate for the decreasing cross-sections of the synthesis of heavy elements, increasing beam currents are needed from the accelerators. In the following efforts are described aiming to improve the facilities for experiments at low-projectile energies at the UNILAC at the end of the 1980s. A new high-charge injector was built including a 14-GHz-ECR (electron cyclotron resonance) CAPRICE-type ion source [22]. The ion source is followed by a radio-frequency quadrupole and an interdigital H-structure accelerator that provide a beam energy of 1.4 MeV/u (MeV per mass unit u) for direct injection into the Alvarez section of the UNILAC. The advantages, compared to the previously used Penning ion source, are:

(1) Low consumption of material ($\approx$ 0.2–4 mg/h).
(2) High-beam intensity provided over extended time periods.
(3) A high-quality beam of low emittance, halo free, and of well-defined energy.

The high-beam quality is a result of the high-ionic charge state attained, e.g., 10^+ in the case of ^{70}Zn [23]. This charge state is maintained throughout the acceleration process as an increase by stripping of electrons is unnecessary. A reduction of the projectile background behind the recoil separator is partially due to the increased beam quality.

The beam energy is variable and defined by a set of single resonators. The relative accuracy of the beam energy is ± 0.003 MeV/u. The absolute energies are accurate to ± 0.01 MeV/u. This high accuracy is sufficient for the measurement of narrow excitation functions, as observed for the one-neutron and two-neutron emission channels in reactions for the production of heavy elements. Beam energies are determined by time-of-flight measurements using pick-up coils. Recently, an independent method based on so-called cusp electrons was developed [24]. The relative accuracy of this method is $\pm 1 \times 10^{-3}$. The measuring equipment is transportable and could thus help to solve uncertainties related with energy measurements at different accelerators.

The beam intensities are rather high. For example, the following values could be obtained at the target: 1.6 particle-µA for ^{48}Ca^{10+}, 4.0 particle-µA for ^{40}Ar^{8+}, 1.2 particle-µA for ^{58}Fe^{8+}, and 0.4 particle-µA for ^{82}Se^{12+} (1 particle-µA = 6.24×10^{12} ions/s). The given values represent mean currents reached on target at a duty factor of 28% (5.5 ms wide pulses at 20 Hz repetition frequency). In the case of rare isotopic abundance, the source material was enriched to reach a concentration higher than 90%.

2.2 Targets

High-beam currents, in turn, demand a high resistance of the targets. The present target technology uses target wheels which rotate with high speed through the beam. In general, the beam intensity is limited by the melting point of the target material. In some cases higher melting point can be achieved by using chemical compounds or alloys. Already successfully tested is a PbS target (melting point 1118°C) produced by depositing the target material on a carbon backing [25]. By heating the backing during evaporation (up to about 300°C), the formation of a crystalline needle structure of PbS, which would result in uncontrolled energy loss of the projectiles, is avoided. Using the 'heated' PbS target, a 1n excitation function was measured, which was identical to the previously measured one obtained with a metallic lead target. These targets were irradiated with ^{54}Cr beam with an intensity of up to 1.2 particle-μA at a 28% duty factor without observable damage. Other examples of high melting-point compound targets which have already been experimentally tested, are BiO_2 and UF_4. In the case of uranium targets, the elementary metallic uranium has a higher melting point, i.e., 1132°C compared to 960°C of UF_4. However, the production of the target by sputtering of the material on a thin carbon foil is more complicated than the evaporation of UF_4 [26]. The advantage of high-temperature targets is the increased radiative cooling which makes the application of more complicated gas cooling superfluous. However, gas cooling has to be used in the case of targets of low-melting point. The cooling medium can be a stream of helium, blown with low pressure (1–10 mbar) from both sides in the direction of the beam spot. The cooling effect of a gas acting on a target is well known from gas-filled separators and helium-jet systems, where the beam currents can be increased by a factor 5–10 compared to targets in vacuum.

The gas-cooled target must be used in experiments where only targets of low-melting point are available or have to be used for experimental reasons. At highest beam intensities, the gas cooling method is also interesting in cases where the target material is not available in gaseous or liquid form or where radioactive, fixed targets are used, e.g., curium or californium.

A crucial item is also the intensity distribution of the beam across the target. Quadrupoles as ion-optical elements allow only for a Gaussian-shaped beam intensity profile. Thus the highest power load still occurs in the center region and tails at the outer areas. Therefore the target most likely melts in the middle, whereas the tail of the intensity distribution causes background when hitting the target frame. The intensity distribution can be optimized using an octupole doublet in the beam line in addition to the quadrupoles. With the use of these magnets an almost rectangular intensity distribution should be achievable [27].

The beam intensity distribution and the resulting temperature distribution across the target can be monitored by an infrared video camera. The target thickness can be controlled on-line by registration of elastically scattered projectiles. A precise two-dimensional thickness measurement was developed by using a narrow beam of electrons of 20 keV energy [28]. This method is based on measuring the reduction of the electron beam intensity due to scattering at the target material.

2.3 Recoil-Separation Techniques

The identification of the first transuranium elements was achieved by chemical means. In the early 1960s separation and transport techniques were developed which allowed for detection of nuclei with lifetimes down to a few tenth of seconds at high sensitivity. A further improvement of the physical methods was obtained with the development of recoil separators and large area position-sensitive detectors. As a prime example for such instruments, we will describe

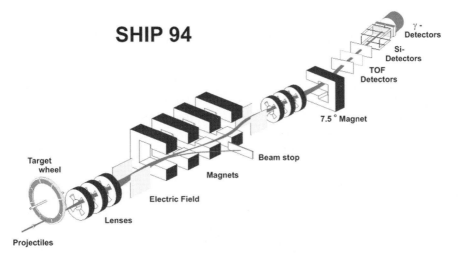

Fig. 1. The velocity filter SHIP (separator for heavy ion reaction products) and its detection system [17–19] as it was used for the study of element 110 in 1994. The drawing is approximately to scale, however, the target wheel and the detectors are enlarged by a factor of 2. The length of SHIP from the target to the detector is 11 m. The target wheel has a radius up to the center of the targets of 155 mm. It rotates synchronously with beam macrostructure at 1,125 rpm [29]. The target thickness is usually $450\,\mu g/cm^2$. The detector system consists of three large area secondary-electron time-of-flight detectors [30] and a position-sensitive silicon-detector array (see text and Fig. 2). The flight time of the reaction products through SHIP is 1–$2\,\mu s$. The filter, consisting of two electric and four magnetic dipole fields plus two quadrupole triplets, was later extended by a fifth deflection magnet, allowing for positioning of the detectors away from the straight beam line and leading to further reduction of the background (See also Plate 15 in the Color Plate Section)

the SHIP and its detector system, which were developed at the UNILAC (see Fig. 1). The principle of separation and detection techniques used in most of the other laboratories is comparable.

In contrast to the recoil-stopping methods, as used in helium-jet systems or mass separators where ion sources are utilized, recoil-separation techniques use the ionic charge and momentum of the recoiling fusion product obtained in the reaction process. Spatial separation from the projectiles and other reaction products is achieved by combined electric and magnetic fields. The separation times are determined by the recoil velocities and the lengths of the separators. They are typically in the range of 1–$2\,\mu s$. Two types of recoil separator have been developed: (1) The gas-filled separators use the different magnetic rigidities of the recoils and projectiles travelling through a low pressure (about 1 mbar) gas-filled volume in a magnetic dipole field [31]. In general, helium or hydrogen is used in order to obtain a maximum difference in the rigidities of slow reaction products and fast projectiles. A mean charge state of the ions is achieved by frequent collisions with the atoms of the gas.

(2) Wien-filter or energy separators use the specific kinematic properties of the fusion products. The latter ones are created with velocities and energies different from the projectiles and other reaction products. Their ionic charge state is determined when they escape from a thin solid-state target into vacuum. A whole charge state distribution is created with a width of about $\pm 10\%$ around the mean value. Therefore, ionic-charge achromaticity is essential for high transmission. It is achieved by additional magnetic fields or symmetric arrangements of electric fields.

2.4 Detectors

Recoil separators are designed to filter out nuclei produced in fusion reactions with a high rate of transmission. Since higher separation yield also leads to increased background level, the transmitted particles have to be identified by detector systems. The detector type to be selected depends on the particle rate, energy, decay mode, and half-life. Experimental as well as theoretical data on the stability of heavy nuclei show that they decay by α emission or electron capture (EC) or SF, with half-lives ranging from microseconds to days. Therefore, silicon semiconductor detectors are well suited for the identification of nuclei and for the measurement of their decay properties.

If the total rate of ions striking the focal plane of the separator is low the particles can be implanted directly into the silicon detectors. Using position-sensitive detectors, one can measure the local distribution of the implanted particles. In this case, the detectors act as diagnostic elements to optimize and control the ion optical properties of the separator.

Given that the implanted nuclei are radioactive, the positions measured for the implantation and all subsequent decay processes are the same. This is the case because the range of implanted nuclei, α particles, recoiling daughter nuclei, and fission products is small compared with the detector thickness. Recording the data event by event allows for the analysis of delayed coincidences. Thus, by inspecting the implantation and decay positions and the time windows between implantation and decay, decay chains can be identified.

This method was developed and tested in experiments investigating neutron-deficient α emitters and proton radioactivity near $N=82$ [18]. Subsequently, the detector system was enlarged [32], and an array of seven position-sensitive silicon detectors was used in the identification of the elements bohrium, hassium, and meitnerium [33]. Finally an even larger system was built at the SHIP in order to search for elements beyond meitnerium [34].

The new detector system is composed of three time-of-flight detectors, seven identical 16-strip silicon wafers, and three germanium detectors. A three-dimensional view of the detector arrangement is shown in Fig. 2, together with a cross-section drawn to scale. In front of the silicon detectors, there is a mechanism for inserting calibration sources and degrader foils. The thickness of the foils (Mylar) can be varied in increments of $0.5\,\mu m$ in order to facilitate

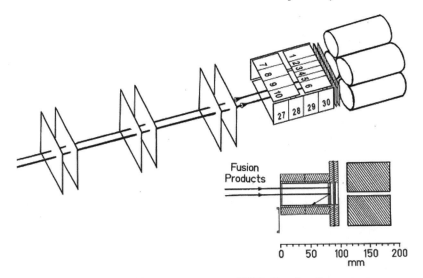

Fig. 2. Present detector set-up at SHIP. For details see text

the absorption of low-energy projectiles and reduce the implantation energy, thus avoiding pile-up effects on the signals.

The active area of each silicon wafer is $35 \times 80 \, \text{mm}^2$. Each of the 16 strips is 5 mm wide and position-sensitive in the vertical direction with a spatial resolution of 150 µm FWHM for α-decay events. For that reason, the stop detector is equivalent to 3,700 single detectors, each of them having an active area of $0.15 \times 5 \, \text{mm}^2$. The energy resolution of new detectors is 14 keV (FWHM) for α particles from a ^{241}Am source or α decays of implanted nuclei. If the resolution declines below a level of approximately 35 keV, which usually happens after about 2 years of operation, the stop detector is replaced. Six wafers are mounted like a box in the back hemisphere facing the stop detector. They measure escaping α particles or fission fragments with a solid angle of 80% of 2π. In the case of the box detectors, neighboring strips are connected galvanically, forming 28 energy sensitive segments. Thus, the direction of the escaping α particle or fission fragment can be roughly retraced. All silicon detectors are cooled to $-15°$C. The energy resolution obtained by summing the energy-loss signal from the stop detector and the residual energy from the box detector is 40 keV for α particles.

In this way lifetimes as short as 20 µs are measured with sufficient position and energy resolution by using fast ADCs (analog-to-digital converters, 3.5 µs conversion time) and a fast ADC multiplexer system (AMUX) with a front-end data buffering in an event queue with a length of 400 events on AMUX modules [35]. The dead time of about 20 µs corresponds to the total widths of the detector signals. These are determined by the shaping-time constants of 0.3 µs for the position signals and 2.0 µs for the energy signals. The energy signals are obtained by summing the two preamplifier signals of each individual strip. The shorter position signals are also used to obtain the energy information for the 3–20 μs range. If two events appear within 20 μs, the signals of the second event (energy, position, and time-to-amplitude converter signal) are shifted beyond the dead-time of the first event by a constant delay of 50 μs.

The set-up is completed by a silicon-veto detector and a germanium-clover detector, both mounted behind the stop detector. The clover detector is separated from the SHIP vacuum by a 1 mm thick aluminium window. The arrangement consists of four germanium crystals, each with a diameter of 50 mm and a length of 70 mm.

The germanium detectors measure X-rays or γ-rays that are coincident with signals from the silicon detectors within a time window of 4 μs. This allows for the detection of α transitions to excited levels in the daughter nucleus, which decay by γ-ray emission. In the case of an electron conversion process characteristic X-rays may be emitted, which would allow for a clear element identification. Although the probability for detecting coincident events is small, the germanium detectors provide useful spectroscopic information if the cross-sections are of the order of nanobarns or higher.

Three secondary-electron time-of-flight foil detectors are mounted at distances of 780, 425, and 245 mm upstream of the silicon detectors and the degrader stack [30]. Two foils made of 30 μg/cm² thick carbon are needed for each detector. Between the foils an electric potential of ≈4 kV is applied in order to accelerate electrons emitted from the first foil during the passage of heavy ions. Perpendicularly, a magnetic field is applied in order to bend the electrons onto a channel plate for further amplification. The foils are self-supporting and the transmission of electrons from the foils to the channel plate is close to 100%. The detector signals are used to distinguish implantation from radioactive decays of previously implanted nuclei. Three detectors are used to increase detection efficiency. Because of the high efficiency of each of these detectors, the background in the decay spectra due to implantation of heavy ions is suppressed by a factor of 1,000, and the time window for measuring generic parent–daughter decays is significantly prolonged. The time resolution of the foil detectors is about 0.8 ns (FWHM), which is small enough that, taking the energy signals from the silicon detector into account, a rough mass assignment with an accuracy of ±10% for the implanted ions is achievable.

The establishment of a generic link of signals from radioactive decays to known daughter decays provides a method for unambiguous identification of the unknown parent isotope. Ghiorso et al. [36] originally applied this method when they identified ^{257}Rf and ^{258}Rf by 'milking' their daughter nuclides, ^{253}No and ^{254}No. In measurements at recoil separators, the event chain already starts with the implantation of the produced nuclide. The measured signals deliver time-of-flight, moment, energy, and position of implantation into the silicon detector. In an ideal case, a sequence of α-decay signals follows. From each decay, the time, energy, and position are again measured, but no signal is obtained from the time-of-flight detectors. The chain ends due to SF or long half-life of the decay products. The longest half-life that can be reached by this method is solely determined by the background conditions or, in the case of high fusion cross-sections, by the rate of implanted evaporation residues.

3 Experimental Results

3.1 Elements Produced in Cold-Fusion Reactions

In this section, we present results dealing with the synthesis of elements 107–113 using cold-fusion reactions based on lead and bismuth targets. A detailed presentation and discussion of the GSI-SHIP results on the decay properties of elements 107–112 was given in previous reviews [19, 33, 37]. The results of experiments at RIKEN on the confirmation of elements 110–112 and the first production of an isotope of element 113 using a cold-fusion reaction were published in [38–42]. Known elements and their position in the periodic table of the elements are shown in Fig. 3. An overview of nuclei in the region of SHEs, which are known to date, is given in the partial chart of nuclides displayed in Fig. 4.

Bohrium, element 107, was the first new element synthesized at SHIP using the method of in-flight recoil separation and generic correlation of parent–daughter nuclei. The reaction used was ^{54}Cr + ^{209}Bi → ^{263}Bh*. Five decay chains of ^{262}Bh were observed [45]. The next lighter isotope, ^{261}Bh, was synthesized at a higher beam energy [46]. Additional data were obtained from the α decay of ^{266}Mt [47], and the isotope ^{264}Bh was identified as granddaughter in the decay chain of ^{272}Rg [48, 49]. The isotopes ^{266}Bh and ^{267}Bh were produced using the hot-fusion reaction ^{22}Ne + ^{249}Bk → ^{271}Bh* [50, 51]. These nuclei were used for a study of the chemical properties of bohrium.

A new isotope of bohrium, ^{265}Bh, was synthesized in 2003 at the Heavy Ion Research Facility (HIRFL) in Lanzhou, China [52]. It was produced in a 4n evaporation channel using the hot-fusion reaction ^{26}Mg + ^{243}Am → ^{269}Bh*. For completeness we add here that the identification of the new isotope ^{259}Db was published by the same group already in 2001. In this case the reaction ^{22}Ne + ^{241}Am → ^{263}Db* was used [53].

The excitation function for the production of ^{262}Bh (Z=107) in the odd-Z-projectile reaction ^{55}Mn + ^{208}Pb → ^{263}Bh* was studied in [54]. The interesting result was a higher cross-section of 540 pb than in the reaction ^{54}Cr + ^{209}Bi → ^{263}Bh* (σ=163 pb) studied at SHIP [46]. However, in the earlier SHIP experiment, the excitation function was measured only rudimentary and the cross-section maximum could have been missed. Only recently the Berkeley group was successful in synthesizing the so far lightest isotope of bohrium, ^{260}Bh [55]. The reaction ^{52}Cr + ^{209}Bi → ^{260}Bh + 1n was used, and a cross-section of 59 pb was measured.

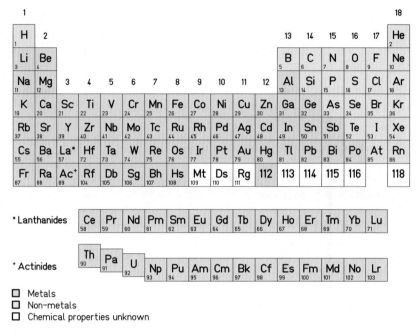

Fig. 3. Periodic table of elements. The known transactinide elements 104–116 and 118 take the positions from below hafnium in group 4 to below radon in group 18. Elements 108, hassium (Hs), and element 112, the heaviest elements chemically investigated, are placed in groups 8 and 12, respectively. The arrangement of the actinides reflects the fact that the first actinide elements still resemble, to a decreasing extent, the chemistry of the other groups: Thorium (group 4 below hafnium), protactinium (group 5 below tantalum), and uranium (group 6 below tungsten) [43]. The name 'roentgenium', symbol 'Rg', was proposed for element 111 and recommended for acceptance by the Inorganic Chemistry Division of IUPAC in 2004 [44] (See also Plate 16 in the Color Plate Section)

On the neutron-rich side isotopes up to ^{272}Bh were measured in experiments using hot-fusion reactions with a ^{48}Ca beam. These experiments and the synthesized isotopes will be discussed in the following Sect. 3.2.

Hassium, element 108, was first synthesized in 1984 using the reaction ^{58}Fe + ^{208}Pb. The identification was based on the observation of three atoms [56]. Only one α decay chain was measured in the irradiation of ^{207}Pb with ^{58}Fe. The measured event was assigned to the even–even isotope ^{264}Hs [57]. The results were confirmed in later works [37, 58], and for the decay of ^{264}Hs an SF branching of 50% was measured. The isotope ^{269}Hs was discovered as a link in the decay chain of ^{277}112 [49, 59], and ^{270}Hs was identified in a recent chemistry experiment using a hot-fusion reaction [60, 61].

Meitnerium, element 109, was first observed in 1982 in the irradiation of ^{209}Bi with ^{58}Fe by a single α decay chain [62, 63]. This result was confirmed

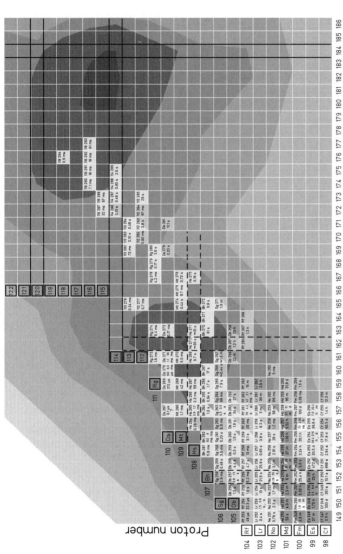

Neutron number

Fig. 4. *Upper end* of the chart of nuclei showing the presently (2007) known nuclei. For each known isotope the element name, mass number, and half-life are given. The relatively neutron-deficient isotopes of the elements up to proton number 113 were created after evaporation of one neutron from the compound nucleus (cold-fusion reactions based on ^{208}Pb and ^{209}Bi targets). The more neutron-rich isotopes from element 112 to 118 were produced in reactions using a ^{48}Ca beam and targets of ^{238}U, ^{237}Np, ^{242}Pu, ^{244}Pu, ^{243}Am, ^{245}Cm, ^{248}Cm, and ^{249}Cf. The magic numbers for protons at element 114 and 120 are emphasized. The *bold dashed lines* mark proton number 108 and neutron numbers 152 and 162. Nuclei with that number of protons or neutrons have increased stability; however, they are deformed contrary to the spherical superheavy nuclei. At $Z=114$ and $N=162$ it is uncertain whether nuclei in that region are deformed or spherical. The background structure in gray shows the calculated shell correction energy according to the macroscopic–microscopic model. See Sect. 4 and Fig. 9 for details (See also Plate 17 in the Color Plate Section)

in [64]. In the most recent experiment [47] 12 atoms of ^{266}Mt were measured, revealing a complicated decay pattern, as concluded from the wide range of α energies from 10.5 to 11.8 MeV. This property seems to be common to many odd and odd–odd nuclides in the region of the heavy elements. The more neutron-rich isotope ^{268}Mt was measured after α decay of ^{272}Rg [48, 49].

Darmstadtium, element 110, was discovered in 1994 using the reaction ^{62}Ni $+ \, ^{208}$Pb $\rightarrow \, ^{269}$Ds $+ 1$n [58]. The main experiment was preceded by an accurate study of the excitation functions for the synthesis of ^{257}Rf and ^{265}Hs using beams of ^{50}Ti and ^{58}Fe in order to determine the optimum beam energy for the production of element 110. New information on the decay pattern of these nuclei was also obtained. The data revealed that the maximum cross-section for the synthesis of element 108 was shifted to a lower excitation energy relative to the peak for production of element 104, different from the predictions of reaction theories. A so-called *extra-push energy* [65, 66] was not measured.

The heavier isotope ^{271}Ds was synthesized with a beam of the more neutron-rich isotope ^{64}Ni [37]. The important result for the further production of elements beyond meitnerium was that the cross-section was enhanced from 2.6 to 15 pb by increasing the neutron number of the projectile by two. This observation gave hope that the cross-sections for the synthesis of heavier elements could decrease less steeply with available stable, more neutron-rich projectiles. However, this expectation was not proven in the case of element 112, see below.

An overview of all data measured at SHIP from the decay chains observed in the reaction ^{64}Ni $+ \, ^{208}$Pb $\rightarrow \, ^{272}$Ds* is given in Fig. 5. We will describe these data in more detail, because they are representative for the detection and assignment of new decay data by correlation to the decay of known nuclei. In the figure the energies and lifetimes of α decays directly succeeding the implantations are shown (describing single event chains it is preferable to use the lifetime τ instead of the half-life $T_{1/2}$, because τ is directly measured as time difference between two subsequent signals). On top of the upper abscissa of Fig. 5, the α spectra deduced from literature are plotted for the decays of ^{255}Md, ^{255}No, ^{259}Rf, and ^{263}Sg, in order to compare the energy and intensity pattern with the measured data assigned to the decay of ^{271}Ds. In case of ^{267}Hs the three α decays observed in [67] are plotted.

A total number of 57 α decays was measured and assigned to the 13 decay chains shown in Fig. 5. Thirty-eight α particles were emitted in-beam direction and stopped in the 300 μm thick detector. These α events are marked by the little filled squares. The width of the squares shows the energy resolution of 13 keV (2σ) obtained for these full energy α signals. An exception are the ^{271}Ds α decays of chain 9 and 11 for which the energy had to be corrected by $+ 60$ and -40 keV, respectively, due to short lifetime and tails of the signals from the preceding implantation. The exact shape of these tails as a function of energy was measured with energy-degraded fission fragments of a ^{252}Cf source. A larger uncertainty, as indicated by the rectangles, was assigned to

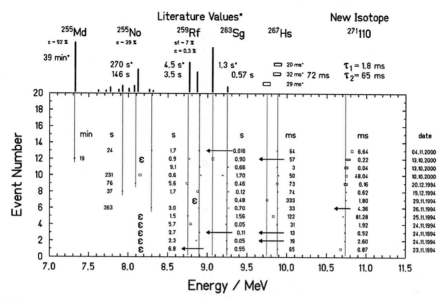

Fig. 5. Energies and lifetimes of the 13 α decay chains resulting from the reaction ^{64}Ni + ^{208}Pb → ^{271}Ds + 1n. Plotted are the data measured in experiments at SHIP. The isotope ^{271}Ds was identified by comparison of the decay properties of the daughter products with literature data marked by an *asterisk* (^{267}Hs [67], ^{263}Sg [68], ^{259}Rf [69], ^{255}No [69], ^{255}Md [69]). New decay data of the isotope ^{263}Sg could be deduced. The 13 decay chains are arranged chronologically with the date of production given at the *righthand ordinate*. For each chain, the time sequence of the α decays is from right to left following the decreasing α energies. The lifetimes of the first α decays at 10.7 MeV are the time differences between implantation and α decay. For description of the symbols see text. Obviously, the α decay energy serves as a finger print for the identification of nuclei. The higher the energy resolution is and the less transitions of different energy for one nucleus occur, the higher is the significance of the assignment. See text for details

these data. The larger, open squares mark escape α particles for which the full energy was summed from ΔE signals between 0.6 and 1.4 MeV in the stop detector and coincident residual energies in the backward crystals. A total of 12 of these α events was observed, for which an energy resolution of 34 keV (2σ), as marked by the width of the squares, was obtained in test reactions at higher statistical significance. In one of these cases, chain number 12, the first α decay was corrected by +60 keV due to the short lifetime after the implantation, and a larger error bar was used. Seven α particles escaped from being recorded in the backward crystals but still yielded a ΔE signal and thus the time and position information. The amplitudes of the signals corresponded to energies between 1.0 and 5.2 MeV, which is characteristic for escaping α particles. They are marked in Fig. 5 by the arrows pointing left. The ratio of the α particles measured with full energy/escape plus residual energy/escape is 38/12/7. It agrees within the respective experimental uncertainties with the ratio 34/17/6 determined by test reactions and normalized to the total number of 57 α particles as measured in the ^{271}Ds experiment.

The lower discriminator level for signals of the stop detector amounted to 260 keV, which is below the lowest ΔE signals of escaping α particles emitted from heavy nuclei implanted 4–6 μm into the active detector material. An estimate of the implantation depths was obtained by the measured implantation energy signals of $\approx$25 MeV. As an important result a detection efficiency of 100% for α decay was obtained, the same as for the much higher energetic SF events. Therefore the reason for non-observation of an α-decay or fission signal is almost certainly that the nucleus, which has been identified as an α decay product, undergoes β decay or EC. In Fig. 5 such missing α or fission decays are marked by 'ϵ'. If the time windows between implantation and disintegration is wide enough to cover the full lifetime range, the detector system allows an unambiguous measurement of β branching ratios.

In agreement with the known branching ratios of ^{255}No (b_ϵ=39% [69]) and ^{255}Md (b_α=8% [69]) is the observation of the α decay of ^{255}Md in one of 13 cases (see chain 12 in Fig. 5). This result allows one to draw two conclusions. Firstly, lifetimes up to 19 min are measurable at low-background rate (in our case the ^{255}Md α-decay occurred during a beam pause of 14.5 ms). Secondly, the agreement of the measured α energies with literature data is much more significant for monoenergetic decays, simple decay patterns, and high-energy resolution of the detector. The widely spread α energies of the ^{255}No decay represent an opposite example.

The decay data of ^{271}Ds and its daughter products were confirmed in an experiment at RIKEN, where the same reaction, ^{64}Ni + ^{208}Pb $\rightarrow$ ^{272}Ds*, was studied and a total of 14 decay chains was measured [38]. The α energy agrees well with the SHIP value of 10.74 MeV and also for the lifetime a long-lived (three events) and a short-lived (11 events) component was measured. Mean values obtained from all five long-lived and 22 short-lived decays are τ=100 and 2.35 ms, corresponding to half-lives of $(69\,^{+56}_{-21})$ and $(1.63\,^{+0.44}_{-0.28})$ ms, respectively. Note that the uncertainties decrease considerably with increasing number of events. A proper application of the Poisson statistics for the determining of uncertainties at low statistics is given in [70].

Obviously, isomeric states are responsible for the two different half-lives. Because the measured α energies are almost identical, an explanation suggested in [38] seems very likely: The isomeric state has the longer half-life and decays dominantly by γ-ray or conversion-electron emission into the shorter-lived ground state. Both states are populated in the reaction, but because the lifetime is measured as interval between implantation and α decay, we observe two different values for one and the same α transition. Theoretically, low-spin and high-spin levels ($1/2^+$ to $13/2^-$) which could result in isomeric states close to the ground state were predicted by Cwiok et al. [71]. The isomeric ratio between population of the 69 and 1.6 ms states is 5/22. Therefore, we may further conclude that the relatively long-lived isomeric state has a low-spin value and is less populated. However, this spin dependence of the production cross-section known from lighter nuclei may be changed for heavy systems due to reduced fission probability of the compound nucleus at high spin. Also the possibility that both levels decay by α emission with almost the same α energy cannot be completely excluded. An indication could be the slightly lower α energy for the longer-lived decay in the case of chain number 5 (see Fig. 5), which was measured with high precision.

Further confirmation of the production of ^{271}Ds in the reaction ^{64}Ni$+^{208}$Pb $\rightarrow$ ^{272}Ds* was reported from experiments performed at the Berkeley gas-filled separator (BGS) [72, 73]. At a beam energy of 309.2, 311.5, and 314.3 MeV at half-target thickness, a total of 2, 5, and 2 decay chains were measured, respectively. Cross-sections of 8.3, 20, and 7.7 pb were deduced. The decay chains are in excellent agreement with the previously measured data [37, 38]. Position and shape of the excitation functions agree within experimental uncertainties.

We conclude that the maximum deviation of beam energy measured at the LBNL 88-inch Cyclotron and the UNILAC is ±2 MeV. A similar deviation is deduced from a comparison of the data measured at RIKEN [38] and at the UNILAC. Possibilities to improve the accuracy of beam-energy measurements were discussed before.

Two more isotopes of darmstadtium have been reported in the literature. The first one is ^{267}Ds, produced at LBNL in the irradiation of ^{209}Bi with ^{59}Co [74]. The second isotope is ^{273}Ds, reported to be observed at JINR in the irradiation of ^{244}Pu with ^{34}S after the evaporation of five neutrons [75]. Both observations need further experimental clarification.

The even–even nucleus ^{270}Ds was synthesized using the reaction ^{64}Ni$+$ ^{207}Pb [76]. A total of eight α decay chains was measured during an irradiation time of 7 days. Decay data were obtained for the ground state and a high-spin K isomer, for which calculations predict spin and parity 9^-, 10^- or 8^+ [77]. The relevant single particle Nilsson levels are $\nu[613]_{7/2+}$ and $\nu[615]_{9/2+}$ below the Fermi level and $\nu[725]_{11/2-}$ above the Fermi level. Configuration and calculated energy of the excited states are $\{\nu[613]_{7/2+} \ \nu[725]_{11/2-}\}_{9-}$ at 1.31 MeV, $\{\nu[615]_{9/2+} \ \nu[725]_{11/2-}\}_{10-}$ at 1.34 MeV, and $\{\nu[613]_{7/2+} \ \nu[615]_{9/2+}\}_{8+}$ at 1.58 MeV.

The new nuclei ^{266}Hs and ^{262}Sg were identified as daughter products of α decay of ^{270}Ds. Spontaneous fission of ^{262}Sg terminates the decay chain. A proposed partial decay scheme of ^{270}Ds is shown in Fig. 6.

Roentgenium, element 111, was synthesized in 1994 using the reaction ^{64}Ni$+$ ^{209}Bi $\rightarrow$ ^{273}Rg*. A total of three α chains of the isotope ^{272}Rg were observed [48]. Another three decay chains were measured in a confirmation experiment in 2000 [49].

The GSI data on ^{272}Rg were confirmed in a 50-days irradiation at RIKEN performed in the period from February 12 to May 12, 2003 [39]. A total of 14 α decay chains were measured. In eight cases the α decays were followed down to the α decay of ^{256}Lr and in three cases down to the α decay of ^{260}Db. In one case, the chain terminated by SF after population of ^{260}Db, and in two

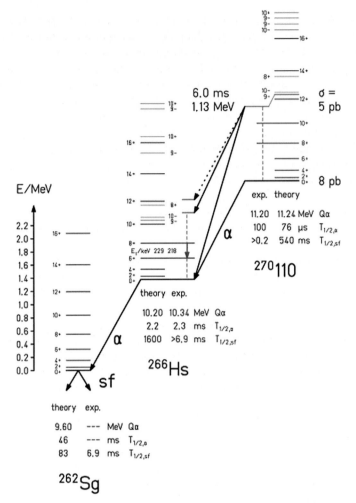

Fig. 6. Assignment of measured α and γ decay and SF data observed in the reaction $^{64}\mathrm{Ni} + {}^{207}\mathrm{Pb} \rightarrow {}^{271}\mathrm{Ds}^{*}$. The data were assigned to the ground-state decays of the new isotopes $^{270}\mathrm{Ds}$, $^{266}\mathrm{Hs}$, and $^{262}\mathrm{Sg}$ and to a high-spin K isomer in $^{270}\mathrm{Ds}$. *Arrows* in bold represent measured α-and γ-rays and SF. The data of the proposed partial level schemes are taken from theoretical studies of Muntian et al. [78] for the rotational levels, of Cwiok et al. [77] for the K isomers and of Smolanczuk [79] and Smolanczuk et al. [80] for the α energies and SF half-life of $^{262}\mathrm{Sg}$, respectively. For a detailed discussion see [76] (See also Plate 18 in the Color Plate Section)

cases by SF after population of $^{264}\mathrm{Bh}$. The resulting SF branching ratios are 9.6 and 15%, respectively. It remains open, whether the two nuclei themselves decay by SF or whether the known spontaneously fissioning even–even nuclei $^{260}\mathrm{Rf}$ and $^{264}\mathrm{Sg}$ are populated by EC decay (see Fig. 4). However, considering the SF half-lives of about 10–100 ms of even–even nuclei in this region, a SF hindrance factor for odd–odd nuclei of about 10^6 [81], and a calculated partial EC half-life of about 10–100 s (see Sect. 4), it seems more likely that ^{260}Db and ^{264}Bh have an EC branching. From the total half-lives of 1.5 and 1.0 s and the measured branching ratios [39] follow reasonable partial EC half-lives of 15 and 10 s, respectively. A purely experimental determination of a possible EC or SF branch of ^{260}Db and ^{264}Bh decay is difficult because the SF half-lives of the daughters ^{260}Rf and ^{264}Sg are short, i.e., 20 and 68 ms, respectively. This means that the measured time difference between α decay of ^{264}Bh and SF of ^{260}Rf is not prolonged significantly by the SF lifetime of ^{260}Rf in the case of an (unobservable) EC decay of ^{260}Db, compared to the time difference between α decay of ^{264}Bh and SF of ^{260}Db. The same argument holds for the chain ^{268}Mt – ^{264}Bh – ^{264}Sg.

Further confirmation of the decay pattern of ^{272}Rg was achieved in an experiment at LBNL [73]. However, for the synthesis the reaction ^{65}Cu + ^{208}Pb → ^{273}Rg* was used.

Element 112 was investigated at SHIP using the reaction ^{70}Zn + ^{208}Pb → ^{278}112* [59]. The irradiation was performed in 1996. Over a period of 24 days, a total of 3.4×10^{18} projectiles were collected. One α-decay chain, shown as the first one from the left in Fig. 7, was observed resulting in a cross-section of 0.5 pb. The chain was assigned to the 1n channel. The experiment was repeated in 2000 with the aim of confirming the synthesis of ^{277}112 [49]. During a similar measuring time, but using slightly higher beam energy, one more decay chain was observed, which is also shown in Fig. 7. The two experiments yield agreement concerning the decay pattern of the first four α decays of the ^{277}112 decay chain.

A new result was the occurrence of fission which terminated the second decay chain at ^{261}Rf. A SF branch of this nucleus was not known, however, it was expected from theoretical calculations. The new results on ^{261}Rf were proven in a recent chemistry experiment [60, 61], in which this isotope was measured as granddaughter in the decay chain of ^{269}Hs and SF of ^{261}Rf was also observed (see Sect. 3.2).

A reanalysis of all relevant results obtained at SHIP since 1994, including a total of 34 decay chains analyzed, revealed that the previously published first decay chain of ^{277}112 [59] (not shown in Fig. 7) and the second one of the originally published four chains of ^{269}Ds [58] represented spurious events. Details of the results of the reanalysis are given in [49].

In 2004, the ^{70}Zn + ^{208}Pb → ^{278}112* irradiation was repeated at RIKEN [40]. Using a beam energy comparable to that used at SHIP, two decay chains were measured, which fully confirmed the SHIP data. For comparison, all four decay chains are shown in Fig. 7. The two chains observed at RIKEN also verified the SF branch of ^{261}Rf. In both of these chains, the α energy of ^{265}Sg produced in the decay chain of ^{277}112 was measured for the first time and is now available for comparison with the value measured in [61]. Also in this case agreement was observed giving further support to the assignments made in [49, 59].

So far, element 113 is the heaviest one produced by a cold-fusion reaction, namely in $^{70}\text{Zn} + {}^{209}\text{Bi} \rightarrow {}^{279}113^*$. A first attempt to synthesize element 113 was made at the GSI-SHIP in 1998, using a net irradiation time of 46 days and a beam dose of 7.5×10^{18} [19]. The experiment was continued in 2003, reaching a net irradiation time of 36 days and the beam dose of 7.4×10^{18} [82]. This experiment was running in parallel to the search for element 113 at RIKEN. Combining both parts of the SHIP experiment, an upper one-event cross-section limit of 160 fb was deduced. The "one-event limit" corresponds to a cross-section in the case that one event would have been observed, statistical fluctuations are not considered. For an estimate of observation limits in the case of negative results including statistical fluctuations, see [70].

The search experiment for element 113 at RIKEN was performed in 2003 and 2004, the net irradiation time being 79 days and the beam dose 17×10^{18} [41]. During this experiment one decay chain was observed and assigned to the isotope $^{278}113$. The chain terminated after four subsequent α decays by SF of the known isotope ^{262}Db. Also known was the last α emitter ^{266}Bh. New were the chain members ^{270}Mt and ^{274}Rg. The measured cross-section of (55^{+150}_{-45}) fb is in agreement with the limit obtained at SHIP. Despite comparable beam doses, the three times lower cross-section value reached at RIKEN is due to the higher efficiency, 80% instead of 50% at SHIP, and the use of thicker targets in the RIKEN experiment.

The $^{70}\text{Zn} + {}^{209}\text{Bi}$ irradiation was continued at RIKEN between January 2005 and May 2006, with several intermissions. The net irradiation time of this experiment was 161 days and the total beam dose amounted to 44.5×10^{18}.

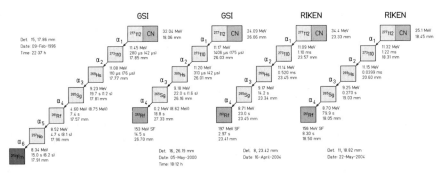

Fig. 7. Decay chains measured in the cold-fusion reaction $^{70}\text{Zn} + {}^{208}\text{Pb} \rightarrow {}^{278}112^*$. In the *left part*, the two chains are shown which were measured in 1996 and 2000 at SHIP [49, 59], in the *right part* those measured in 2004 at RIKEN [40]. The chains were assigned to the isotope $^{277}112$ produced by evaporation of one neutron from the compound nucleus. The lifetimes given in brackets were calculated using the measured α energies. In the case of escaped α particles the α energies, given in brackets, were determined using the measured lifetimes (See also Plate 19 in the Color Plate Section)

During this period, a second decay chain was measured, which confirmed the previous decay chain and its assignment to $^{278}113$ [42]. The cross-section determined from the RIKEN experiments is (31^{+40}_{-20}) fb. This value is the lowest one ever measured for a heavy-ion fusion reaction.

Cold fusion was also applied to search for elements 116 and 118. In both cases ^{208}Pb targets were irradiated. The beams were ^{82}Se and ^{86}Kr, respectively. The 116 experiment was performed at SHIP in order to search for the radiative capture (0n) channel. At five different excitation energies between 0 and 11 MeV, cross-section limits of about 5 pb were reached [19].

Subsequent to reports on positive results of the synthesis of element 118 in 1999 [83], confirmation experiments were performed. However, only cross-section limits of about 1 pb were reached at various laboratories [19, 84, 85]. Eventually, the first announcement was retracted in 2001 [86] after additional experiments and after a re-analysis of the data of the first experiment.

3.2 Elements Produced in Hot Fusion Reactions

Superheavy-element studies based on hot fusion reactions involve targets made from actinide elements. A number of differences exist compared with reactions using lead or bismuth targets. Probably the most significant one is the excitation energy of the compound nucleus at the lowest beam energies necessary to initiate a fusion reaction. Values are of the order of 10–20 MeV in reactions with lead targets and 35–45 MeV in reactions with actinide targets, which led to the widely used terminology of 'cold' and 'hot' fusion reactions. Due to the lack of targets between bismuth and thorium, a gradual change from cold to hot fusion cannot be studied experimentally. A second important difference of actinide-target based reactions is the synthesis of more neutron-rich isotopes compared with a cold-fusion reaction leading to the same element, e.g., ^{270}Hs from a ^{248}Cm target and ^{265}Hs from a ^{208}Pb target using beams of ^{26}Mg and ^{58}Fe, respectively.

Actinides served already as targets when neutron capture and subsequent β^- decay were used for the first synthesis of transuranium elements. Later, up to the synthesis of seaborgium [68], actinides were irradiated with light-ion beams from accelerators. Even later cold-fusion reactions were used with lead and bismuth targets, which resulted in higher yield for the synthesis of heavy nuclei with proton number greater than 106.

The argumentation changed again when elements 110–112 had been discovered in cold-fusion reactions and continuously decreasing cross-sections were measured. The combination of actinide targets with beams as heavy as ^{48}Ca became promising to study more neutron-rich isotopes, which are closer to the region of spherical SHEs and for which also longer half-lives were expected. In addition the lowest excitation energies of compound nuclei from fusion with actinide targets are obtained with beams of ^{48}Ca.

The experimental difficulty with the use of a ^{48}Ca beam is the low natural abundance of only 0.19% of this isotope, which makes enrichment very expensive. Therefore, the development of an intense ^{48}Ca beam at low consumption of material in the ion source and high transmission through the accelerator was the aim of the work accomplished at the FLNR during a period of about 2 years until 1998 [87]. Till now (2008), this ^{48}Ca beam is successfully used for irradiation of various actinide targets, which aim at the the synthesis of SHEs up to Z=118. All results of these Dubna experiments are described and discussed in a recent review article [88].

The experiments at the Dubna U400 cyclotron were performed at two different recoil separators, which had been built during the 1980s. The separators had been upgraded in order to improve the background suppression and detector efficiency. The energy-dispersive electrostatic separator VASSILISSA was equipped with an additional deflection magnet [89, 90]. The Dubna gas-filled recoil separator (DGFRS) was tuned for the use of very asymmetric reactions with emphasis on the irradiation of highly radioactive targets [91]. A specific characteristic of the DGFRS is the hydrogen gas used in the separator, which enables better suppression of projectile- and target-like recoils at the focal plane than the filling with helium gas [88]. Both separators are equipped with time-of-flight detectors and with an array of position-sensitive silicon detectors in an arrangement similar to the one shown in Figs. 1 and 2.

At VASSILISSA attempts were undertaken to search for new isotopes of element 112 by irradiation of ^{238}U with ^{48}Ca ions in 1998 [92]. Two SF events were measured resulting in the relatively high cross-section of 5.0 pb. The two events were tentatively assigned to the residue ^{283}112 after 3n evaporation. The measured half-life was $\left(81\,^{+147}_{-32}\right)$ s.

The experiments were continued in 1999. The reaction ^{48}Ca + ^{242}Pu $\rightarrow$ ^{290}114* was investigated [93]. It was expected that, after evaporation of three neutrons, the nucleus ^{287}114 would be produced and would decay by α emission into the previously investigated ^{283}112. Over a period of 21 days, a total of four SF events were detected. Two of them could be assigned to short-lived fission isomers. The other two fission signals were preceded by signals from α particles (one was an escape event with an α energy of 2.31 MeV) and implantations. A cross-section of 2.5 pb was obtained for the two events. They were assigned to the nuclide ^{287}114. The four SF events preceded by α decay, observed when irradiating ^{238}U and ^{242}Pu with ^{48}Ca and interpreted as decay of an isotope of element 112 and 114, respectively, are consistent with each other. The fission lifetimes are within the limits given by statistical fluctuations. Fission was measured again after α decay when the target was changed from ^{238}U to ^{242}Pu. The low-background rate in the focal plane of VASSILISSA makes chance coincidences unlikely. However, further investigation was needed for an unambiguous assignment.

At the DGFRS a search for element 114 was started in 1998. The experiments were performed in collaboration between the FLNR and the Lawrence Livermore National Laboratory (LLNL). A ^{244}Pu target was irradiated with a ^{48}Ca beam for a period of 34 days. One decay chain was extracted from the data. The chain was claimed to be a candidate for the decay of ^{289}114. The measured cross-section was 1 pb [94].

The ^{48}Ca + ^{244}Pu experiment was repeated in 1999. During a period of 3.5 months, two more α decay sequences terminating in SF, were observed [95]. The two chains were identical within statistical fluctuations and detector-energy resolution, but differed from the first chain measured in 1998. The two new events were assigned to the decay of ^{288}114, the 4n evaporation channel. The cross-section was 0.5 pb.

An investigation of element 116 was started in 2000. Using a ^{248}Cm target, the previously detected isotopes ^{289}114 or ^{288}114 were expected to be observed as daughter products from the decay of the corresponding parent isotope of element 116 produced after evaporation of three or four neutrons. The first decay chain which was assigned to ^{292}116 was measured after an irradiation of 35 days [96]. The irradiation was continued later, and two more decay chains were measured in 2001 [97]. The cross-section was deduced to be about 0.6 pb from a total beam dose of 22.5×10^{18}.

The newly measured chains are of high significance. The data reveal internal redundancy and the lifetimes are relatively short, making an origin by chance events extremely unlikely. In particular, all further decays in the chain following observation of a parent decay, were measured during a beam free period. This was achieved by switching off the beam, using as a trigger the time-of-flight and energy signals from the implantation and the α decay from the parent. The assignment to the 4n channel was likely, but remained subject to further investigation until an unambiguous identification would become possible. As the chains end at ^{280}Ds by SF, generic relations to known nuclei cannot be used. Other possible procedures which could help to establish a unique assignment, could be measurements of excitation functions, further cross bombardments, direct mass measurements and chemical analysis of parent or daughter isotopes. Also a systematic investigation of the nuclei in the gap between those studied with cold fusion and those measured in Dubna using hot fusion would be useful.

Several of the confirmation studies suggested before were performed during the following years. However, before starting this enormous work, e.g., the measurement of an excitation function means about five to ten times more irradiation time than the synthesis of a few atoms at one energy, an attempt was undertaken to search for element 118. The reaction ^{48}Ca + ^{249}Cf $\rightarrow$ ^{297}118* was studied in 2002 [98, 99]. Two events were measured, one involving a time correlation between implantation and SF and the other one characterized by a correlation between implantation of the evaporation residue, two α decays and SF (ER–α1–α2–SF), the α energies being 11.65 and 10.71 MeV, respectively. The latter α energy had a relatively large uncertainty of 0.17 MeV because the energy had to be determined from the sum of signals in the stop and box detector. The two events were assigned to the even–even nucleus ^{294}118.

In order to confirm this assignment, the daughter isotope ^{290}116 was produced directly in the reaction ^{48}Ca + ^{245}Cm $\rightarrow$ ^{293}116* in 2003 [100]. Only one beam energy was used. Events with α energies of 10.74 and 10.88 MeV were measured and assigned to the decay of ^{291}116 (2n channel) and ^{290}116 (3n channel), respectively.

In the same year, three more reactions were studied and a fourth one was started, which was continued in 2004. Moreover, an attempt was made to study excitation functions for a ^{244}Pu [100] and a ^{242}Pu [101] target using projectiles of three and four different energies, respectively.

The measurement of the excitation function with the ^{244}Pu target allowed for a correction of the previously made xn assignments and a consistent interpretation was given of all so far observed decay chains from isotopes of the elements 114, 116, and 118 [100]. These corrected assignments are shown in Fig. 4. The results from the irradiation of ^{242}Pu are especially important because the isotopes ^{288}114 and ^{287}114, produced in 4n and 5n channels with a ^{244}Pu target, were now observed at lower excitation energy also in 2n and 3n channels, respectively [100, 101]. Finally, in the 4n channel with the ^{242}Pu target, the granddaughter ^{286}114 from the decay of ^{294}118 was also produced.

As discussed in Sect. 3.1, the RIKEN group started an attempt to produce element 113 using the cold fusion reaction ^{70}Zn + ^{209}Bi → ^{279}113* in 2003. From a previous SHIP experiment the low cross-section limit of 600 fb [19] was known at that time. Nevertheless, taking into account the long beam times available at RIKEN, this endeavor seemed to be justified.

Concerning production of element 113 using hot-fusion reactions, cross-section estimates were made on the basis of the great number of reactions with ^{48}Ca beams already studied in Dubna. Reasonable estimates resulted in values of about 1 pb. A possible target for the synthesis of element 113 would be ^{237}Np. However, the extrapolation of experimental data as well as cross-section calculations [102, 103] indicated that production of element 115 using a ^{243}Am target could have even a higher yield. Such an experiment was performed in Dubna in 2003 [104, 105]. At two beam energies of 248 and 253 MeV (values calculated for projectiles at half of the target thickness) three and one, respectively, α decay chains were measured consisting of five subsequent α particles (in one case the last α particle was probably missed). The chains were assigned to the isotopes ^{288}115 and ^{287}115 of the new element 115. The isotopes ^{284}113 and ^{283}113 of the also new element 113 were thus produced as daughter products. The chains ended by SF after α decay of the new isotopes ^{280}Rg, ^{276}Mt, ^{272}Bh and ^{279}Rg, ^{275}Mt, respectively, was detected.

The interesting question which arises actually is the fissioning nucleus at the end of the chain (see also [104]). Taking into account the theoretical and systematic Q values for α and EC decay, the corresponding half-lives as well as calculated SF half-lives and odd particle hindrance factors, it seems most likely that the odd–odd isotope ^{268}Db populated by the measured α decay of ^{272}Bh decays by EC with the measured half-life of 16 h to ^{268}Rf which decays by SF with a short half-life of 1.4 s. The latter value is the theoretical one taken from [80]. In the case of the single decay chain of ^{287}115 it was concluded that the α decay of ^{271}Bh was most likely missed for technical reasons and that the daughter ^{267}Db decays by SF.

As mentioned in Sect. 3.1, the first decay chain assigned to ^{278}113 at RIKEN was measured in 2004. The period, when an experiment was performed and when events like decay chains are measured, are not necessarily decisive for assigning priority of discovery. However, the knowledge of these dates is certainly interesting for historical reasons. Decisive for assigning priority of discovery is the date when the publication on the results of an experiment or a theoretical study was received by the editor of a journal, preferably a scientific journal publishing refereed articles.

In Dubna, the series of experiments using the ^{48}Ca beam was continued in 2003 and 2004. The reaction with a ^{238}U target was studied with the intention to produce ^{282}112 (4n channel) which is the grand-granddaughter of ^{294}118. In addition, previous data measured at VASSILISSA [92, 93] for the isotope ^{283}112 and the consistency of the decay data should be checked, when this isotope is produced as daughter in the α decay of ^{287}114. The result was α emission of ^{283}112 with an energy of 9.54 MeV and a half-life of 4.0 s [101]. The daughter isotope ^{279}Ds decays by SF with $T_{1/2}$=0.18 s. Later, in 2005–2007, the study of ^{283}112 moved again into the center of interest, when attempts were made to produce this nucleus in independent experiments.

In 2004, a ^{233}U target was irradiated in an attempt to synthesize ^{277}112 through the 4n channel. This was the isotope previously studied in cold-fusion reactions with a ^{208}Pb target [49, 59]. In the case of 3n evaporation it was expected that the known nuclei ^{270}Hs and ^{266}Sg would be produced in the decay chain of ^{278}112. The irradiation was performed at an excitation energy of 34.9 MeV, however, only an upper cross-section limit of 0.6 pb was achieved [101].

Up to summer 2004 there was still no news about a positive result from the ongoing search for element 113 at RIKEN. However, for confirmation purposes of previous results, an experiment was prepared in Dubna to separate the long-lived isotope ^{268}Db $[T_{1/2} = (16^{+19}_{-6})\,\mathrm{h}]$, which terminates the decay chain of ^{288}115 by SF, by chemical means and to measure its decay. The exact dates of this experiment as given in the publication were from June 11 to 22, 2004. A total of 15 SF events were detected [105–107]. A half-life of $(32^{+11}_{-7})\,\mathrm{h}$ was measured in agreement with the previous data obtained at the DGFRS. The astonishingly high yield of 15 SF events in the chemistry experiment compared to three events measured at the DGFRS at a comparable beam dose, is due to the fact that thicker targets could be used, 1.2 instead of 0.36 mg/cm^2, and that the efficiency was 80% in the chemistry experiment instead of 35% at the DGFRS.

By 2005, the amount of data on cross-sections as well as decay properties of nuclei up to element 116 had increased considerably. This wide basis of new data for nuclei which would be members of decay chains from element 118, suggested a repetition of the ^{48}Ca + ^{249}Cf → ^{297}118* irradiation first performed in 2002 [98, 99]. The corresponding experiment was carried out in

2005, yielding two more chains of the type ER–α1– α2–SF and ER–α1–α2–α3–SF [108]. The first of the new chains agreed with that measured in the 2002 experiment whereas the second one was different: The granddaughter, $^{286}114$, decayed by α emission and not by SF. However, in the 2003 ^{48}Ca + ^{245}Cm → $^{293}116^*$ experiment [100], two of three measured chains were assigned only tentatively to $^{290}116$ (the daughter of $^{294}118$) under the assumption that the first α particles populating $^{286}114$ were missed. Although the chains and α/SF ratios could be reasonably well explained, a more precise determination of the α/SF ratio of $^{286}114$ was desirable. A second SF decay of $^{294}118$ itself, the first one being measured in the 2002 experiment, was not observed.

Therefore, in 2005 the ^{48}Ca + ^{245}Cm → $^{293}116^*$ irradiation was repeated in order to study carefully the yield and the decay chains of element 116 isotopes expected as α-decay daughters of isotopes of element 118. In this experiment nine decay chains of $^{290}116$ were measured, which fully confirms the previous assignment of the chains measured in the 118 experiment [98, 99] to the isotope $^{294}118$. Also branching ratios for α and SF decay of $^{286}114$ were established.

A remarkably long decay chain was measured in the ^{48}Ca + ^{245}Cm experiment at an excitation energy of 37.9 MeV [108]. The chain consists of six consecutive α decays and terminated by a SF. The total decay time of all nuclei in this chain was about 0.4 h. This sequence of decays was assigned to the parent isotope $^{291}116$ produced via the 2n evaporation channel. The decay properties of $^{291}116$ were determined in a previous experiment [100]. In addition, the daughter isotope, $^{287}114$, was observed in two reactions, ^{48}Ca + ^{242}Pu → $^{287}114 + 3n$ and ^{48}Ca + ^{244}Pu → $^{287}114 + 5n$, and, finally, the granddaughter isotope, $^{283}112$, was produced in the ^{48}Ca + ^{238}U → $^{283}112 + 3n$ reaction [100, 101].

The decay chains of the isotopes $^{291}116$, $^{287}114$, and $^{283}112$ usually end in SF of ^{279}Ds ($T_{1/2}=0.2$ s). However, in three cases out of 26 observed, ^{279}Ds underwent α decay ($b_\alpha=10\%$), which was followed by α decay of ^{275}Hs and terminated in one case by SF of ^{271}Sg ($T_{1/2}=1.9$ min) and in two other cases by another α decay and SF of ^{267}Rf ($T_{1/2}=1.3$ h) [88, 101]. The long decay chain of the even–odd nucleus $^{291}116$ is an interesting case of a transition from the region of heaviest nuclei ($^{291}116$ and $^{287}114$) which are stabilized by the influence of a spherical shell closure at $Z=114$ and $N=184$, to isotopes (^{271}Sg or ^{267}Rf) which are located in the region of deformed nuclei owing their stability to single particle energy gaps at $Z=108$ and $N=162$.

Recently, element 113 which was first observed as daughter after α decay of element 115 [104], was also produced directly in a hot fusion–evaporation reaction. In this case the target ^{237}Np was irradiated for a period of 32 days in 2006 [88, 109]. Two decay chains were measured, which were assigned to the new isotope $^{282}113$. Subsequent to the α decay of $^{282}113$, three α emissions were observed and assigned to the new isotopes ^{278}Rg, ^{274}Mt, and ^{270}Bh, respectively. In one case SF was detected after the α decay of ^{270}Bh, in the other case the chain could not be measured beyond the α decay of ^{274}Mt.

Like in the case of the decay chains of ^{287}115 and ^{288}115 the most probable ending of the new chains of ^{282}113 was discussed [109]. The observed SF event of the first chain was assigned to the even–even isotope ^{266}Rf populated in EC of ^{266}Db which was produced in the measured α decay of ^{270}Bh. However, a SF branching of ^{266}Db itself is not completely excluded. In agreement with the non-observation of SF in the second chain, α decay of ^{266}Db populating the known isotope ^{262}Lr would be possible. However, this disintegration was not observed. A possible α decay branch of ^{262}Lr would lead to the also known long-lived isotope ^{258}Md ($T_{1/2}$=51.5 d), which would explain the non-observation of SF in this chain. If this explanation would be confirmed this chain would be the first candidate for connecting the new Dubna chains to known and well-established nuclei.

Among the many different reactions studied in Dubna, the technically least problematic is the irradiation of a ^{238}U target. The reason is the relatively low radioactivity of this material. In addition, in reactions with ^{48}Ca a relatively high maximum cross-section of $(2.5^{+1.8}_{-1.1})$ pb was measured for production of the isotope ^{283}112 in a 3n evaporation channel [101]. However, despite the high level of experimental standards in heavy-ion laboratories worldwide, only negative or ambiguous results were obtained in several repetition experiments [110–114]. Most reliable were the negative results obtained in two experiments performed at the gas-filled separator BGS of LBNL in 2002 and 2004 [110, 112]. The irradiations were performed at two different beam energies resulting in excitation energies of 31.9 and 36.3 MeV. One event upper cross-section limits of 0.80 and 0.96 pb were obtained, respectively.

Negative results were also obtained in a chemistry experiment performed at GSI [113]. Under the assumption that element 112 behaves like mercury, a cross-section limit of 2.3 pb at 95.45% confidence was deduced. In the case of radon-like behavior, the limit was at 2.7 pb.

Successful, however, was a chemistry experiment performed at the cyclotron U400 in Dubna in 2006. In this experiment the reaction ^{48}Ca + ^{242}Pu $\rightarrow$ ^{287}114 +3n was used to study the adsorption properties of the relatively long-lived daughter isotope ^{283}112 on a cooled, gold covered detector surface [114, 115]. Two events were observed, which had decay properties in agreement with the results obtained at the DGFRS. By extracting the activity from a stopping chamber, transporting it through a capillary, and determining its adsorption enthalpy it was concluded that the α decaying isotopes belong to element 112. The reason is that lighter elements from Group 11 down to Group 3 in the periodic table (Fig. 3) are expected to be not sufficiently volatile for being transported through the capillary to the cooled detector. This experiment represents the first independent confirmation of results on SHEs that have been obtained at the DGFRS by physical means.

Recently, Dubna data measured in the reaction ^{48}Ca + ^{238}U $\rightarrow$ ^{283}112 + 3n were confirmed independently at the velocity filter SHIP [116]. The experiment was performed in three parts in 2005, 2006, and 2007. The reaction was studied at three different beam energies resulting in excitation energies of 37.2,

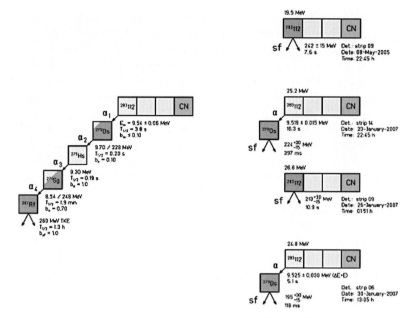

Fig. 8. Events observed for the reaction ^{48}Ca + ^{238}U → ^{286}112* at SHIP (*right panel*) [116] and at the DGFRS (*left panel*) [101]. The latter data represent mean values obtained from a total of 22 decay chains measured at the DGFRS including those produced by α decay of heavier elements [88]. The two implantation–α–SF events from the SHIP work completely agree with the data assigned to the decay of ^{283}112 in [101]. A new result obtained at SHIP was the observation of two implantation–SF events which were assigned to a 50% SF branch of ^{283}112 (See also Plate 20 in the Color Plate Section)

34.6 (mean value of 34.8 and 34.4.), and 32.3 MeV. At 34.6 MeV two implantation–SF events and two implantation–α–SF events were measured. The events are plotted in Fig. 8 together with the mean values of data obtained at Dubna from a total of 22 decay chains including those produced by α decay of heavier elements [88]. No events were measured at the other two beam energies. The two implantation–α–SF events are especially important for comparison with each other and with results from the Dubna experiments, because the α energy is characteristic for the particular isotope and can be determined with high accuracy. The two events agreed with each other within statistical uncertainties and detector resolution. The weighted mean value of the α energy is (9.520 ± 0.015) MeV. Half-lives of $(7.4^{+13.5}_{-2.9})$ s and $(0.18^{+0.32}_{-0.07})$ s were obtained for ^{283}112 and its spontaneously fissioning daughter ^{279}Ds, respectively. The corresponding data from the Dubna experiments are (9.54 ± 0.06) MeV, and $(3.8^{+1.2}_{-0.7})$ s and $(0.17^{+0.81}_{-0.08})$ s, respectively.

A new result was the observation of the two implantation–SF events with a half-life of $(6.4^{+11.6}_{-2.5})$ s. They were assigned to a 50% SF branch of ^{283}112.

From all four events a mean half-life of $(6.9^{+6.9}_{-2.3})$ s was obtained. The relatively long SF half-lives measured earlier in the reaction ^{48}Ca + ^{238}U [92, 111, 117] were not observed in the SHIP experiment.

Particular importance of the recent SHIP experiment lies in the fact that its separation properties differ from that of the DGFRS. For example, nuclei produced by α evaporation or pre-compound α emission are strongly suppressed in velocity filters. In the case of reactions with actinide targets an α particle emitted from a compound nucleus at a Coulomb barrier energy of 24 MeV would result in a change of the velocity of the evaporation residue by $\pm 9\%$ in-beam direction or of the deflection angle by $\pm 5°$ if the α particle is emitted perpendicular to the beam direction. As SHIP has a velocity window of $\pm 5\%$ and an acceptance angle of $\pm 2°$, reaction products including α evaporation are strongly suppressed. Therefore, the observation of products from the reaction ^{48}Ca + ^{238}U at SHIP supports the previous assignment to a neutron evaporation channel. An independent determination of the isotopic assignment cannot be given by the SHIP results alone. However, the systematic measurement of excitations functions at the DGFRS revealed two spontaneously fissioning nuclei with short half-lives of 0.82 and 97 ms, which were assigned to the even–even nuclei ^{282}112 and ^{284}112, respectively [88]. Therefore, the assignment of the four long-lived events measured at SHIP to the 3n channel and thus to the isotope ^{283}112 is reasonable and fully in agreement with the results obtained in Dubna.

How well chemical properties can be used for the separation and identification of even single atoms was recently demonstrated in an experimental study of hassium [60, 61]. Using the hot-fusion reaction ^{26}Mg + ^{248}Cm $\rightarrow$ ^{274}Hs*, the isotope ^{269}Hs was produced after evaporation of five neutrons. On the basis of three decay chains measured, it was concluded that ^{269}Hs atoms react with oxygen to form the volatile compound HsO_4. Thus it was proven independently by chemical means that the produced atom belongs, like osmium which also forms a volatile tetroxide, to group 8 and thus to element 108 in the periodic table of elements (Fig. 3). The measured decay properties of the chemically separated atoms fully confirms the data obtained for ^{269}Hs from the decay chain of ^{277}112 [40, 49].

Hot-fusion reactions applied to synthesize long-lived nuclides of elements 104 through 108 and 112 for chemical studies are summarized in Table 1. Cross-sections vary from about 10 nb to a few pb [50, 51, 60, 118–120]. With typical beam intensities of 3×10^{12} atoms/s on targets of about 0.8 mg/cm^2 thickness, production yields range from a few atoms/min for rutherfordium and dubnium isotopes to five atoms/h for ^{265}Sg and even less for ^{267}Bh and heavier nuclides. Therefore, all chemical separations are performed with single atoms on an 'atom-at-a-time' scale. Similar to the experiments with recoil separators, characteristic α decays and time correlated α–α decay chains are used after chemical separation to identify these nuclides in specific fractions or at characteristic depositions, i.e., surfaces of special properties and tem-

Table 1. Nuclides from hot-fusion reactions used in chemical investigations. Part of the data was taken from [43]

Nuclide	$T_{1/2}$ / s	Beam	Target	Channel	Cross-section / pb	Yield
^{261m}Rf	78	^{18}O	^{248}Cm	5n	$\approx$10,000	2 min^{-1}
		^{22}Ne	^{244}Pu	5n	4,000	1 min^{-1}
^{262}Db	34	^{18}O	^{249}Bk	5n	6,000	2 min^{-1}
		^{19}F	^{248}Cm	5n	1,000	0.5 min^{-1}
^{263}Db	27	^{18}O	^{249}Bk	4n	10,000	3 min^{-1}
^{265}Sg	7.4	^{22}Ne	^{248}Cm	5n	$\approx$240	5 h^{-1}
^{266}Sg	21	^{22}Ne	^{248}Cm	4n	$\approx$25	0.5 h^{-1}
^{267}Bh	17	^{22}Ne	^{249}Bk	5n	$\approx$70	1.5 h^{-1}
^{269}Hs	14	^{26}Mg	^{248}Cm	5n	$\approx$6	3 d^{-1}
^{270}Hs	2–7	^{26}Mg	^{248}Cm	4n	4	2 d^{-1}
^{283}112[a]	3.8	^{48}Ca	^{242}Pu	3n	3.6	0.7 d^{-1}

[a] The α decay of the parent nucleus ^{287}114 was not observed

perature. Overviews on research related to the chemical properties of SHEs are given in [121–123].

4 Nuclear Structure and Decay Properties

The calculation of the ground-state binding energy provides the basic step to determine the stability of SHEs. In macroscopic–microscopic models the binding energy is calculated as sum of a predominating macroscopic part (derived from the liquid-drop model of the atomic nucleus) and a microscopic part (derived from the nuclear shell model). In this way, more accurate values for the binding energy are obtained than in the cases of using only the liquid-drop model or the shell model. The shell correction energies of the ground state of nuclei near closed shells are negative, which results in further decreased values of the negative binding energy from the liquid-drop model – and thus increased stability. An experimental signature for the shell-correction energy is obtained by subtracting a calculated smooth macroscopic part from the measured total binding energy.

The shell-correction energy is plotted in Fig. 9a using data from [124]. Two equally deep minima are obtained, one at $Z=108$ and $N=162$ for deformed nuclei with deformation parameters $\beta_2 \approx 0.22$, $\beta_4 \approx -0.07$ and the other one at $Z=114$ and $N=184$ for spherical SHEs. Different results are obtained from self-consistent Hartree–Fock–Bogoliubov calculations and relativistic mean-field models [125–129]. They predict for spherical nuclei shells at $Z=114$, 120, or 126 (indicated as *dashed lines* in Fig. 9a) and $N=172$ or 184.

The knowledge of ground-state binding energies, however, is not sufficient for the calculation of partial SF half-lives. Here it is necessary to determine the size of the fission barrier over a wide range of deformation. The most

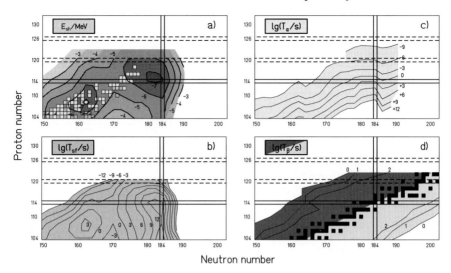

Fig. 9. Shell-correction energy (**a**) and partial half-lives for SF, α and β decay (**b**)–(**d**). The calculated values in (**a**)–(**c**) were taken from [80, 124] and in (**d**) from [131]. The *squares* in (**a**) mark the nuclei presently known, the *filled squares* in (**d**) indicate the β stable nuclei (See also Plate 21 in the Color Plate Section)

accurate data have been obtained for even–even nuclei using a macroscopic–microscopic model [80]. Partial SF half-lives are plotted in Fig. 9b. The landscape of fission half-lives reflects the landscape of shell-correction energies, because in the region of SHEs the height of the fission barrier is, firstly, mainly determined by the ground-state shell correction energy, while the contribution from the macroscopic liquid-drop part approaches zero for $Z=104$ and above, and, secondly, the shell correction energy at the saddle point is small [130]. Nevertheless, the SF half-life is predicted to significantly increase from 10^3 s for deformed nuclei to 10^{12} s for spherical SHEs. This difference originates from an increasing width of the fission barrier when going from deformed to spherical nuclei.

Partial α half-lives decrease almost monotonically from 10^{12} s down to 10^{-9} s near $Z=126$ (Fig. 9c). However, as pointed out in [131], non-smooth changes in Q_α values and associated half-lives can occur for some very proton-rich nuclei e.g., the calculated partial α half-life for $^{318}128$ is $10^{-8.15}$ s, whereas the half-life for $^{319}128$ is $>10^{20}$ s. These extreme differences are due to multiple minima with extremely different quadrupole deformation in the potential energy surface.

The valley of β-stable nuclei passes through $Z=114$, $N=184$. At a distance of about 20 neutrons away from the bottom of the valley, β half-lives of isotopes have dropped down to values of 1 s [131] (Fig. 9d).

Combining the results from the individual decay modes, one obtains the dominating partial half-life as shown in Fig. 10a for even–even nuclei. The

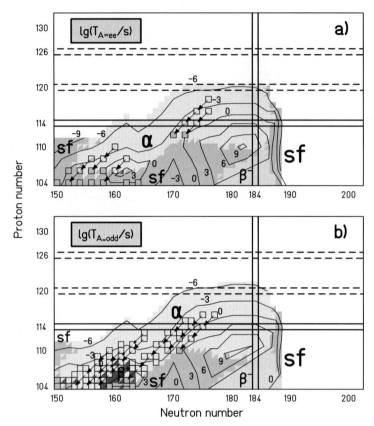

Fig. 10. Dominating partial half-lives for α decay , β^+ decay/EC, β^- decay, and SF: (**a**) for even–even nuclei; (**b**) for odd-A nuclei. Nuclei and decay chains known at present are marked in (**a**) and, in the latter case (**b**) also the known odd–odd nuclei are included (See also Plate 22 in the Color Plate Section)

two regions of deformed heavy nuclei near N=162 and spherical SHEs merge and form a region of α emitters surrounded by spontaneously fissioning nuclei. The longest half-lives are 1,000 s for deformed heavy nuclei and 30 years for spherical SHEs. It is interesting to note that the longest half-lives are not reached for the doubly magic nucleus 114, but for Z=110 and N=182. This is a result of continuously increasing Q_α values with increasing atomic number. Therefore, α decay becomes the dominant decay mode beyond darmstadtium with continuously decreasing half-lives. For nuclei at N=184 and $Z < 110$ half-lives are determined by β^- decay.

The four member α decay chain of ^{294}118, the heaviest even–even nucleus, observed in the recent experiment in Dubna [108], is also displayed in Fig. 10a. The *arrows* follow approximately the 1–10 ms contour line down to ^{282}112. This is in agreement with the experimental observation. The nucleus

^{282}112 is close to the region where theory predicts SF. The average values of the measured half-lives of the nuclei along the decay chain are 0.9–7.1–130–0.82 ms, respectively (see Sect. 3.2). Only the predicted half-life of ^{286}114 deviates from experiment by more than a factor of ten. However, this deviation is well within the accuracy limits of the calculation, e.g., a change of the α energy of ^{286}114 by 350 keV only changes the half-life by a factor of ten. Similar agreement with the half-life predictions exists for the decay chain of ^{292}116 [101].

The decay chain of ^{264}Hs [37, 57] and those of the two recently synthesized even–even nuclei, ^{270}Ds [76] and ^{270}Hs [60], are also shown in Figs. 9 and 10. In these cases the decay chains end by SF at ^{256}Rf, ^{262}Sg, and ^{262}Rf, respectively.

For the odd nuclei displayed in Fig. 10(b), the partial α and SF half-lives calculated in [124] have to be multiplied by a factor of 10 and 1,000, respectively, thus making provisions for the odd-particle hindrance factors. However, we have to keep in mind that fission hindrance factors show a wide distribution from 10^1–10^5, which is mainly a result of the specific levels occupied by the odd nucleon [81]. For odd–odd nuclei, the fission hindrance factors from both the odd proton and the odd neutron have to be taken into account. For odd and odd–odd nuclei, the island character of α emitters disappears and for nuclei with neutron numbers 150–160 α decay prevails down to rutherfordium and beyond. In the allegorical representation, where the stability of SHEs is seen as an island in a sea of instability, even–even nuclei portray the situation at high-tide and odd nuclei at low-tide, when the island is connected to the mainland.

The interesting question arises, if and to which extent uncertainties related to the location of proton and neutron shell closures change the half-lives of SHEs. Partial α and β half-lives are only insignificantly modified by shell effects because their decay process occurs between neighboring nuclei. This is different for fission half-lives which are primarily determined by shell effects. However, the uncertainty related to the location of nuclei with the strongest shell-effects, and thus longest partial SF half-life at Z=114, 120, or 126 and N=172 or 184, is irrelevant concerning the longest 'total' half-life of SHEs. The decays of all of these SHEs are dominated by α decay. Alpha-decay half-lives are only modified by a factor of up to approximately 100 if the double shell closure is not located at Z=114 and N=184. Only if shell effects are as strong as in the double magic ^{208}Pb, the half-lives could become significantly shorter for nuclei above the shell closure and longer for the nuclei below.

The line of reasoning is, however, different concerning the production cross-section. The survival probability of the compound nucleus is mainly determined by the fission barrier. Therefore, for reliably estimating the production cross-section, the knowledge of the location of minimal negative shell-correction energy is highly important. However, it may also turn out that shell effects in the region of SHEs are distributed across a number of subshell closures, e.g., for the proton numbers 114, 120, and 126. In that case, a wider

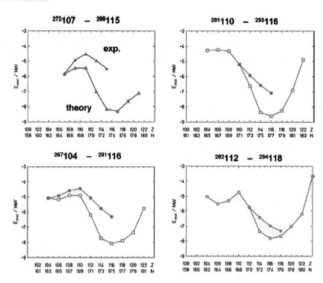

Fig. 11. Comparison of shell correction energies as calculated in [132] to 'experimental' ones obtained as difference between the calculated liquid-drop binding energy taken from [132] and the 'experimental' total binding energy obtained by using the measured Q_α values of nuclei along a decay chain [88]. The unknown shell correction energy of the nucleus at the end of a chain is normalized to the theoretical value (See also Plate 23 in the Color Plate Section)

region of less deep shell-correction energy would exist with corresponding modification of stability and production yield of SHEs.

Experimental shell-correction energies, calculated as difference of binding energies from the liquid-drop part of macroscopic–microscopic models, e.g., from [132, 133], and measured binding energies cannot be determined for the new neutron-rich nuclei discovered in Dubna, because the endpoints of the α decay chains are not connected to known nuclei. However, it is possible to extract a trend of shell correction energies for nuclei along the measured decay chains by normalizing the binding energies to theoretical values at the end points. In Fig. 11, these data are plotted for the four chains starting at $^{288}115$, $^{291}116$, $^{293}116$, and $^{294}118$. The theoretical values were taken from [132].

The 'experimental' data deduced in this way reveal a monotonic decrease of the (negative) shell correction energies when neutron number 184 is approached. No discontinuity is observed, when the predicted closed shell at proton number 114 is crossed at neutron numbers 172–176. This result could mean that proton number 114 as a closed shell or subshell has less influence on the stability of SHEs than the closed neutron shell at $N=184$. When planning future experimental work this aspect should be taken into account. The aim will be to reach $N=184$, which is possible in various reactions using actinide targets and the neutron-richest, but still stable projectiles. These reactions lead into a region of elements as heavy as $Z=120$ and beyond, where, however, increasing destructive Coulomb effects could further reduce half-lives and cross-sections. If, however, the predicted subshell closure at Z=120 has a stronger influence on the stability of SHEs than the subshell closure at Z=114, isotopes near Z=120 and N=184 might gain from increased stability and thus become accessible to experimental studies. These considerations are supported by results from recent theoretical investigations based on self-consistent nuclear models, which obtain increased stability for nuclei in the region of Z=120 and N=184 with certain parameter sets for the nuclear forces [126, 128].

Two questions that had remained unanswered for a long time were recently tackled experimentally: Which is the maximum angular momentum that heavy nuclei can bear during the formation process, and what is the degree of deformation in the ground state? The observation of high-spin states ($I \geq 10$) gives information on the fission barrier of heavy nuclei at high-angular momentum. This information is important for understanding the mechanism of the synthesis, since the fission barrier governs the survival probability. In fact, it is a priori not obvious that high-spin states of shell-stabilized nuclei will even survive against fission. Secondly, the stability of the isotopes that have so far been discovered of elements beyond fermium is predicted to arise from the ability of the nucleus to deform. However, a direct proof and a measurement of the degree of deformation was still missing.

The standard method for identifying high-spin states is in-beam γ spectroscopy, but it is rarely used for studying very heavy nuclei because of the overwhelming fission background. This problem was overcome in recent experiments at the Argonne superconducting linear accelerator ATLAS and at the cyclotron of the University of Jyväskylä by using the recoil-detection technique [134, 135] described in Sect. 2.4.

The first reaction studied was ^{48}Ca + ^{208}Pb → ^{254}No + 2n, which has a relatively high cross-section of $3\,\mu$b. The important result was the observation of the ground-state rotational band up to spin 18. The experiments demonstrated that shell effects stabilize heavy nuclei up to such high-spin values and that ^{254}No is indeed a deformed nucleus. From the energies of the transitions, a quadrupole deformation parameter β_2=0.27 $\pm$ 0.02 was deduced, which is in excellent agreement with theoretical predictions [71, 78, 125, 126, 132, 136–139]. The nucleus ^{254}No is a prolate spheroid with an axis ratio of 4:3. A review on more recent in-beam studies of rotational bands of nuclei in the region of ^{252}No is given in [140].

The reason for the increased stability of nuclei near Z=100 and N=152 at relatively large deformation is the existence of large gaps between the single-particle Nilsson levels at deformation, which results in compression of levels below and above the gaps. If the Fermi level is within such a band of close-lying levels and levels of both high and low spin exist, relatively low-lying high-spin K isomers can be formed. A detailed knowledge of the spectroscopic properties of these levels is particularly interesting since levels could be involved which are relevant for the location of the shell closures in the region of spherical SHEs.

Experimentally, the first K isomers in the region of interest were measured already 34 years ago [141]. These were the $T_{1/2}=$ 1.8 s isomer in ^{250}Fm and the 0.28 s isomer in ^{254}No. A revival of more detailed studies of isomeric states in the region of heavy elements started in 2001 with the observation of a $T_{1/2}=$ 6.0 ms isomer at an energy of 1.13 MeV in ^{270}Ds and its interpretation as a K isomer [76], see Fig. 6. For the deformation of this nucleus the level gaps at $Z=108$ and $N=162$ are responsible. Single particle energies in the region of interest were calculated, using, e.g., the macroscopic–microscopic model [71, 137] or a self-consistent mean field approach [129]. Extensive theoretical studies of K isomers in the region of heavy and superheavy elements were published in [142, 143]. Recently, experiments were performed for detailed investigation of K isomers in ^{254}No, ^{252}No, ^{250}Fm, and ^{256}Fm [144–148].

5 Nuclear Reactions

The main features which determine the fusion process of heavy ions are (1) the fusion barrier and the related beam energy and excitation energy, (2) the ratio of surface tension versus Coulomb repulsion which determines the fusion probability and which strongly depends on the asymmetry of the reaction partners (the product $Z_1 Z_2$ at fixed $Z_1 + Z_2$), (3) the impact parameter (centrality of collision) and related angular momentum, and (4) the ratio of neutron evaporation and of γ emission versus the fission of the compound nucleus.

In fusion reactions toward SHEs the product $Z_1 Z_2$ reaches extremely large and the fission barrier extremely small values. In addition, the fission barrier itself is 'fragile', because it is solely built up from shell effects (see Sect. 4). For these reasons the fusion of interest for production of SHEs is hampered twofold: (1) in the entrance channel by a high probability for re-separation and (2) in the exit channel by a high probability for fission. In contrast, the fusion of lighter elements proceeds unhindered through the contracting effect of the surface tension and the evaporation of neutrons instead of fission.

Cross-section data obtained in reactions for production of heavy elements are plotted in Fig. 12. As already described in Sects. 3.1 and 3.2, 1n and 2n reactions with ^{208}Pb and ^{209}Bi targets lead to low-excitation energies of about 10–15 and 15–25 MeV, respectively (therefore named cold fusion) whereas reactions with actinide targets yield excitation energies of 35–45 MeV (hot fusion) with the 3n and 4n channels being particular interesting. The lowest cross-section value was measured to be 31 fb for the production of element 113 in cold fusion [42]. This is the extreme limit presently set by experimental constraints. Considering the already long irradiation time of $\approx$2 weeks to reach a cross-section of 1 pb, it seems difficult to perform systematic studies at this cross-section level or even below. Further improvement of the experimental conditions is mandatory. Note in this context that the experimental sensitivity increased by three orders of magnitude since the 1982–1983 LBNL–GSI search experiment for element 116 [149] using a hot-fusion reaction, see Fig. 12(d).

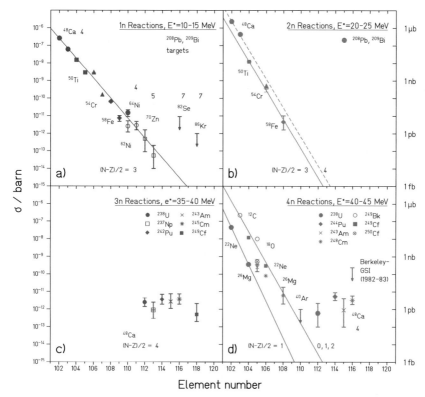

Fig. 12. Measured cross-sections for fusion reactions with ^{208}Pb and ^{209}Bi targets and evaporation of one (**a**) and two (**b**) neutrons and for fusion reactions with actinide targets and evaporation of three (**c**) and four neutrons (**d**). Values $(N-Z)/2$ characterize the neutron excess of the projectile (See also Plate 24 in the Color Plate Section)

The cross-sections for elements up to 113 decrease by a factor of 4 per element in the case of cold fusion (1n channel) and those for elements lighter than 110 by a factor of 10 in the case of hot fusion (4n channel). The decrease is explained as a combined effect of increasing probability for re-separation of projectile and target nucleus and fission of the compound nucleus. Theoretical considerations and studies, see, e.g., [19, 37, 102, 150–154] suggest that the steep decrease of cross-sections for cold-fusion reactions with increasing Z may be strongly linked to increasing re-separation probability at high values of $Z_1 Z_2$ while hot fusion cross-sections mainly drop because of strong fission losses at high-excitation energies and decreasing fission barrier already in the ground state.

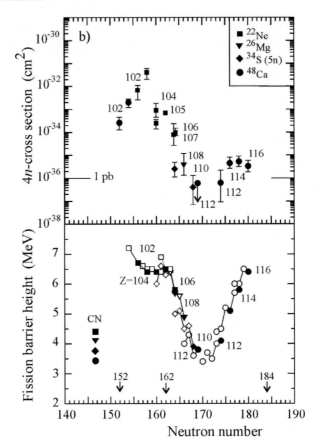

Fig. 13. Measured cross-sections of hot-fusion reactions (*upper panel*) and calculated fission barriers [80, 155] (*lower panel*), revealing the mutual relation between these two quantities (see text for details). The figure is taken from [101]

Extremely small cross-section values result by extrapolating these data into the region of element 114 and beyond. However, the experimental data in the case of hot fusion reveal an opposite trend. The cross-sections increase again and reach maximum values of about 5 pb at element 114 and 116 for both 3n and 4n evaporation channels, Fig. 12(c) and (d). Only relatively high cross-section limits exist for element 116 and 118 in the case of cold fusion.

The mutual relation between measured cross-sections and calculated fission barriers is plotted in Fig. 13. As discussed before, cross-sections and fission barriers decrease up to element 110 whereas both values increase again beyond. The reason that the cross-sections increase less than the fission barriers could be that the latter quantities are actually smaller than predicted or, which seems more likely, that increasing Coulomb repulsion in the entrance channel leads to a reduced fusion probability.

The height of the fission barriers is approximately the negative of the ground-state shell correction energy plotted in Fig. 9a. This is so because the fission barriers are mainly determined by the ground-state shell correction energy in the region of heavy elements, where the liquid-drop barrier vanishes, e.g., for $^{296}116$ the fission-barrier height is 6.5 MeV (see Fig. 13) and the ground-state shell correction energy is -7 MeV (see Fig. 9a). The small difference of 0.5 MeV is due to the shell correction energy at the saddle point. This value is in agreement with the result of a recent theoretical study of saddle-point shell correction energies [130]. Therefore, the negative values of the shell correction energy plotted in Fig. 9a represent a good approximation to the fission-barrier heights in the whole region of heavy elements. As can be seen from Fig. 13, minimum values are obtained at $Z=110$ and $N=170$ while maximum values result for elements $Z=114$–116 and $N=180$–184. In the region of deformed nuclei, which is the domain of cold-fusion reactions, the maximum values are at $Z=108$ and $N=162$. Beyond, fission barriers decrease with increasing element number. This implies that in the case of cold fusion at least up to element 114 decreasing fission barriers further reduce the cross-sections in addition to the increasing probability for re-separation of projectile and target in the entrance channel.

Locally, an increase of the cross-section by a factor of 5.8 was measured for darmstadtium in cold-fusion reaction when the beam was changed from ^{62}Ni to ^{64}Ni. It was speculated that this increase could be due to the increased value of the projectile neutron number. However, the assumption was not confirmed in the case of element 112 which was synthesized using the most neutron-rich stable zinc isotope with mass number 70. The isotopic dependence of fusion cross-sections, including the possible use of radioactive neutron-rich beams, was discussed in several recent publications [160–167].

A number of excitation functions was measured for the synthesis of elements from nobelium to darmstadtium using lead and bismuth targets [19]. For the even elements these data are shown together with the two data points measured for $^{278}112$ in Fig. 14. The maximum evaporation residue cross-section (1n channel) was measured at beam energies well below a one-dimensional fusion barrier [156]. At the optimum beam energy projectile and target are just reaching the contact configuration in a central collision. The relatively simple fusion barrier based on the Bass model [156] is too high and a tunnelling process through this barrier cannot explain the measured cross-section. Various processes may result in a lowering of the fusion barrier. Among these processes transfer of nucleons and an excitation of vibrational degrees of freedom are most important [157, 158, 168].

Target nuclei of actinide targets are strongly deformed and the height of the Coulomb barrier depends on the orientation of the deformation axes. Two of the measured excitation functions for production of $^{292}114$ and $^{283}112$ [101] after evaporation of four and three neutrons, respectively, are shown in Fig. 14. A comparison with the cold fusion data reveals that the element 114 excitation function is located completely above the Bass contact configuration which was

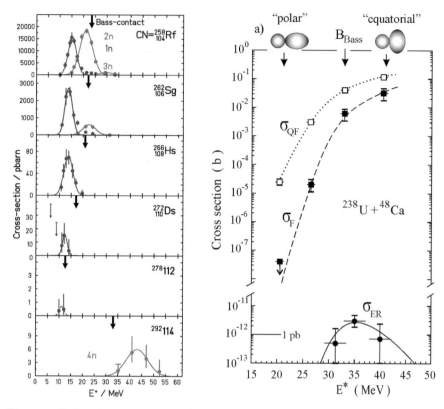

Fig. 14. *Left side*: Measured excitation functions of even elements from ruther-fordium to element 112 produced in reactions with ^{208}Pb targets and beams from ^{50}Ti to ^{70}Zn. The data were measured in experiments at SHIP. For comparison the excitation function for synthesis of element 114 in the reaction ^{48}Ca $+$ ^{244}Pu is plotted in the *bottom panel* [101]. The *arrows* mark the energy at reaching a contact configuration using the model by Bass [156]. *Right panel*: Comparison of the cross-sections as function of the excitation energy E^* for quasifission (σ_{QF}), compound-nucleus fission (σ_F), and evaporation residues (σ_{ER}) for the reaction ^{48}Ca $+$ ^{238}U. The figure on the *right side* is taken from [101] (See also Plate 25 in the Color Plate Section)

calculated for a mean radius of the deformed target nucleus. In addition, the curves for the hot-fusion reactions are significantly broader, e.g., 10.6 instead of 4.6 MeV (FWHM) as measured for ^{265}Hs.

In the *right side* of Fig. 14 the fusion–evaporation excitation function is compared with the yield of quasi-fission and compound nucleus fission [169]. At low-projectile energies nuclear reactions can occur only at polar orientation of the deformed target nucleus. In the *upper part* of the figure the orientation at the touching point and the corresponding Coulomb energy is indicated. Although the excitation energy of the compound nucleus would be low, about

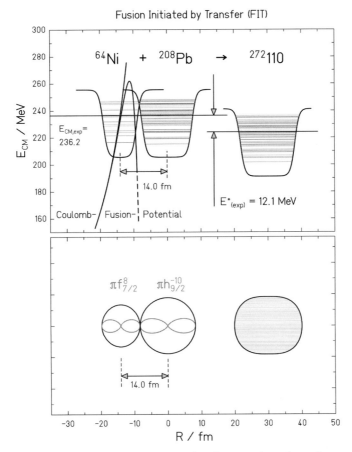

Fig. 15. Energy-against-distance diagram for the reaction of an almost spherical ^{64}Ni projectile with a spherical ^{208}Pb target nucleus resulting in the deformed fusion product ^{271}110 after emission of one neutron. At the center-of-mass energy of 236.2 MeV, the maximum cross-section was measured. In the *top panel*, the reaction partners are represented by their nuclear potentials (Woods–Saxon) at the contact configuration where the initial kinetic energy is exhausted by the Coulomb potential. At this configuration projectile and target nuclei are 14 fm apart from each other. This distance is 2 fm larger than the Bass contact configuration [156], where the mean radii of projectile and target nucleus are in contact. In the *bottom panel*, the outermost proton orbitals are shown at the contact point. For the projectile ^{64}Ni, an occupied $1f_{7/2}$ orbit is drawn, and for the target ^{208}Pb an empty $1h_{9/2}$ orbit. The protons circulate in a plane perpendicular to the drawing. The Coulomb repulsion, and thus the probability for separation, is reduced by the transfer of protons. In this model, the fusion is initiated by transfer (see also [157, 158]). The figure is taken from [159] (See also Plate 26 in the Color Plate Section)

20–25 MeV resulting in reduced fission probability of the compound nucleus, the nuclei do not fuse but re-separate with high probability at this elongated configuration. Only the compact configuration in the case of equatorial collisions results in fusion despite the fact that the excitation energy of the compound nucleus is considerably higher, i.e., about 35–40 MeV.

It has been pointed out [170–172] that projectile and target nuclei with a closed-shell configuration are favorable for synthesizing SHEs. The reason is not only a low (negative) reaction Q-value and thus a low-excitation energy, but also that fusion of such systems is connected with a minimum of energy dissipation. The fusion path proceeds along cold-fusion valleys, where the reaction partners maintain kinetic energy up to the closest possible distance. In the case of cold fusion with spherical targets the maximum fusion yield is obtained at projectile energies just enough high so that projectile and target nucleus come to rest when just the outer orbits are in contact. The configuration at this point is plotted in Fig. 15. From there on the fusion process occurs well ordered along paths of minimum dissipation of energy. Empty orbits above the closed shell nucleus ^{208}Pb favor a transfer of nucleons from the projectile to the target and thus initiate the fusion process. An adequate theoretical description of this process is the application of the two-center shell model [173–176].

On a first glance the situation seems to be different in the case of hot fusion. The maximum of the excitation function is located at the higher energy side of the value needed to reach the contact configuration according to the Bass model [156], see Fig. 14. However, taking into account the deformation of the target nucleus and considering fusion at equatorial orientation, it is evident that the projectile and target nuclei come to rest when just the outer orbits are in contact. This distance corresponds to the energy, where the maximum yield is measured, see Fig. 14. It is, like in cold fusion located on the left side of the 'Bass' contact configuration for deformed nuclei at equatorial collisions. Also in this case the empty orbits in the equatorial plane of the prolate deformed target nucleus favor the transfer of nucleons.

At low-projectile energies, where a contact configuration at zero kinetic energy in the center of mass system is reached only in polar collisions, the nuclei do not fuse. The reasons are, firstly, that at the elongated configuration re-separation is enhanced due to the unfavorable ratio of Coulomb repulsion and surface attraction and, secondly, due to the occupied Nilsson levels originating from orbits of high spin in the target nucleus, which hinders a transfer of nucleons from the projectile to the target. Recent experimental work is aiming to study the transition from fusion with deformed actinide target nuclei and light projectiles, e.g., ^{12}C or ^{16}O, where polar collisions at low-beam energy result in enhanced sub-barrier fusion, to heavier projectiles like ^{48}Ca just described, as function of the projectile mass and charge [177].

Figure 16 illustrates the two different scenarios of reactions starting from an elongated and a compact configuration. Triggered by the recent experimental success of heavy element synthesis, a number of theoretical studies were performed or are in progress, aiming to obtain a detailed quantitative understanding of the reaction processes involved in heavy element synthesis [153, 161, 168, 178–187].

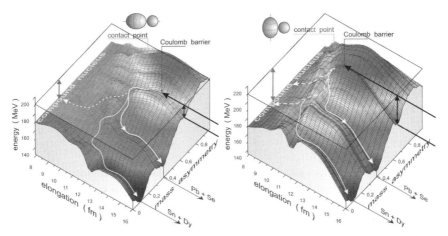

Fig. 16. Three dimensional potential energy surface as function of elongation and mass asymmetry for the reaction ^{48}Ca $+$ ^{248}Cm. Two scenarios are shown: On the *left side* for a reaction at low-projectile energy so that contact can occur only at polar orientation. On the *right side* the projectile energy is high enough so that contact occurs also at equatorial orientation. In both cases the projectile energies are chosen so that the kinetic energy in the center of mass system is zero at the contact configuration. The *rectangular plane* marks the total energy. It is drawn through the potential energy at contact. The difference between this plane and the curved potential energy surface is mainly kinetic energy in the entrance channel, after contact mainly intrinsic excitation energy of the system. The valleys in the potential energy surface are due to shell effects of the reaction partners. After contact the paths of the configuration in the plane of total kinetic energy is influenced by the structure of the potential energy surface. Trajectories above the valleys are favored due to enhanced conversion of potential energy into kinetic energy, which results in re-separation dominantly with the double magic ^{208}Pb as one of the reaction partners. Due to the longer distance from contact to the compound system (*boarder on the left*), the probability for re-separation is higher in the case of the elongated configuration. Therefore, the fusion cross-section is considerably smaller as in the case of the compact configuration, although the excitation energy of the compound nucleus, marked by the length of the *double arrow on the left*, is significantly smaller, about 20 MeV instead of 40 MeV. The main part of the figure has been provided by courtesy of V.I. Zagrebaev (2007) (See also Plate 27 in the Color Plate Section)

Due to the great uncertainty concerning the influence of the various steps in the fusion of heavy elements, more and more precise experimental data are needed. It is especially important that various combinations of projectile and target be investigated, from very asymmetric systems to symmetric ones, studied for both fusion and transfer reactions, the latter for systems as heavy as ^{238}U + ^{248}Cm [189], and that excitation functions are measured. This provides information on how fast the cross-section decreases with increasing energy due to fission of the compound nucleus, and how fast cross-sections decrease on the low-energy side due to the fusion barrier and re-separation of projectile and target nuclei. From both slopes, information about the shape of the fission and the fusion barriers can be obtained. At a high enough cross-section, these measurements can be complemented by in-beam γ-ray spectroscopy using the RDT method in order to study the influence of angular momentum on the fusion and survival probability [134, 190, 191].

6 Summary and Outlook

The experimental work of the last three decades has shown that cross-sections for the synthesis of the heaviest elements do not decrease continuously as it was measured up to the production of element 112 using a cold-fusion re-action. Recent data on the synthesis of elements 113–116 and 118 in Dubna using hot fusion break this trend when the region of spherical SHEs is reached. Some of the results originally obtained in Dubna were confirmed in independent experiments and with different methods, including the use of chemical, element-specific properties. Due to the systematic study of different reactions for the production of SHEs and the measurement of excitation functions, the present assignment of the measured decay chains to new relatively neutron-rich isotopes of SHEs is highly reliable. Results of independent experiments supported the assignment and the correctness of the published data. The region of the predicted spherical SHEs has finally been reached and the exploration of the 'island' has started and can be performed even on a relatively high cross-section level.

The progress towards the exploration of the island of spherical SHEs is difficult to predict. However, despite the exciting new results, many questions of more general character are still awaiting an answer. New developments will not only make it possible to perform experiments aimed at synthe-sizing new elements in reasonable measuring times, but will also allow for a number of various other investigations covering reaction physics and spectroscopy.

One can hope that, during the coming years, more data will be measured in order to promote a better understanding of the stability of the heaviest elements taking into account the fundamental properties of the nuclear forces. A microscopic description of the fusion process will be needed for an effective explanation of all measured phenomena in the case of low-dissipative energies. Then, the relationships between fusion probability and stability of the fusion products will result in accurate predictions of the production yield of SHEs using various reaction mechanisms.

An opportunity for the continuation of experiments in the region of SHEs at low cross-sections afford, among others, further accelerator developments [134, 135]. High-current beams and radioactive beams are options for the future. At increased beam currents, values of tens of particle-μA's may become accessible, the cross-section level for the performance of experiments can be shifted down into the region of tens of femtobarns, and excitation functions can be measured on the level of tenths of picobarns. High currents, in turn, call for the development of new targets and separator improvements. Radioactive ion beams, not as intense as the ones with stable isotopes, will allow for approaching the closed neutron shell $N=184$ already at lighter elements. The study of the fusion process using radioactive neutron-rich beams will be highly interesting.

The half-lives of spherical SHEs are expected to be relatively long. Based on nuclear models, which are effective predictors of half-lives in the region of the heaviest elements, values from microseconds to years have been calculated for various isotopes. This wide range of half-lives encourages the application of a wide variety of experimental methods in the investigation of SHEs, from the safe identification of short-lived isotopes by recoil-separation techniques to atomic physics experiments on trapped ions, and to the investigation of chemical properties of SHEs using long-lived isotopes. Finally, on the basis of a broader knowledge of the properties of SHEs and the mechanisms of their synthesis, it will become possible to answer reliably the questions, if SHEs could be produced in nature and where the highest probability may exists for finding them.

Acknowledgement

Various experiments at SHIP were performed in collaboration with D. Ackermann, P. Armbruster, H.G. Burkhard, H. Folger, F.P. Heßberger, S. Heinz, B. Kindler, J. Khuyagbaatar, B. Lommel, R. Mann, G. Münzenberg, W. Reisdorf, K.H. Schmidt, H.J. Schött, B. Sulignano (GSI Darmstadt), A.N. Andreyev, A.Yu. Lavrentev, A.G. Popeko, A.V. Yeremin (FLNR-JINR Dubna), S. Antalic, P. Cagarda, R. Janik, S. Šaro, B. Streicher, M. Venhart (University Bratislava), K. Nishio (JAEA, Tokai), P. Kuusiniemi, M. Leino, J. Uusitalo (University Jyväskylä), V.F. Comas, J.A. Heredia (InSTEC, Habana), C. Stodel (GANIL, Caen), and R. Dressler (PSI, Villigen). The most important prerequisite for realizing the experimental program was a stable and high-current beam of accurate energy from the UNILAC. The excellent performance of the beam, the targets and the data acquisition, and analysis were due to the efforts of the people from the ion-source group, the accelerator, the target laboratory and the department of experimental electronics and data acquisition. Many of the technical problems were treated and solved by the designers and mechanics of the GSI. For many stimulating and informative discussions, I wish to express my appreciation to Y. Abe, G. Adamian, N.V.

Antonenko, E.A. Cherepanov, H.W. Gäggeler, W. Greiner, R.K. Gupta, M.G. Itkis, C. Kozhuharov, J. Maruhn, P. Möller, K. Morita, A.K. Nasirov, W. Nazarewicz, Yu.Ts. Oganessian, W. von Oertzen, W. Scheid, A. Sobiczewski, W.J. Swiatecki, V.V. Volkov, and V.I. Zagrebaev.

References

1. O. Hahn, F. Straßmann, Naturwissenschaften **27**, 11 (1939)
2. L. Meitner, O.R. Frisch, Nature **143**, 239 (1939)
3. G. Gamov, Proc. R. Soc. London A **126**, 632 (1930)
4. C.F. von Weizsäcker, Z. Phys. **96**, 431 (1935)
5. M. Göppert-Mayer, Phys. Rev. **74**, 235 (1948)
6. O. Haxel et al., Phys. Rev. **75**, 1769 (1949)
7. H. Grawe, Lect. Notes Phys. **651**, 33 (2004)
8. W.D. Myers, W.J. Swiatecki, Nucl. Phys. **81**, 1 (1966)
9. H. Meldner, Ark. Fys. **36**, 593 (1967)
10. S.G. Nilsson et al., Nucl. Phys. A **115**, 545 (1968)
11. U. Mosel, W. Greiner, Z. Phys. A **222**, 261 (1969)
12. J. Grumann et al., Z. Phys. A **228**, 371 (1969)
13. E.O. Fiset, J.R. Nix, Nucl. Phys. A **193**, 647 (1972)
14. J. Randrup et al., Phys. Rev. C **13**, 229 (1976)
15. W. Grimm et al., Phys. Rev. Lett. **26**, 1040 (1971)
16. G.N. Flerov, G.M. Ter-Akopian, Rep. Prog. Phys. **46**, 817 (1983)
17. G. Münzenberg et al., Nucl. Instr. Meth. **161**, 65 (1979)
18. S. Hofmann et al., Z. Phys. A **291**, 53 (1979)
19. S. Hofmann, G. Münzenberg, Rev. Mod. Phys. **72**, 733 (2000)
20. K.H. Schmidt, et al., Phys. Lett. B **168**, 39 (1986)
21. N. Angert et al., GSI Report **89-1**, 372 (1989)
22. R. Geller et al., Rev. Sci. Instrum. **63**, 2795 (1992)
23. J. Bossler et al., GSI Rep. **97-1**, 155 (1997)
24. R. Mann et al., GSI Report **2006-1**, Scientific Report 2005, 198 (2006)
25. B. Lommel et al., Nucl. Instr. Meth. A **561**, 100 (2006)
26. B. Kindler et al., Nucl. Inst. Meth. A **590**, 126 (2008)
27. E. Kashy, B. Sherrill, Nucl. Instr. Meth. B **26**, 610 (1987)
28. R. Mann et al., GSI Report **2004-1**, Scientific Report 2003, 224 (2004)
29. H. Folger et al., Nucl. Instr. Meth. A **362**, 64 (1995)
30. S. Saro et al., Nucl. Instr. Meth. A **381**, 520 (1996)
31. P. Armbruster et al., Nucl. Instr. Meth. **91**, 499 (1971)
32. S. Hofmann et al., Nucl. Instr. Meth. **223**, 312, 1984
33. G. Münzenberg, Rep. Prog. Phys. **51**, 57 (1988)
34. S. Hofmann, J. Alloys Compd. **213/214**, 74 (1994)
35. J. Hoffmann et al., GSI Report **2005-1**, Scientific Report 2004, 340 (2005)
36. A. Ghiorso et al., Phys. Rev. Lett. **22**, 1317 (1969)
37. S. Hofmann, Rep. Prog. Phys. **61**, 639 (1998)
38. K. Morita et al., Eur. Phys. J. A **21**, 257 (2004)
39. K. Morita et al., J. Phys. Soc. Jpn. **73**, 1738 (2004)
40. K. Morita et al., J. Phys. Soc. Jpn. **76**, 043201 (2007)

41. K. Morita et al., J. Phys. Soc. Jpn. **73**, 2593 (2004)
42. K. Morita et al., J. Phys. Soc. Jpn. **76**, 045001 (2007)
43. M. Schädel, J. Nucl. Radiochem. Sci. **3**, 113 (2002)
44. J. Corish, G.M. Rosenblatt, Pure Appl. Chem., **76**, 2101 (2004)
45. G. Münzenberg et al., Z. Phys. A **300**, 107 (1981)
46. G. Münzenberg et al., Z. Phys. A **333**, 163 (1989)
47. S. Hofmann et al., Z. Phys. A **358**, 377 (1997)
48. S. Hofmann et al., Z. Phys. A **350**, 281 (1995)
49. S. Hofmann et al., Eur. Phys. J. A **14**, 147 (2002)
50. P.A. Wilk et al., Phys. Rev Lett. **85**, 2697 (2000)
51. R. Eichler et al., Nature **407**, 63 (2000)
52. Z.G. Gan et al., Eur. Phys. J. A **20**, 385 (2004)
53. Z.G. Gan et al., Eur. Phys. J. A **10**, 21 (2001)
54. C.M. Folden III et al., Phys. Rev. C **73**, 014611 (2006)
55. S.L. Nelson et al., Phys. Rev. Lett. **100**, 022501 (2008)
56. G. Münzenberg et al., Z. Phys. A **317**, 235 (1984)
57. G. Münzenberg et al., Z. Phys. A **324**, 489 (1986)
58. S. Hofmann et al., Z. Phys. A **350**, 277 (1995)
59. S. Hofmann et al., Z. Phys. A **354**, 229 (1996)
60. Ch.E. Düllmann et al., Nature **418**, 859 (2002)
61. A. Türler et al., Eur. Phys. J. A **17**, 505 (2003)
62. G. Münzenberg et al., Z. Phys. A **309**, 89 (1982)
63. G. Münzenberg et al., Z. Phys. A **315**, 145 (1984)
64. G. Münzenberg et al., Z. Phys. A **330**, 435 (1988)
65. S. Bjørnholm, W.J. Swiatecki, Nucl. Phys. A **391**, 471 (1982)
66. P. Fröbrich, Phys. Lett. B **215**, 36 (1988)
67. Yu.A. Lazarev et al., Phys. Rev. Lett. **75**, 1903 (1995)
68. A. Ghiorso et al., Phys. Rev. Lett. **33**, 1490 (1974)
69. M.R. Schmorak, Nuclear Data Sheets **59**, 507 (1990)
70. K.H. Schmidt et al., Z. Phys. A **316**, 19 (1984)
71. S. Cwiok et al., Nucl. Phys. A **573**, 356 (1994)
72. T.N. Ginter et al., Phys. Rev. C **67**, 064609, 2003
73. C.M. Folden III et al., Phys. Rev. Lett. **93**, 212702 (2004)
74. A. Ghiorso et al., Phys. Rev. C **51**, R2293 (1995)
75. Yu.A. Lazarev et al., Phys. Rev. C **54**, 620 (1996)
76. S. Hofmann et al., Eur. Phys. J. A **10**, 5 (2001)
77. S. Cwiok et al., Phys. Rev. Lett. **83**, 1108 (1999)
78. I. Muntian et al., Phys. Ref. C **60**, 041302 (1999)
79. R. Smolanczuk, Phys. Rev. Lett. **83**, 4705 (1999)
80. R. Smolanczuk et al., Phys. Rev. C **52**, 1871 (1995)
81. D.C. Hoffman, Nucl. Phys. A **502**, 21c (1989)
82. S. Hofmann et al., GSI Report **2004-1**, Scientific Report 2003, 1 (2004)
83. V. Ninov et al., Phys. Rev. Lett. **83**, 1104 (1999)
84. C. Stodel et al., *Proceedings of the Tours Symposium on Nuclear Physics IV*, edited by M. Arnould et al., AIP Conference Proceedings 561 (AIP, New York, 2001) p. 344.
85. K. Morimoto et al., *Proceedings of the Tours Symposium on Nuclear Physics IV*, edited by M. Arnould et al., AIP Conference Proceedings 561 (AIP, New York, 2001) p. 354.

86. V. Ninov et al., Phys. Rev. Lett. **89**, 039901 (2002)
87. V.B. Kutner et al., *Proceedings of the 15th International Conference on Cyclotrons and their Applications*, edited by E. Baron and M. Lieuvin (IOP, Bristol, 1998), p. 405
88. Yu.Ts. Oganessian, J. Phys. G, Nucl. Part. Phys. **34**, R165 (2007)
89. A.V. Yeremin et al., Nucl. Instr. Meth. A **350**, 608 (1994)
90. A.V. Yeremin et al., Nucl. Instr. Meth. B **126**, 329 (1997)
91. Yu.A. Lazarev et al., *Proceedings of the International School Seminar on Heavy Ion Physics*, edited by Yu.Ts. Oganessian et al.. (Joint Institute for Nuclear Research, Dubna, 1993), Vol. II, p. 497
92. Yu.Ts. Oganessian et al., Eur. Phys. J. A **5**, 63 (1999)
93. Yu.Ts. Oganessian et al., Nature **400**, 242 (1999)
94. Yu.Ts. Oganessian et al., Phys. Rev. Lett. **83**, 3154 (1999)
95. Yu.Ts. Oganessian et al., Phys. Rev. C **62**, 041604 (2000)
96. Yu.Ts. Oganessian et al., Phys. Rev. C **63**, 011301 (2000)
97. Yu.Ts. Oganessian et al., Phys. At. Nucl. **64**, 1349 (2001)
98. Yu.Ts. Oganessian et al., Preprint JINR E7-2002-287
99. Yu.Ts. Oganessian et al., Nucl. Phys. A **734**, 109 (2004)
100. Yu.Ts. Oganessian et al., Phys. Rev. C **69**, 054607 (2004)
101. Yu.Ts. Oganessian et al., Phys. Rev. C **70**, 064609 (2004)
102. I.V. Zagrebaev et al., Phys. At. Nucl. **66**, 1033 (2003)
103. V.I. Zagrebaev, Nucl. Phys. A **734**, 164 (2004)
104. Yu.Ts. Oganessian et al., Phys. Rev. C **69**, 021601 (2004)
105. Yu.Ts. Oganessian et al., Phys. Rev. C **72**, 034611 (2005)
106. S.N. Dmitriev et al., Int. Symp. on Exotic Nuclei, Edts. Yu.E. Penionzhkevich and E.A. Cherepanov, Peterhof, Russia, July 5–12, 2004, World Scientific Publishing, Singapore, 2005, p. 285
107. D. Schumann et al., Radiochim. Acta **93**, 727 (2005)
108. Yu.Ts. Oganessian et al., Phys. Rev. C **74**, 044602 (2006)
109. Yu.Ts. Oganessian et al., Phys. Rev. C **76**, 11601(R) (2007)
110. W. Loveland et al., Phys. Rev. C **66**, 044617 (2002)
111. A.B. Yakushev et al., Rad. Chem. Acta **91**, 433 (2003)
112. K.E. Gregorich et al., Phys. Rev. C **72**, 014605 (2005)
113. R. Eichler et al., Radiochim. Acta **94**,181 (2006)
114. R. Eichler et al., Nucl. Phys. A **787**, 373c (2007)
115. R. Eichler et al., Nature **447**, 72 (2007)
116. S. Hofmann et al., Eur. Phys. J. A **32**, 251 (2007)
117. Yu.Ts. Oganessian et al., Nucl. Phys. A **734**, 196 (2004)
118. B. Kadkhodayan et al., Radiochim. Acta **72**, 169 (1996)
119. J.V. Kratz et al., Phys. Rev. C **45**, 1064 (1992)
120. A. Türler et al., Phys. Rev. C **57**, 1648 (1998)
121. G.T. Seaborg W.D. Loveland, *The elements beyond uranium* (John Wiley and Sons, Inc., New York, 1990)
122. J.V. Kratz, 'Chemsitry of Transactinides', in *Handbook of Nuclear Chemsitry, Vol. 2.* ed. by A. Vertes, S. Nagy, and Z. Klencsar (Kluwer Academic Publishers, Dordrecht, 2003) pp. 323–395
123. M. Schädel, *The Chemistry of Superheavy Elements.* (Kluwer Academic Publishers, Dordrecht 2003)

124. R. Smolanczuk, A. Sobiczewski, *Proceedings of the XV. Nuclear Physics Divisional Conference on Low Energy Nuclear Dynamics*, ed. by Yu.Ts. Oganessian et al.. (World Scientific, Singapore, 1995), p. 313.
125. S. Cwiok et al., Nucl. Phys. A **611**, 211 (1996)
126. K. Rutz et al., Phys. Rev. C **56**, 238 (1997)
127. A.T. Kruppa et al., Phys. Rev. C **61**, 034313 (2000)
128. M. Bender et al., Phys. Lett. B **515**, 42 (2001)
129. M. Bender et al., Nucl. Phys. A **723**, 354 (2003)
130. W.J. Swiatecki et al., Acta Phys. Pol. B **38**, 1565 (2007)
131. P. Möller et al., At. Data Nucl. Data Tables **66**, 131 (1997)
132. P. Möller et al., At. Data Nucl. Data Tables **59**, 185 (1995)
133. I. Muntian et al., Acta Phys. Pol. B **32**, 691 (2001)
134. P. Reiter et al., Phys. Rev. Lett. **82**, 509 (1999)
135. M. Leino et al., Acta Phys. Pol. B **30**, 635 (1999)
136. G.A. Lalazissis et al., Nucl. Phys. A **608**, 202 (1996)
137. Z. Patyk, A. Sobiczewski, Nucl. Phys. A **533**, 132 (1991)
138. I. Muntian et al., Phys. Lett. A **500**, 241 (2001)
139. A. Sobiczewski et al., Phys. At. Nucl. **64**, 1105 (2001)
140. R.-D. Herzberg, J. Phys. G, Nucl. Part. Phys., **30**, R123 (2004)
141. A. Ghiorso et al., Phys. Rev. C **7**, 2032 (1973)
142. F.R. Xu et al., Phys. Rev. Lett. **92**, 252501 (2004)
143. J.-P. Delaroche et al., Nucl. Phys. A **771**, 103 (2006)
144. S. Eeckhaudt et al., Eur. Phys. J. A **26**, 227 (2005)
145. C.N. Davids et al., Phys. Rev. C **74**, 014316 (2006)
146. S.K. Tandel et al., Phys. Rev. Lett. **97**, 082502 (2006)
147. R.-D. Herzberg et al., Nature **442**, 896 (2006)
148. B. Sulignano et al., Eur. Phys. J. A **33**, 327 (2007)
149. P. Armbruster et al., Phys. Rev. Lett. **54**, 406 (1985)
150. W. Reisdorf, M. Schädel, Z. Phys. A **343**, 47 (1992)
151. M. Schädel, S. Hofmann, J. Radioanal. Nucl. Chem. **203**, 283 (1996)
152. W.J. Swiatecki et al., Phys. Rev. C **71**, 014602 (2005)
153. V.I. Zagrebaev, Phys. Rev. C **64**, 034606 (2001)
154. V.I. Zagrebaev et al., Phys. Rev. C **65**, 014607 (2002)
155. I. Muntian et al., Acta Phys. Pol. B **34**, 2141 (2003)
156. R. Bass, Nucl. Phys. A **231**, 45 (1974)
157. W. von Oertzen, Z. Phys. A **342**, 177 (1992)
158. V.V. Volkov, Phys. Part. Nuclei **35**, 425 (2004)
159. S. Hofmann, 'Production and stability of new elements'. In, *Proceedings of the XV. Nuclear Physics Divisional Conference on Low Energy Nuclear Dynamics, St.Petersburg, Russia, April 18–22, 1995*, ed. by Yu.Ts. Oganessian, R. Kalpakchieva, and W. von Oertzen. (World Scientific, Singapore 1995) pp. 305–312
160. S. Hofmann, Prog. Part. Nucl. Phys. **46**, 293 (2001)
161. G.G. Adamian et al., Nucl. Phys. A **678**, 24 (2000)
162. G.G. Adamian et al., Phys. Rev. C **69**, 11601 (2004)
163. G.G. Adamian et al., Phys. Rev. C **69**, 14607 (2004)
164. G.G. Adamian et al., Phys. Rev. C **69**, 44601 (2004)
165. W. Loveland et al., Phys. Rev. C **74**, 44607 (2006)
166. W. Loveland, Phys. Rev. C **76**, 14612 (2007)

167. J.F. Liang et al., Phys. Rev. C **75**, 54607 (2000)
168. V.Yu. Denisov, S. Hofmann, Phys. Rev. C **61**, 034606 (2000)
169. M.G. Itkis et al., Phys. Rev. C **65**, 044602 (2002)
170. J. Maruhn, W. Greiner, Z. Phys. A **251**, 431 (1972)
171. A. Sandulescu et al., Phys. Lett. B **60**, 225 (1976)
172. R.K. Gupta et al., Z. Phys. A **283**, 217 (1977)
173. D. Scharnweber et al., Phys. Rev. Lett. **24**, 601 (1970)
174. U. Mosel et al., Phys. Lett. B **34**, 587 (1971)
175. A. Sandulescu, W. Greiner, Rep. Prog. Phys. **55**, 1423 (1992)
176. A. Sandulescu et al., Int. J. Mod. Phys. E **1**, 379 (1992)
177. K. Nishio et al., Eur. Phys. J. A **29**, 281 (2006)
178. V.Yu. Denisov, W. Nörenberg, Eur. Phys. J. A **15**, 375 (2002)
179. Y. Aritomo et al., Phys. Rev. C **59**, 796 (1999)
180. G. Giardina et al., Eur. Phys. J. A **8**, 205 (2000)
181. R. Smolanczuk, Phys. Rev. C **63**, 044607 (2001)
182. G. Fazio et al., Eur. Phys. J. A **20**, 89 (2004)
183. G. Fazio et al., Mod. Phys. Lett. A **20**, 391 (2005)
184. G.G. Adamian et al., Phys. Rev. C **71**, 34603 (2005)
185. G.G. Adamian, N.V. Antonenko, Phys. Rev. C **72**, 64617 (2005)
186. T. Ichikawa et al., Phys. Rev. C **71**, 44608 (2005)
187. V.Yu. Denisov, N.A. Pilipenko, Phys. Rev. C **76**, 14602 (2007)
188. I.V. Zagrebaev, W. Greiner, J. Phys. G **31**, 825 (2005)
189. V.I. Zagrebaev et al., Phys. Rev. C **73**, 31602(R) (2006)
190. P. Reiter et al., Phys. Rev. Lett. **84**, 3542 (2000)
191. R.-D. Herzberg et al., Phys. Rev. C **65**, 014303 (2001)

Experimental Tools for Nuclear Astrophysics

C. Angulo

Université catholique de Louvain, Louvain-la-Neuve, Belgium

Abstract This chapter concentrates on experimental techniques currently used to investigate nuclear reactions of astrophysical interest. After a brief introduction, I shall present the basic quantities and equations governing thermonuclear reaction rates in stellar plasma. The various astrophysical scenarios, from hydrostatic to advanced burning stages up to/and including explosive mechanisms, as well as some key reactions, are briefly presented. I will concentrate on the experimental approaches to study nuclear reactions involved in both quiescent and explosive stellar burning. Particular emphasis is given to the use of radioactive ion beams and their importance for characterizing explosive nucleosynthesis in novae and X-ray bursts. A few recent examples will be shown in more detail to illustrate these techniques. Some key open questions will be discussed in the context of future facilities.

1 Understanding the Universe

The present configuration of the Universe is the result of the evolution of the primordial matter which was mainly composed, a few minutes after the Big Bang,[1] of the two lightest elements, hydrogen and helium. It is well-known at present that all other chemical elements, that make up, for instance, all living beings and celestial objects have been produced by nuclear reactions in the heart of stars (see, for example, the review papers on stellar nucleosynthesis [1–5]). These nuclear reactions produce the energy that powers the stars and this balances the gravitational force to prevent their collapse. In turn, the production of energy and elements explain the structure and evolution of the Universe, with stars that will progressively cool down and die and stars that will explode producing cataclysmic events.

To understand the structure and the evolution of the Universe it is essential to understand the synthesis of the elements [6]. Most of the questions about

---

[1] The theory of the Big Bang established 80 years ago, is largely accepted at present as the theory of the origin of the Universe.

Angulo, C.: *Experimental Tools for Nuclear Astrophysics.* Lect. Notes Phys. **764**, 253–282 (2009)
DOI 10.1007/978-3-540-85839-3_7

the nucleosynthesis have been answered by nuclear physics in the twentieth century, with remarkable contributions from, among others, H. Bethe [7–9] and F. Hoyle [10]. The reader will find a concise but rather complete historical review in the book of D.D. Clayton [6]. Just 50 years ago, in a famous review article [11], E.M. Burbidge, G.R. Burbidge, W.A. Fowler and F. Hoyle gave a complete explanation of the synthesis of elements by different nuclear mechanisms. They explained hydrogen burning in the Big Bang and in the main sequence stars (such as our sun), helium burning and the triple-α process in red giant stars and in the asymptotic giant branch stars, as well as the production of elements beyond iron by the s-process (slow neutron capture reactions), the r-process (rapid neutron capture reactions) and the p-process (photodisintegration and proton capture reactions).

Modern nuclear astrophysics has thus been born on the crossroad of nuclear physics, astrophysics and astronomy. It involves careful and dedicated experimental and theoretical studies of a large variety of nuclear processes (nuclear physics) as indispensable tools for the modelling of stellar evolution and nucleosynthesis (astrophysics). Nuclear reactions cross sections, nuclear masses, β-decay rates, and other nuclear properties are fundamental inputs to understand the structure, evolution and composition of a large variety of cosmic objects, including the Solar System. Figure 1 summarizes the link between the different fields.

The roles of nuclear physics and of astrophysics are well defined. Nuclear physics is dedicated to measure (nuclear experiments) or to calculate (nuclear theory) the fundamental quantities playing a role in the stellar processes, to extrapolate the cross section data down to astrophysical energies, and to interpret the results using theoretical models. It also has the role of calculating the reaction rates that will allow the astrophysicists to model the different stellar environments and to study the evolution of the stars. Reciprocally, by modelling the impact of the nuclear uncertainties one obtains important indications on the key reactions and the key quantities that have to be studied in the laboratory. Finally, by the astronomy, the experimental data face the observations (γ-radioactivity of the galaxy, element abundances obtained by spectroscopy, supernova light-curves, ...) as well as the results of the analysis of meteorites and grains. The interaction nuclear physics–astrophysics–astronomy must be considered to yield an entire comprehension of the Universe (see Fig. 1).

Whatever the process considered is (primordial, stellar or explosive nucleosynthesis), the calculation of element abundances and the codes of stellar evolution request a huge number of nuclear reaction cross sections. The stellar environment being considered determines the energy region within which the nuclear reaction processes need to be determined. This region is known as the Gamow window [6, 13], and the specific nuclear properties within this energy domain can play a vital role in determining the nucleosynthesis that occurs. Thus, a reaction that may play an important role in a certain stellar environment can be completely negligible under different temperature and density

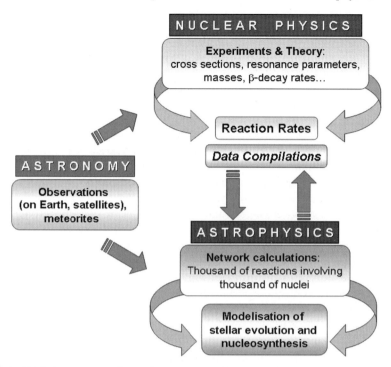

Fig. 1. Link between nuclear physics, astrophysics and astronomy and their respective tasks within nuclear astrophysics [12] (See also Plate 28 in the Color Plate Section)

conditions. Therefore, each nuclear reaction must be treated as a unique process [14]. Most of these reactions involve charged particles whose relative energy, that depends on the temperature of the star, is in general much smaller than the Coulomb barrier of the nuclear systems. Because the probability to tunnel the barrier decreases rapidly with energy, the cross sections of astrophysical interest are among the smallest ever studied in the laboratory (of the order of 10^{-9} b and less) and remain largely uncertain in many cases [15].

Among the many astrophysical sites where nuclear reactions occur, explosive environments such as novae, supernovae and X-ray bursts are particularly fascinating cases. They represent dramatic events that are characterized by high temperatures and densities and produce energy at a rate greater than in almost any other astrophysical phenomena. Under such conditions the interaction times are short enough to be of the order of the lifetimes of the β-radioactive nuclei that will be largely involved in the reaction networks [14, 16]. The sequence of reactions involving such loosely bound nuclei is determined by a balance between proton and α capture rates and rates for β decay or photo dissociation. Hundreds of different reactions involving radioactive nuclei may lie along the reaction path, and unfortunately our current knowledge concerning the properties of these nuclei and reaction rates is typically very incomplete. The information needed includes nuclear masses, excited state properties, decay properties and lifetimes, electron capture rates, neutrino and photon interaction rates, as well as light particle reaction rates [16]. To gain such information for unstable nuclei is experimentally very challenging as, for instance, typically weak intensities of radioactive ion beams require the use of efficient and highly selective experimental techniques. Knowledge of individual reaction rates is of critical importance. In many situations, there is no alternative other than to measure the properties of individual resonances.

Indeed, the production and acceleration of radioactive beams have completely changed nuclear astrophysics studies in recent years (see, for example, [17] for a review on the techniques of production of radioactive beams). However, in spite of numerous efforts there are still very few laboratories capable of delivering radioactive beams with characteristics (energy range, intensity, stability, isobaric purity) that make them useful for studies of nuclear reactions of astrophysical interest. For some nuclei, especially those at the high-Z end of the r-process path, this information will remain inaccessible for many years to come. In all cases, a rigorous theoretical treatment is necessary to extrapolate the cross sections from the experimentally measured values to the energies characterizing the astrophysical processes [18] and to calculate quantities not accessible in the laboratory.

In the following, the main quantities used in nuclear astrophysics (cross section, S-factor, resonance strength, Gamow window) are introduced. The principal burning cycles are schematically presented and some key reactions are discussed. Some promising experimental techniques are presented and some recent important results are highlighted. Complementary information may be found in recent reviews (see, for example [16, 19, 20]).

2 Relevant Quantities at Stellar Energies

Fusion reactions in stellar plasma involve charged particles and neutrons.[2] In addition to the nuclear force, the interaction implies the electrostatic repulsive force between the nuclei described by the Coulomb barrier E_C. For energies well-below the Coulomb barrier, which is always the case in the energy range relevant for astrophysics, the probability for penetrating the barrier, $\mathcal{P}_\ell$, can be approximated by the so-called Gamow factor for $\ell = 0$ (usually dominating at low energies) [6]:

$$\mathcal{P}_0 \approx \exp(-2\pi\eta(E)), \qquad (1)$$

where E is the energy in the centre-of-mass reaction system and $\eta(E)$ is the Sommerfeld parameter,

---

[2] Only charged-particle induced reactions are discussed here, see, for example, the review article [4] for neutron-induced reactions.

$$\eta(E) = \frac{Z_1 Z_2 e^2}{\hbar v}, \tag{2}$$

with Z_1, Z_2 being the charge number of the interacting nuclei, e the electron charge, $\hbar = h/2\pi$ (h is the Planck constant), and v the relative velocity of the nuclei. Numerically, $\eta(E) = 0.1575 Z_1 Z_2 (\mu/E)^{1/2}$, with $\mu = M_1 M_2/(M_1 + M_2)$ being the reduced mass of the system, M_1 and M_2 the masses of the nuclei, and E the energy in MeV. On the other hand, the cross section $\sigma(E)$ is proportional to a geometrical factor $\pi/k^2 \propto 1/E$ (k is the wavenumber). These two factors, $\mathcal{P}_0$ and π/k^2, explicitly represent the non-nuclear dependence of $\sigma(E)$. One can thus write the cross section, for energies $E \ll E_C$, as the product of three factors [21]:

$$\sigma(E) = S(E) \frac{1}{E} \exp(-2\pi\eta(E)). \tag{3}$$

This equation defines the so-called astrophysical S-factor, $S(E)$, that contains all the information related to the nuclear properties of the interacting nuclei (resonances, subthreshold states, resonance interferences, ...). If the $\ell = 0$ partial wave is dominant, the S-factor for non-resonant reactions is nearly independent of the energy. If the contribution of other partial waves ($\ell > 0$) are important, the energy dependence of the S-factor cannot be neglected at astrophysical energies. This is the case, for example, of the reactions d(p,γ)^{3}He reaction, d(d,γ)^{4}He and d(α,γ)^{6}Li [15]. Bound states near the reaction threshold (negative energy) can strongly influence the low-energy cross section. The most famous case is the ^{12}C(α,γ)^{16}O reaction that determines the ratio ^{12}C/^{16}O after helium burning and, thus, the evolution of massive stars [6]. Two subthreshold states dominate its cross section at the astrophysical energies ($E \simeq 300$ keV). In addition, interference with states above threshold strongly complicate the determination of its reaction rate that should be known to better than 20% in order to allow one to draw significant conclusions. But, because its cross section is of the order of 10^{-27} b, it cannot be experimentally determined at the astrophysically relevant energies (the cross section values presently accessible in the laboratory are of the order of 10^{-12} b [22]).

The quantity used in nucleosynthesis calculations is the thermonuclear reaction rate, which is a function of the density of the interacting nuclei, their relative velocity and the reaction cross section. Most of its lifetime, a star is in hydrostatic equilibrium: the gravitational pressure is compensated by the thermal pressure due to nuclear burning, the gas is non-degenerated and the particles are non-relativistic. Under these conditions the velocity distribution is given by the Maxwell–Boltzmann distribution. By using Eq. (2), the reaction rate for a pair of projectile and target nuclei is given by [6]:

$$\langle \sigma v \rangle = \left(\frac{8}{\pi\mu} \right)^{1/2} \frac{1}{k_B T}^{3/2} \int_0^\infty S(E) \exp\left(-\frac{E}{k_B T} - 2\pi\eta(E) \right), \tag{4}$$

where T is the temperature of the stellar interior and k_B the Bolztmann constant. This definition implies the evaluation of an integral from zero to infinity.

Table 1. Values of $E_0 \pm \Delta E_0/2$ and E_C/E_0 for some reactions of the proton–proton chains at $T_9 = 0.015$ (see text)

Reaction	$E_0 \pm \Delta E_0/2$ (keV)	E_C/E_0
p(p,νe+)d	5.9 ± 3.2	46
d(p,γ)^{3}He reaction	6.5 ± 3.4	40
^{3}He(^{3}He,2p)^{4}He	21.4 ± 6.0	43
^{3}He(α,γ)^{7}Be	22.4 ± 6.2	39
^{7}Li(p,α)^{4}He reaction	14.8 ± 5.0	46
^{7}Be(p,γ)^{8}B	17.9 ± 5.5	50

However, because the factor $\exp(-E/k_BT)$ decreases rapidly with energy and the factor $\exp(-2\pi\eta(E))$ increases rapidly with energy, the integrand needs to be evaluated only in a relative narrow energy range called the Gamow window [6, 13] centred around an energy $E_0 = 0.122\mu^{1/3}(Z_1Z_2T_9)^{2/3}$ MeV (T_9 is the temperature in units of 10^9 K). By approximating this energy region by a Gaussian function, one gets a FWHM value given by $\Delta E_0 = 0.2368(Z_1^2Z_2^2\mu)^{1/6}T_9^{5/6}$ MeV. In Table 1, the values of the Gamow window, $E_0 \pm \Delta E_0/2$, as well as the ratio E_C/E_0 are given for some relevant reactions of the proton–proton chains at a typical temperature $T_9 = 0.015$ (centre of the Sun). E_C is calculated as in [23].

Because the cross section does not show details, the advantage of introducing the S-factor is obvious. Figure 2 shows the cross section and the S-factor of the ^{3}He(^{3}He,2p)^{4}He reaction, which is a typical non-resonant reaction. It has been the first reaction to be investigated in the laboratory at energies within the Gamow window [22, 24, 25].

For a resonant reaction, the cross section can be approximated by a Breit–Wigner expression. Thus the reaction rate depends exponentially on the resonant energy E_R and on the resonance strength, $\omega\gamma$ [6]:

$$\langle\sigma v\rangle \cong \left(\frac{2\pi}{\mu k_BT}\right)^{3/2} \hbar^2(\omega\gamma) \exp\left(-\frac{E_R}{k_BT}\right),$$ (5)

where $\omega\gamma$ is defined by:

$$\omega\gamma = (1+\delta_{12})\frac{2J+1}{(2I_1+1)(2I_2+1)}\frac{\Gamma_i\Gamma_f}{\Gamma_{tot}},$$ (6)

with J being the resonance spin, I_1 and I_2, the spin of the nuclei, Γ_i (Γ_f) the initial (final) width, and $\Gamma_{tot} = \Gamma_i + \Gamma_f + ...$ the total width of the state that should contain all open channels. The spin J is the result of the coupling $\vec{J} = \vec{I_1} + \vec{I_2} + \vec{\ell}$. Equation (6) is valid only if the resonance is narrow ($\Gamma_{tot} << \Delta E_0$). For broad resonances the calculation of the reaction rate must be performed numerically [18]. At low energies, the resonant rate essentially

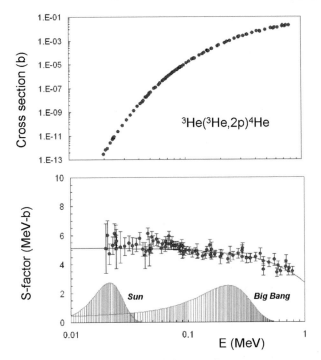

Fig. 2. Cross section and S-factor of ^{3}He(^{3}He,2p)^{4}He (data are taken from [15]). The *solid curve* is a theoretical extrapolation. The Gamow window for $T_9 = 0.015$ (Sun) and $T_9 = 0.5$ (Big Bang) are indicated (See also Plate 29 in the Color Plate Section)

depends on the resonances with lower kinetic moments, generally $\ell = 0$. For example, the reaction rate of ^{13}N(p,γ)^{14}O, which is the first reaction of the hot CNO cycle, is dominated (at $T_9 < 1$) by a $J^\pi = 1^-$ ($\ell = 0$) resonance at $E_R = 528.4$ keV [15]. On the contrary, the $J^\pi = 1^+$ ($\ell = 1$) resonance of the ^{7}Be(p,γ)^{8}B reaction is negligible at energies $E < 300$ keV [26].

Resonances with different ℓ values do not interfere among each other as far as the integrated cross section is concerned [18]. In general, if the angular momentum of the resonance ℓ_R is different from the lower value ($\ell = 0$), the contribution of the resonance is added to the non-resonant contribution, the latter being essentially that obtained for $\ell = 0$, $\sigma(E) \approx \sigma_{\ell_R}(E) + \sigma_{0,NR}(E)$. This is the case for ^{7}Be(p,γ)^{8}B, where the S-factor is dominated by the asymptotic behaviour of the Coulomb functions at $E = 0$. On the contrary, if the resonance angular momentum ℓ_R is zero, the total cross section depends on that resonance and thus, $\sigma(E) \approx \sigma_{\ell_R}(E)$, with the contribution of the other angular momenta $\ell \neq 0$ being strongly attenuated. Reactions such as ^{12}C(p,γ)^{13}N and ^{13}N(p,γ)^{14}O are examples of such a behaviour.

3 Stellar Cycles and Some Key Reactions

The first burning stage of a star is hydrogen burning that occurs typically at a temperature of the order of 10^7 K. From all the nuclear reactions involved in the proton–proton chain, $^3\text{He}(\alpha,\gamma)^7\text{Be}$ is especially interesting as it is the main source of uncertainty in determining the solar neutrino flux at higher energies [27] which results from the β decay $^8\text{B}(e^+\nu)^8\text{Be}$ following the reaction $^7\text{Be}(p,\gamma)^8\text{B}$. It also plays a role determining the primordial ^7Li abundance, although the $^3\text{He}(\alpha,\gamma)^7\text{Be}$ uncertainties do not explain the ^7Li problem [28]. Data on the $^3\text{He}(\alpha,\gamma)^7\text{Be}$ cross section were obtained so far by using two different methods, i.e. direct γ-ray detection and detection of ^7Be radioactivity [15]. The results of recent experiments [29, 30] are in agreement with an earlier data compilation [31]. New studies are underway at the LUNA Gran Sasso Laboratory.

The CNO cycle is the main energy source for stars that are somewhat more massive than the Sun (at $T \simeq 2\text{--}5.5 \times 10^7$ K). However, all stars produce energy via the CNO cycle at the end of their main sequence lifetimes, and while on the red-giant branch. At present, the reactions that play the more important role in the CNO cycle are: $^{14}\text{N}(p,\gamma)^{15}\text{O}$, $^{17}\text{O}(p,\gamma)^{18}\text{F}$, and $^{17}\text{O}(p,\alpha)^{14}\text{N}$. The latter two reactions on ^{17}O are also of interest in explosive burning, as will be shown below. The $^{14}\text{N}(p,\gamma)^{15}\text{O}$ reaction is the slowest one in the CNO cycle and thus regulates the rate of nuclear energy generation. The power liberated by the CNO cycle and the amount of helium produced are related to the luminosity observed at the transition between the main-sequence and the red-giant branch, and to the luminosity of the horizontal branch. Both of these quantities play a role in determining the ages of globular clusters [32–34].[3] Moreover, since it helps to constrain the temperature and density profiles in the hydrogen-burning shell, $^{14}\text{N}(p,\gamma)^{15}\text{O}$ will affect nucleosynthesis beyond the CNO cycle during the red-giant stage [35, 36]. Two recent, direct and independent studies [35, 37] are presented in more detail in Sect. 4. Important astrophysical consequences are also discussed there.

Hydrogen burning explains the nucleosynthesis of elements with $A \leq 4$ (the elements with $A = 7$ are not produced in sufficient amount by the p–p chain to survive hydrogen extinction). Therefore the more plausible explanation of the ratio $\text{He/H} \sim 0.2$ comes from hydrogen burning produced at the primordial Universe about 13 billions of years ago. Hydrogen burning is followed by a gravitational contraction until the centre of the star reaches a temperature sufficiently high for the ignition of helium burning (typically at about 2×10^8 K) [6, 13]. After hydrogen and helium, the more abundant elements are carbon and oxygen. Because ^8Be is not stable, the fusion process $\alpha + \alpha$ only represent an intermediate state in the ^{12}C synthesis. Because ^{12}C and ^{16}O

---

[3] The globular clusters are old stellar objects presenting heavy-element abundances much lower than those of the main sequence stars. They have presumably been formed from primordial matter, before galaxy formation. They are thus an unique "stellar laboratory" to determine the age of the Universe.

are composed of a number of protons and neutrons equivalent to three and four α particles, respectively, they can be synthesized by three-body reactions, i.e. $3\alpha \longrightarrow {}^{12}C + \gamma$. Only because the $\alpha + {}^8Be$ reaction is resonant (the Hoyle state at an energy of 278 keV above threshold [38]), ${}^{12}C$ is produced in sufficient amount to account for the stellar abundances. Another state at about 10 MeV in ${}^{12}C$ has been also observed [39] but its influence in the triple-α process remains to be firmly established [40]. The triple-α reaction dominates helium burning in the more evolved burning phases [36].

After the triple-α process, helium burning continues through the chain of reactions ${}^{12}C(\alpha,\gamma){}^{16}O(\alpha,\gamma){}^{20}Ne$. The last one is negligible except for very massive stars (more than 30 times the solar mass) [6]. ${}^{12}C(\alpha,\gamma){}^{16}O$ is one of the more important reactions in astrophysics, its cross section at 0.3 MeV (position of the Gamow window for a typical temperature of 0.25 GK) is of the order of 10^{-27} b, comparable to that of the weak interaction. Contrary to the triple-α process, ${}^{12}C(\alpha,\gamma){}^{16}O$ is practically a non-resonant reaction at that energy and its cross section is given by the tails of interfering resonance and subthreshold states. A lot of experimental efforts has been dedicated to the study of this reaction. For a detailed review, see, for example, the recent paper of Buchmann and Barnes [41].

In parallel to the main chain, ${}^{14}N$, the main ash of the CNO cycle, is transformed into ${}^{22}Ne$ by a chain of two α-capture reactions and one β decay, namely ${}^{14}N(\alpha,\gamma){}^{18}F(\beta^+){}^{18}O(\alpha,\gamma){}^{22}Ne$. ${}^{22}Ne$ is one of the main neutron sources due to the ${}^{22}Ne(\alpha,n){}^{25}Mg$ reaction and thus an important path to the s-process. The ${}^{22}Ne(\alpha,n){}^{25}Mg$ reaction cross section remains largely uncertain [15]. The other important neutron source in massive stars is ${}^{13}C(\alpha,n){}^{16}O$ [42]. Its cross section, measured at energies above 0.3 MeV [43, 44], strongly depends on a $1/2^+$ state situated at 3 keV below threshold [45]. The extrapolation to astrophysical energies remains very uncertain.

A sequence of further reactions gradually transform the hearth of stars into heavier and heavier nuclei. The energy produced balances the gravitational contraction. However, once the thermal pressure is not sufficient to balance the gravitational force the hearth contracts and the temperature increases enough to ignite the ashes of previous burning phases. Thus, the helium burning ashes in massive stars are the fuel of successive nuclear processes. After helium burning, and depending on the mass of the star, phases burning carbon, oxygen, neon and silicon will successively take place to produce iron. The last possible step is explosive burning [13].

Explosive burning takes place at much higher temperatures and densities during events such as novae, supernovae and X-ray bursts. Nucleosynthesis in novae (temperatures $T \simeq 2-3 \times 10^8$ K, densities $\rho \simeq 10^3$ g/cm^3) involves about 100 stable and radioactive nuclei ($A < 40$) and a few hundred reactions [46], mainly (p,γ) and (p,α) reactions with the nucleosynthesis path located at the border of nuclear stability. Data needed for nova models are essentially rates of reactions involving stable and radioactive nuclei and β-decay half lives. The situation is more complex for X-ray bursts

$(T \simeq 10^9$ K, $\rho \simeq 10^6$ g/cm^3) as the main sequence of reactions is far from stability, reaching the proton drip line above $A = 38$ [47]. The data needed when modelling these stellar systems are the cross sections for proton and α-induced reactions on stable and radioactive nuclei and for photodissociation as well as the β-decay half lives, for a few hundred nuclei with masses $A \le 110$ [48]. This means that several thousand reactions are involved [49]. Recent experiments are dedicated to a series of important reactions in nova nucleosynthesis: ^{17}O(p,γ)^{18}F and ^{17}O(p,α)^{14}N (direct and indirect studies at TUNL and CSNSM Orsay) [50–52], ^{21}Na(p,γ)^{22}Mg (direct studies at TRIUMF and indirect studies at KVI) [53–55], ^{22}Na(p,γ)^{23}Mg (indirect studies at ANL) [56], ^{23}Na(p,γ)^{24}Mg and ^{23}Na(p,α)^{20}Ne (direct studies at TUNL) [57], and ^{30}P(p,γ)^{31}S (indirect studies at ANL) [58]. A lot of effort has been dedicated to the ^{18}F(p,α)^{15}O reaction, which is the main destruction reaction of ^{18}F ($T_{1/2} = 110$ min) during nova outbursts and thus related to detection of γ-rays from novae by future satellite missions. Its cross section needs to be known at energies as low as 200 keV (see, for example, [59]). This reaction will be discussed in detail in Sect. 4.5.

Figure 3 shows schematically the relation between stellar sites and nuclear processes. A detailed discussion can be found in Chap. 5 of [6].

4 Experimental Techniques in Nuclear Astrophysics

The experimental techniques used to investigate nuclear quantities of astrophysical interest are very varied and depend on the stellar environment under investigation (quiescent or explosive burning). Some relevant issues are discussed in the following.[4] To illustrate the experimental techniques typically used in nuclear astrophysics studies, three selected examples of reaction studies are described in more detail.

4.1 Targets

The measurement of reaction cross sections at very low energies with intense stable-isotope beams (of the order of mA) and water-cooled thick solid targets requests the monitoring of the stability and of the stoichiometry of the target as a function of the beam dose. This is typically performed by measuring from time to time the cross section at a reference energy (target stability) and by nuclear reaction analysis before and after the target has been employed (target stoichiometry). The appropriate corrections must be applied to the experimental results [13].

The study of reactions involved in explosive astrophysical process requires the development of radioactive beams [17]. Reactions involving hydrogen and helium are the most important ones in explosive burning. Because

---

[4] The choice here is exclusively based on the author's preferences and experience.

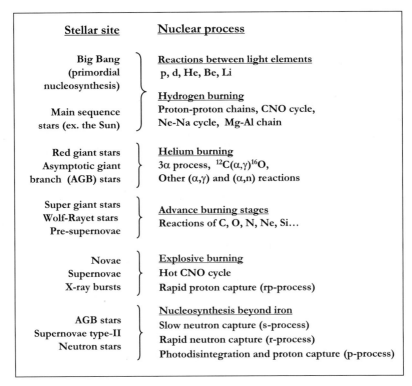

Fig. 3. Relation between astrophysical sites and nuclear processes [12]

the lifetime of the interacting nuclei are too short to allow one to produce targets, inverse kinematics methods using radioactive beams require the use of hydrogen-rich or helium-rich targets. The choice of the target must be adapted to the physics goal and the other experimental conditions (such as beam energy and intensity, detection system, etc). Polyethylene foils $[(CH_2)_n]$ are easy to handle and have been one of the most popular and successful targets for investigation of hydrogen burning reactions. Foils with thicknesses between $40\,\mu g/cm^2$ and several mg/cm^2 have been used with beam intensities as high as $10^9\ s^{-1}$ [60, 61] without significant degradation, though care must be taken to distribute the beam power by, for example, rotating the target. Solid targets containing helium are produced by implantation. Implanted helium targets have been developed at Louvain-la-Neuve with helium thicknesses up to 10^{18} atoms/cm^2, sufficient for measurement of elastic scattering and some reactions with radioactive ion beams [62]. Gas targets are an obvious alternative to foils. Gas cells with thin windows are easy to handle, but the windows produce similar challenges as with foil targets, degrading the beam energy and inducing background reactions. Windowless gas targets eliminate the problems associated with windows. However, many pumping stations are required to decrease the pressure to the 10^{-7} mbar range required when performing on-line experiments at accelerators. Hence the gas targets are large and costly, and the target thickness is limited [63].

4.2 Detectors

Studies of capture reactions of interest in quiescent burning are mainly limited by the cosmic background of the γ detectors. One can build a lead wall around the detectors but the interaction of the cosmic rays with the material will produce γ-rays and neutrons that will also affect the measurements. Another possibility to partially reduce the activation problem is an active shielding using, for example, plastic scintillators operated in anti-coincidence with the γ detectors. The best (but not always the easiest) solution is going underground [64]. The pioneers of underground laboratories for nuclear astrophysics reaction measurements is the LUNA laboratory, situated under the Gran Sasso mountain in Italy. Its unique character is a suppression of the cosmic rays equivalent to 4,000 m of water. Two linear accelerators that are installed at LUNA (50 and 400 keV) have allowed to measure cross sections of the order of 0.01 pb [22]. An example of these measurements at the limits of the technical possibilities is the study of the $^{14}N(p,\gamma)^{15}O$ reaction [37]. In the United States, a new project for the construction of an underground laboratory has been recently launched by a collaboration of several universities and laboratories from US and Europe.

In studies involving radioactive beams, the low-beam intensities (at present, typically of the order of $10^4 - 10^8$ s^{-1} on target) require the use of very efficient detection systems. Arrays of γ, neutron and charged-particle detectors have been constructed at many facilities in order to maximize the detection efficiency [16]. The advent of large-area silicon strip detectors covering a large solid angle has played a crucial role. These detectors may be segmented in one or two dimensions (doubled-sided detectors) to any practical level of pixelation. The shape of the strips can also be tailored to experimental requirements. For example, strips are curved in a circular pattern in many annular detector designs to allow better reaction angle resolution in a strip. This approach was used in the Louvain–Edinburgh detector array (LEDA), one of the pioneering charged-particle arrays used with radioactive ion beams for nuclear astrophysics [65]. LEDA is composed of independent 16-strip sectors. Its typical electronic resolution is of the order of 10 keV, the energy resolution for 5.5 MeV α particles is about 20 keV, while the time resolution is of about 1 ns. Figure 4 shows a schematic drawing of one of the LEDA sectors and a typical experimental setup using two LEDA arrays. A broader range of detector thicknesses has recently become available, and detectors between 50 μm and 1 mm are common. This broad range of thicknesses allows Z identification of a broad range of charged particles through $\Delta E - E$ techniques. These detectors require new associated electronic modules and new data acquisition systems capable to work with a large number of signal chan-

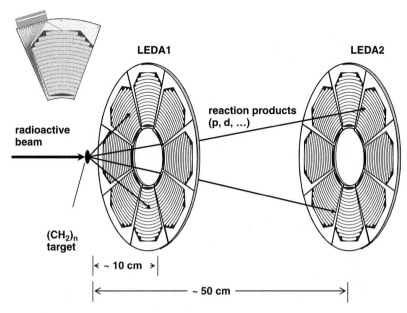

Fig. 4. Typical experimental setup for the measurement of a transfer reaction by means of the LEDA detector [65]. The insert on the *top left* is a schematic drawing of one LEDA sector (See also Plate 30 in the Color Plate Section)

nels and a small deadtime. Also because of the very low-beam intensities it is not possible to use the conventional methods to measure the accumulated beam dose with sufficient precision. In the measurement of cross sections or of resonance strengths, the absolute normalization is always one of the most important tasks. One of the techniques successfully employed is to evaporate a very thin gold layer on the target (when using a plastic foil as a target) and normalize to the measurement of the Rutherford cross section [66].

Arrays of γ-ray detectors have played an important role in several new approaches using both stable-isotope and radioactive ion beams. For instance, the high-total efficiency of the Gammasphere array [67] allowed $\gamma - \gamma$ coincidence measurements. They accurately determined excitation energies of levels in proton-rich nuclei that are of astrophysical importance. Some of the problems with direct measurements, mainly the low efficiency of γ-detectors, the radioactivity of the target material, and the background sources can be solved by performing measurements in inverse kinematics and detecting the recoiling reaction products in recoil separators. This technique is briefly described in Sect. 4.3.

4.3 Recoil Separators

Recoil separators are devices which separate the nuclear reaction products (recoils) leaving the target from the primary beam and focus the former onto a detector system [68]. The experimental challenge is to maintain a high transmission of the heavy reaction recoils while maximizing the rejection of the primary beam. This is difficult because of the small mass and momentum difference between the projectiles and recoils. It is also difficult because all the projectiles enter the separator, their intensity being typically $10^{10} - 10^{15}$ times larger than of the recoils. To obtain the maximum separation, the primary beam is blocked at an early stage of the separator. Some recoil separators have the additional property of dispersing the reaction products at the focal plane according to their mass-to-charge ratio. The scattered beam rejection is enhanced by filtering out particles on the basis of both velocity and ratio between both mass and ionic charge, which necessitates the use of either velocity filters combined with magnetic dipoles or a combination of electric and magnetic dipole elements.

A recoil separator suitable for (p,γ) and (α,γ) studies in astrophysics should have the following specifications [68]:

(i) High-transport efficiency for a relative small solid angle (typically less than 5 msr): due to the inverse kinematics, the maximum angles of the ejectiles should peak near $0°$.
(ii) High-beam rejection over a broad mass range of beams.
(iii) Relative low-mass resolution ($\delta M/M \leq 0.5\%$).
(iv) Target chamber capable of accommodating a variety of detector arrays and gas targets (both jet and extended targets).
(v) Capability of running with different ion optical modes for reactions with different kinematics.
(vi) Incorporation of careful beam handling upstream of the separator (e.g. clean recoil beam with small dispersion and no beam halo).

Several laboratories have developed recoil separators that are designed to collect heavy reaction products and disperse them by their mass-to-ionic-charge ratio. Such separators for astrophysical studies are, for example, DRAGON at ISAC (TRIUMF) [63], ERNA at Bochum [69], the Daresbury Recoil Separator at the HRIBF (Oak Ridge) [70], and the FMA at ANL (Argonne) [61]. A new separator dedicated to nuclear astrophysics studies using stable-isotope beams is under construction at the University of Notre-Dame [71].

4.4 Ground State Properties

The β-decay half lives and masses of nuclei are important for understanding explosive processes. Half lives for β decay can be long compared to the time scale for nuclear reactions, and thus the decays of nuclei near the proton drip line can govern energy generation and nucleosynthesis in the explosion. As the rates of nuclear reactions depend exponentially on the reaction Q value, mass measurements are a crucial first step towards determining these rates. The development of highly selective spectrometers, traps, detectors, and other instrumentation at laboratories around the world has allowed nearly all isotopes of interest for explosive hydrogen burning to be produced and identified in recent years. Half lives and masses, which can be measured with relatively few atoms, are often the first quantities determined experimentally.

Thanks to this impressive world-wide efforts [72–74], now there remain only a few particle-stable neutron-deficient isotopes with $Z < 53$ whose half lives have not been measured with reasonable accuracy. Most important for the rp-process are the last unobserved even–even nuclei below ^{100}Sn, i.e. ^{74}Sr, ^{78}Zr ($N = Z - 2$) and ^{96}Cd ($N = Z$) which could have reasonably long half lives. Although the general progress in studying half lives in this region of the chart of nuclei is indeed impressive, there still exist many isotopes heavier than nickel that lack accurate mass measurements or other experimental information. The techniques developed for capturing and cooling nuclei in traps and storage rings should allow the situation with nuclear masses to be much improved in the near future. However, the additional structure information necessary to extract reaction rates will require substantially more effort.

4.5 Resonances Properties

Some reaction rates at the temperatures of explosive burning are totally or partially dominated by the contribution of resonances. It is therefore important to study the properties of the resonant states using, e.g. elastic and/or inelastic scattering, transfer reactions populating the mirror states, and fusion evaporation reactions. These techniques are discussed in the following subsections.

Elastic and Inelastic Scattering

Applied since the early 1950s [75], elastic scattering is a well-known method to study resonant states. As a natural extension [76, 77], the elastic scattering technique in *inverse kinematics*, used normally to investigate reactions involving radioactive species, makes use of the sensitivity of the protons (or α particles) to the presence of a resonant state in the compound nucleus. Figure 5 shows the principle of the method. The method is based on the fact that the energy loss of heavy ions in a target is significantly larger than the energy loss of protons (which is normally negligible for typical target thicknesses of less than 1–2 mg/cm^2). The resulting recoil proton spectrum can be compared to a "snapshot" of a certain energy range in the level scheme of the compound nucleus. The method can also be applied to α scattering, although being less sensitive. Another advantage is that the laboratory energy of the recoil particles (protons or α particles) are rather high and given by:

$$E_{\text{lab}} = E_{\text{cm}} \frac{4A_{\text{p}}}{A_{\text{p}} + A_{\text{t}}} \cos \phi_{\text{lab}}^2, \tag{7}$$

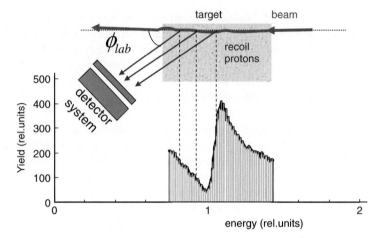

Fig. 5. Principles of the elastic scattering method in inverse kinematics. The spectrum is a typical interference pattern for a $\ell = 0$ resonance (see text) (See also Plate 31 in the Color Plate Section)

where A_p, A_t are the mass number of the projectile and the target nuclei, respectively, and ϕ_{lab} the recoil angle. On the other hand, the cross section for elastic scattering is proportional to the square of two contributions, the Coulomb amplitude and the nuclear amplitude [78]:

$$\left(\frac{d\sigma}{d\Omega}\right)_{\mathrm{elas}} = |f_{\mathrm{N}} + f_{\mathrm{C}}|^2 \tag{8}$$

The nuclear amplitude depends on the collision matrix $U_{\ell I}^{J\pi}$ that can be written as a function of the phase shift [78]:

$$U_{\ell I}^{J\pi} = \exp\left(2i\delta_{\ell I}^{J\pi}\right), \tag{9}$$

where J and π are the spin and parity of the resonant state, respectively, I the channel spin and ℓ the angular momentum. When a resonant state is "scanned" with the appropriate combination projectile–target, the recoil proton spectra show an interference pattern that is indicative of the resonance energy, angular momentum and width (see Fig. 5). All experimental effects, mainly the beam energy resolution inside the target and the angular resolution of the detectors, must be properly taken into account to precisely extract the resonant properties. Fitting procedures, such as the R-matrix method [78], are typically used to evaluate these quantities and to obtain the resonance properties.

Elastic scattering measurements with radioactive beams have been widely applied to determine resonant properties that are important for nuclear astrophysics and nuclear structure [66, 79–81]. One of the main advantages is the large cross section for elastic scattering that allows measurements with radioactive beam intensities as low as 10^4 s^{-1}. Another advantage is that the energy of the recoil protons is usually sufficiently high to be detected readily in standard silicon detectors or in more sophisticated strip detector arrays [65]. The β radioactivity of the beam species may induce background in the particle detectors at low energies. This background can limit the lowest energies measurable, but this time-uncorrelated background can be distinguished from the events of interest by time-of-flight techniques.

A recent example of an elastic scattering measurement of astrophysical interest is the study ^{21}Na+p at the ISAC facility at TRIUMF [82]. An intense $(5 \times 10^7$ s$^{-1})$ ^{21}Na beam at laboratory energies below 1.5 MeV/nucleon bombarded 50–250 μg/cm^2 thick $(CH_2)_n$ targets. The recoil protons were detected using the TUDA strip detector array [65]. Three strong resonances corresponding to states in ^{22}Mg have been identified at energies of 830, 1115 and 1311 keV, respectively. These states dominate high-temperature burning of ^{21}Na via ^{21}Na(p,γ)^{22}Mg and probably influence the low-temperature stellar rate of this reaction. However, to determine the γ strengths and other properties of these states, additional radiative capture measurements are needed [83].

In an inelastic scattering process, the energy of the collision is sufficiently high to excite one of the interacting nuclei. This method becomes particularly important when one attempts to extract a reaction rate from cross section measurements of the inverse reaction. For example, the ^{17}F(p, α)^{14}O cross section has been measured at energies that allow the inverse ^{14}O(α,p$_0$)^{17}F$_{gs}$ reaction rate to be determined by detailed balance [84, 85]. However, the contribution of the ^{14}O(α,p$_1$)^{17}F* reaction branch to the 495 keV first excited state of ^{17}F is not determined by the inverse reaction measured on nuclei in their ground state. Both elastic and inelastic scattering of ^{17}F+p were measured at Argonne National Laboratory [86] and at the Holifield Radioactive Ion Beam Facility (HRIBF) [87] to study states in ^{18}Ne that are important for the ^{14}O(α,p)^{17}F reaction.

Another example is the measurement of ^{19}Ne(p,p$_1$)^{19}Ne* recently performed by a GANIL–Louvain–Edinburgh collaboration at the CRC in Louvain-la-Neuve. The goal was to determine the properties of the ^{19}Ne states near the ^{18}F+p threshold (see Sect. 4.5). Data analysis is ongoing.

Transfer Reactions

Much of our understanding of nuclear spectroscopy has been shaped by the study of transfer reactions in the last decades. These have proved to be a particularly powerful tool for characterizing energy levels of importance for astrophysical reactions. One advantage of these measurements is that states covering a broad region of excitation energy are populated. The main disadvantage from the experimental point of view is that this method needs high-resolution particle detection in order to resolve the states of interest. In addition, difficult targets and problematic kinematics make studies of resonant states via proton transfer in inverse kinematics particularly challenging.

On the theoretical side, many of these reactions can be described by the distorted-wave born approximation (DWBA) which predicts that the shape of the angular distribution of the differential cross section is distinctive of the transferred angular momentum, and the magnitude of the cross section reflects the single-particle character of the state [88]. However, the results are usually model-dependent. Proton-transfer reactions, for example (^{3}He,d), are the best surrogate for proton-induced reactions like (p, γ). The (d, p) neutron-transfer reaction has also been used to indirectly obtain information on single-particle resonances. The properties of neutron single-particle states are studied by the (d, p) reaction on the mirror nucleus, and the properties of proton resonances are determined under the assumption of mirror symmetry. This technique was first applied with a radioactive ion beam to ^{56}Ni at Argonne National Laboratory [89] and it has been used more recently to study the ^{18}F$(p,\alpha)^{15}$O at Louvain-la-Neuve [90, 91] and at the HRIBF [92, 93] (see Sect. 4.5).

Reactions like (p, t) [94], (^{3}He,n) [95, 96] and (^{3}He,^{6}He) [97, 98] have been extensively studied. The distinctive Q-values for these reactions typically allow for a high selectivity for charged particle detection in high-resolution magnetic spectrographs. Comparable resolution is also possible with the (^{3}He,n) reaction using time-of-flight techniques with only a modest flight path. For example, states in ^{26}Si that are important for the ^{25}Al$(p, \gamma)^{26}$Si reaction were studied at the Edwards Accelerator Laboratory at Ohio University using the ^{24}Mg(^{3}He,n)^{26}Si reaction [99]. Neutrons were detected with a flight path of 10 m, and a resolution of about 16 keV was achieved in the region of interest.

More exotic reactions, such as (^{3}He,^{8}Li) [100], (^{4}He,^{8}He) [101], (^{7}Li,^{8}He) [102, 103] and (^{12}C,^{6}He) [96], have also been used to study states of nuclei further away from stability by means of stable-isotope beams and targets. The mechanisms of such reactions are complex, and the cross sections are typically small. However, in some cases high-intensity stable-isotope beams can be used to achieve reasonable reaction yields. States of unnatural parity are sometimes populated with comparable yields to natural parity states, allowing states to be studied that are weakly populated in direct reactions. For example, states of unnatural parity in ^{26}Si, that dominate the rate of the ^{25}Al$(p, \gamma)^{26}$Si reaction and are thus important for understanding the production of ^{26}Al in novae, were studied using the ^{29}Si(^{3}He,^{6}He)^{26}Si reaction at Wright Nuclear Structure Laboratory (WNSL) at Yale University [104].

Nuclei heavier than nickel in the rp-process are too far away from stable nuclei to be produced by transfer reactions involving stable-isotope beams and targets. Radioactive ion beams are thus required to access these nuclei by transfer reactions. Fragmentation facilities like that at the Michigan State University (MSU) produce nuclei near the rp-process path as beams with sufficient intensity to allow transfer reaction studies. Nucleon knock-out reactions like (p, d) have much more favourable cross sections than stripping reactions at the beam energies available from fragmentation facilities. A set of measurements proposed for MSU will use the (p, d) reaction induced by a radioactive ion beam to populate proton-unbound states of nuclei near the path of the *rp*-process. The emitted protons and residual heavy nuclei being detected in coincidence in order to construct excitation energy spectra.

4.6 Selected Examples of Reaction Measurements

The ^{14}N(p,γ)^{15}O Reaction and the Age of the Globular Cluster

The capture reaction ^{14}N(p,γ)^{15}O ($Q = 7.297$ MeV) is the slowest process in the hydrogen burning CNO cycle and thus of high-astrophysical interest. This reaction plays a role in setting the energy production and neutrino spectrum of the sun [27] as well as determining the age of globular clusters cluster [32–34]. Below 2 MeV, several states in ^{15}O contribute to the ^{14}N(p,γ)^{15}O cross section: a $3/2^+$ subthreshold state at -504 keV and three resonant states, $1/2^+$ at 259 keV, $3/2^+$ at 987 keV, and $3/2^+$ at 2187 keV, respectively. Figure 6 shows the level scheme of ^{15}O. Above the ^{14}N+p threshold only the above-mentioned resonant states are indicated.

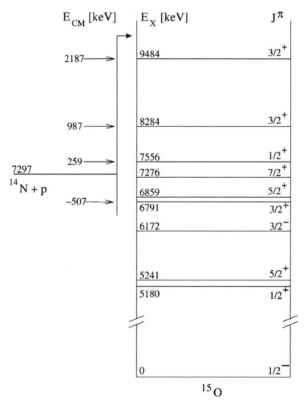

Fig. 6. Level scheme of ^{15}O (from [37]). States relevant for the ^{14}N(p,γ)^{15}O reaction are indicated by *arrows*

A theoretical analysis of previous data [105], using the R-matrix model [106], yielded a difference of a factor of 2 in the rates with respect to those adopted in compilations [15] at the relevant temperatures. This difference is due to the different contribution of the subthreshold state to the capture reaction leading to the ^{15}O ground state, which was found to be negligible in [106]. This result was supported by a lifetime measurement of the subthreshold state via the Doppler-shift [107] and the Coulomb excitation [108] methods. In view of this and of the astrophysical importance of the ^{14}N(p,γ)^{15}O reaction, a new more precise measurement, extending the lower-energy limit below that reached by previous work ($E < 0.24$ MeV), was highly desirable [106].

Two experimental groups have undertaken the study of the ^{14}N(p,γ)^{15}O cross section at low energies, one at the LUNA laboratory at Gran Sasso [37] and the other one at the LENA laboratory at TUNL [35]. The results of the two measurements analysed by using the R-matrix method are in agreement. The result obtained by a simultaneous R-matrix analysis of all exiting data sets [109] are shown in Fig. 7. With the present ^{14}N(p,γ)^{15}O rates, the age at the main-sequence turnoff[5] is 0.5–1.0 Gy older than that deduced from the previous rates. This value depends on the method used to determine the luminosity and the metalicity of the globular clusters even if all other parameters are assumed to be fixed (e.g. distance to globular clusters, time between the

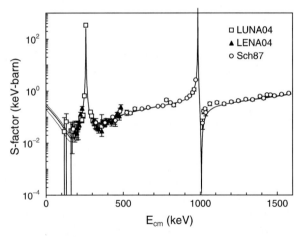

Fig. 7. ^{14}N(p,γ)^{15}O data for the transitions to the ground state in ^{15}O. The experimental results marked by squares, triangle and circles stem from the LUNA [37] and LENA [35] facilities and from the early work of Schöder et al. [105], respectively. The solid curve is the best R-matrix fit [109]. Different curves, corresponding to different R-matrix parameters, are undistinguishable above 0.2 MeV

---

[5] The turnoff point is the moment at which the star evolves from the main sequence (central burning phase) to the asymptotic branch of red giants (shell burning phase).

Big Bang and the star formation, etc.) [12]. In addition, the solar neutrino flux from the CNO cycle is reduced by a factor of 2 [27].

For pedagogical purposes, the measurement of the ^{14}N(p,γ)^{15}O cross section at the LENA laboratory [35] (a typical "above-ground" low-energy accelerator) is briefly described in the following. The measurement was performed using a 1 MV Van der Graaff accelerator providing proton beams at laboratory energies between 155 and 524 keV, with beam currents of 100–150 µA. The ^{14}N targets were fabricated by implanting nitrogen ions into a 0.5 mm-thick tantalum backings, that etched in an acid solution to remove surface impurities. The target composition and thickness remained stable over the accumulated doses of 20–25 C. This was regularly checked by measuring the yield curve for the 0.259 MeV resonance (see Fig. 6). A total of 32 independent measurements were performed to obtain the resonance strength of this resonance, yielding good agreement with previous values. Branching ratios for the decay of the 0.259 MeV resonance were also measured. The Ta/N stoichiometry was measured by Rutherford backscattering. Gamma rays were detected using a 135% HPGe detector placed at 0° close to the target. The energy calibration and the absolute photopeak efficiency were obtained using radioactive sources and the decays from well-known resonances in several capture reactions. An important quantity that allows to correct for summing of coincidence γ rays in the HPGe detector is its total efficiency. The latter quantity was calculated using numerical codes and normalized to source data. Finally, a large long-annulus NaI detector enclosing both the target and the germanium crystal, was used as a cosmic-ray veto while also suppressing events arising from γ-ray cascades. Data for different ^{14}N(p,γ)^{15}O transitions were obtained, whereas in Fig. 7 only the results for the transition to the ^{15}O ground state are shown.

The ^{18}F(p,α)^{15}O Reaction and γ-ray Emission from Novae

Gamma-ray emission from novae is dominated by positron annihilation following β decay of radioactive nuclei [59]. The principal contribution to this emission comes from the decay of newly synthesized ^{18}F and it is therefore of great importance to understand the production and destruction rates of ^{18}F. Moreover, the relatively long half-life of ^{18}F (110 min) means that this annihilation radiation will be present after the expanding envelope of the nova becomes transparent. For this reason, measurements of the ^{18}F abundance are amongst the principal objectives of current and planned γ-ray observatories. The destruction of ^{18}F is determined by the rates of the proton capture reactions ^{18}F(p,α)^{15}O and ^{18}F(p,γ)^{19}Ne, at astrophysically relevant temperatures of 0.1–0.4 GK. The ^{18}F(p,α)^{15}O reaction is dominant. At these temperatures, properties of states of ^{19}Ne in the vicinity of the ^{18}F+p threshold determine the astrophysical S-factor and therefore the ^{18}F destruction rate. Despite an extensive series of measurements of (p,p) [110–113] and (p,α) [111, 113–117] reactions using radioactive ^{18}F beams, and transfer reactions to populate states in ^{19}Ne and ^{19}F [90–93, 118–120], significant uncertainties remain, particularly concerning states near the ^{18}F+p threshold.

The direct measurement of ^{18}F(p,α)^{15}O reaction at the energies corresponding to nova temperature (centre-of-mass energies of about 0.2 MeV) requires a ^{18}F beam of less than 4 MeV with an intensity of at least 10^{12} s^{-1} on target, which is a million times more than the one achievable at present. To overcome the technical difficulty of the direct measurement, indirect methods have been applied, for example, the ^{18}F(d,p)^{19}F(α)^{15}N reaction transfer [90–93]. The aim of these measurements was to determine neutron spectroscopic factors for states in ^{19}F that are mirrors to ^{19}Ne states important for the ^{18}F(p, α)^{15}O reaction. Under the assumption of mirror symmetry, these spectroscopic factor can constrain the rate of the ^{18}F(p, α)^{15}O reaction. In the Louvain-la-Neuve experiment [90, 91], a 14-MeV ^{18}F radioactive beam of about 2×10^6 s^{-1} bombarded a CD$_2$ target. Two LEDA silicon strip detector arrays were used to detect protons in coincidence with α particles or ^{15}N ions from the breakup of α-unbound states in ^{19}F. Differential cross sections for the ^{18}F(d,p)^{19}F reaction were fit by theoretical DWBA distributions to extract neutron spectroscopic factors for excited states in ^{19}F. A similar approach was used in the HRIBF measurement [92, 93], except that better energy resolution was achieved by using a higher bombarding energy (108 MeV). Cross sections for transfer to bound states (with well-known excitation energies) were also measured by detecting ^{19}F ions in the daresbury recoil separator. Figure 8 schematically shows the experimental setup used at HRIBF.

Spectroscopic factors extracted from both of these measurements are consistent and place important new limits on contributions of low-energy resonances to the rate of the ^{18}F(p,α)^{15}O reaction. Although the uncertainty was considerably reduced, the reaction rates at nova temperatures are still largely uncertain. Figure 9 shows the present S-factor estimates at nova temperatures. The main uncertainties arise from the unknown interference signs between the several $3/2^+$ states near the ^{18}F+p threshold.

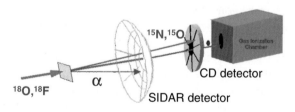

Fig. 8. Experimental setup used at HRIBF to study the ^{18}F(d,p)^{19}F(α)^{15}N reaction [92, 93]. The SIDAR and the CD detectors are composed of independent sixteen-strip silicon sectors (See also Plate 32 in the Color Plate Section)

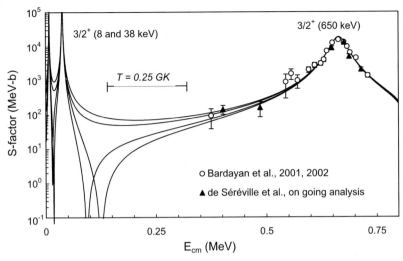

Fig. 9. Astrophysical S-factor of the ^{18}F(p,α)^{15}O reaction. The curves are R-matrix calculations (for a channel radius $a = 5.5$ fm) using different interference sign for the three $3/2^+$ states situated at 8, 38 and 650 keV above threshold. The Gamow window window for a typical nova temperature of 0.25 GK is indicated. The data are from [111, 117] (*open circles*) and some preliminary results from a recent experiment at Louvain-la-Neuve (*triangles*)

The ^{7}Be(d,p)^{8}Be Reaction and the Primordial ^{7}Li Problem

The reaction ^{7}Be(d,p)^{8}Be, one of the destruction channels of ^{7}Be (in competition with the electronic capture ^{7}Be(e$^-\nu$)^{7}Li) is related to primordial ^{7}Li abundance. At present, there is a large difference (about a factor of 2–3) between the ^{7}Li abundances obtained from nucleosynthesis calculations using reaction rates [28, 31] and the ^{7}Li abundances observed in halo stars of the Galaxy [121], if one considers the very precise value of the baryonic content of the Universe, $\Omega_b h^2 = 0.0224 \pm 0.0009$,[6] recently obtained by WMAP [122].

Before suggesting that new physics may be needed to solve this puzzle, effects related to uncertainties in reaction rates involved in the Big Bang nucleosynthesis (BBN) have to be excluded. For high-baryon density, the ^{7}Li abundance from BBN models arises principally from ^{7}Be that further decays to ^{7}Li. Hence reconciliation of BBN, WMAP and ^{7}Li observations by nuclear physics effects can only come from ^{7}Be production and destruction rates. In BBN, the main reactions are ^{3}He(α, γ)^{7}Be and ^{7}Be(n,p)^{7}Li which are sufficiently well known [31] to exclude this option. However, other reactions have to be considered.

---

[6] Ω_b is the ratio of the baryonic density to the critical density, h is the Hubble constant in units of 100 km s^{-1}·Mpc^{-1}

This is the case of ^{7}Be(d,p)2α reaction. Its reaction rate came from an estimate by Parker [123] based on partial experimental data above the centre-of-mass energy of 0.6 MeV from an early work of Kavanagh [124]. In [124], protons corresponding to the 0^+ ground state and the first excited state (3.06 MeV, 2^+) in ^{8}Be were detected at $90°$ using a NaI(Tl) detector. The estimate of the ^{7}Be(d,p)2α cross sections at the Gamow window ($T = 0.1-1$ GK, $E = 0.11-0.56$ MeV) implies an extrapolation of about two orders of magnitude. If the actual ^{7}Be(d,p)2α reaction rate were a factor of about 100 larger at these low energies, where no data existed, the ^{7}Li disagreement would vanish. Figure 10 shows the level scheme of ^{9}B and ^{8}Be, and the ^{7}Be(d,p)2α reaction threshold ($Q = 16.490$ MeV) [125]. In addition to the ground state and the first excited state in ^{8}Be, investigated so far [124], higher-lying states in ^{8}Be can have a non-negligible contribution to the ^{7}Be(d,p)2α reaction cross section. This reaction has been recently investigated using a radioactive ^{7}Be beam at c.m. energies $E = 0.96-1.2$ MeV ($E_{\text{lab}} = 5.55$ MeV) and $E = 0.15-0.38$ MeV ($E_{\text{lab}} = 1.71$ MeV) and a 200 μg/cm^2 (CD$_2$)$_n$ target [126].

Because of the high Q-value, the expected laboratory energies of the reaction products (protons, α-particles, ...) are high and thus particle identification is needed, e.g. by using a ΔE–E telescope. For example, protons produced by an incoming 5.55 MeV ^{7}Be beam over the target thickness, and feeding all ^{8}Be states below the ^{7}Be+d threshold, have energies ranging from 2.5 to 22 MeV for the covered angles. Hence, to distinguish the protons coming from the ^{7}Be+d reaction from those arising from reactions on the carbon content of the target, a stack of two silicon strip detector LEDA arrays [65] were used. They covered laboratory angles from $7.6°$ to $17.4°$. Two energy-loss measurements were performed by detector arrays abbreviated as ΔE_1 and ΔE_2. The former consisted of eight detectors of 0.3 mm thickness, while the latter included four of 0.3 mm and four of 0.5 mm thickness. Alpha particles, recoil and scattered particles from ^{7}Be+^{12}C reactions were completely stopped in ΔE_1. From all the open reaction channels, only protons from the ^{7}Be+d reaction were able to pass through both ΔE_1 and ΔE_2 detector arrays. High-energy protons corresponding to the ground state and the first excited state in ^{8}Be were not completely stopped in the $\Delta E_1 - \Delta E_2$ telescope, while protons corresponding to other higher-lying excited states in ^{8}Be were stopped in ΔE_2.

Figure 11 shows a typical $\Delta E_1 - \Delta E_2$ calibrated spectrum obtained at ^{7}Be beam energies of 5.55 MeV. This spectrum corresponds to the total number of counts integrated in the entire detector [126]. The proton signals are well separated from the uncorrelated background ($\Delta E_2 < 1$ MeV). The most strongly populated regions correspond to the feeding of the 0^+ ground state and 2^+ excited state in ^{8}Be. The two levels are unresolved. The two cluster of events observed are due to the different silicon wafer thicknesses (0.3 and 0.5 mm, respectively) used for ΔE_2. The regions of interest were calculated by taking into account the kinematics of the reaction populating these states, the width of the states, the straggling of the beam in the target, and the energy loss

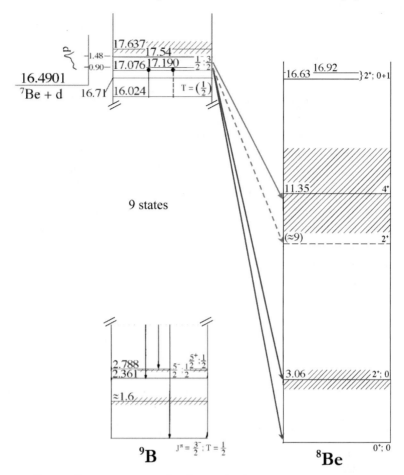

Fig. 10. ^{9}B and ^{8}Be level scheme [125]. The (d,p) reactions populating the ground state and first excited state of ^{8}Be were investigated by Kavanagh [124]. Other ^{8}Be states of interest are also indicated (See also Plate 33 in the Color Plate Section)

of protons in ΔE_1 and ΔE_2 [127]. Other events correspond to the feeding of the higher-lying ^{8}Be levels for which part of the protons are stopped in ΔE_2, hence characterized by a different shape in the $\Delta E_1 - \Delta E_2$ plot.

Figure 12 shows the results as the ^{7}Be(d,p)2α reaction astrophysical S-factor. The full triangles are the contribution from the ground state and the first excited state in ^{8}Be (about 65% of the total), in good agreement with the data from [124] (open circles). The full circles include the contribution of a large 4^+ state at 11.35 MeV in ^{8}Be (see Fig. 10). These results show that the states not observed by [124] account for about 35% of the total S-factor and not of a factor of 3 as previously estimated [123]. This means that the

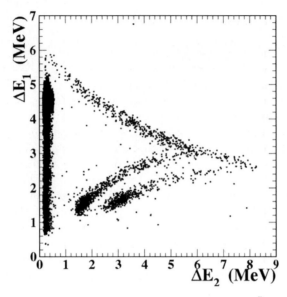

Fig. 11. Typical $\Delta E_1 - \Delta E_2$ proton spectrum obtained for a ^{7}Be beam energy of 5.55 MeV and a 200 μm thick $(CD_2)_n$ target [126]. See text for details

^{7}Be(d,p)2α reaction cross section is about ten times smaller than previously estimated at Big Bang energies, thus excluding a solution via nuclear physics of the primordial ^{7}Li problem.

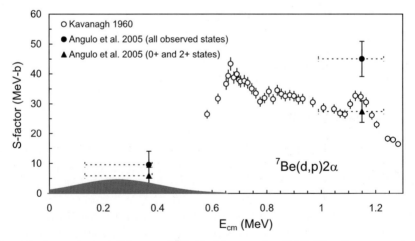

Fig. 12. Astrophysical S-factor of ^{7}Be(d,p)2α reaction [126]. The Gamow peak for a typical Big Bang temperature of 0.8 GK is shown in grey

5 Future Challenges and Conclusions

The understanding of stellar burning confronts physicists with many challenges across a wide range of nuclear physics topics, the corresponding experimental programs being characterized by many technical difficulties. After more than 30 years of research using stable-isotope beams, there are still reactions involved in quiescent burning whose cross sections are unknown in the energy range of interest and, consequently, the astrophysical predictions are still too uncertain for stellar systems that are, in principle, not very exotic, like the sun. Background sources in the laboratory are the main challenge for such measurements. Underground facilities seems to offer a solution to overcome the experimental difficulties, although they face other technical problems.

Our understanding of the energy source and nucleosynthesis in explosive events is reflected in our knowledge of the properties of the related unstable nuclei and the reaction rates involving them. Since about one decade, the development of radioactive beams has made it possible to investigate some reactions involved in such explosive events. Experimental approaches used traditionally with stable-isotope beams (elastic and inelastic scattering, transfer reactions) have been successfully used to obtain information on the properties of astrophysically important states of some radioactive nuclei. Indirect technique using, e.g., transfer reactions are particularly appealing for measurements with low-intensity radioactive beams. However, such approaches are model-dependent, and uncertainties related to the model will be reflected in the results. In spite of the progress, many key reactions and key quantities, which are the most challenging ones, remain largely unknown. This is specially true for the r-process path. More intense and new radioactive beams and sophisticated detection systems are therefore required. On the other hand, nuclear theory must be exploited to avoid misinterpretation of experimental data and a close collaboration with astronomical observations and astrophysical modelling is as essential as ever.

Acknowledgement

The author thanks Ernst Roeckl and Jim Al-Khalili for constructive comments and a careful reading of the manuscript. This work is supported by the European Commission within the Sixth Framework Programme through I3-EURONS (contract no. RII3-CT-2004-506065).

References

1. G. Wallerstein et al.: Rev. Mod. Phys. **69**, 995 (1997)
2. K. Käppeler, F.-K. Thielemann, M. Wiescher: Ann. Rev. Nucl. Part. Sc. **48**, 175 (1998)

3. M. Arnould, T. Takahashi: Rep. Prog. Phys. **62**, 393 (1999)
4. K. Käppeler: Prog. Part. Nucl. Phys. **43**, 419 (1999)
5. Langanke, K., Thielemann, K.-F., Wiecher, M.: Nuclear astrophysics and nuclear far from stability. In: Al-Khalili, J.S., Roeckl E. (eds.) Lecture Notes in Physics, Vol. 651, pp. 383–467. Springer, Heildelberg (2004)
6. D.D. Clayton: *Principles of stellar evolution and nucleosynthesis*, 2nd edn. (University of Chicago Press, 1968)
7. H.A. Bethe, C.L. Critchfield: Phys. Rev. **54**, 248 (1938)
8. H.A. Bethe: Phys. Rev. **55**, 434 (1939)
9. H.A. Bethe: Science **161**, 541 (1968) (Nobel Lecture).
10. F. Hoyle: Mon. Not. R. Astron. Soc. **106**, 343 (1946)
11. E.M. Burbidge, G.R. Burbidge, W.A. Fowler, F. Hoyle: Rev. Mod. Phys. **29**, 547 (1957)
12. C. Angulo: Contributions récents à l'astrophysique nucléaire. Habilitation Thesis, Université catholique de Louvain, Louvain-la-Neuve (2006).
13. C. Iliadis: *Nuclear Physics of Stars* (Wiley-VCH Verlag GmbH, 2007)
14. C. Angulo: 'Experimental approach to nuclear reactions of astrophysical interest involving radioactive nuclei'. In: *Exotic Nuclei and Nuclear/Particle Astrophysics, Carpathian Summer School of Physics 2005 at Mamaia, Constanta, Romania, June 13–24, 2005*, ed. by S. Stoica, L. Trache, R.E. Tribble (World Scientific, Singapore 2006) pp. 264–266
15. C. Angulo et al.: Nucl. Phys. A **656**, 3 (1999)
16. J.C. Blackmon, C. Angulo, A.C. Shotter: Nucl. Phys. A **777**, 531 (2006)
17. Huyse, M.: The why and how of radioactive-beam research. In: Al-Khalili, J.S., Roeckl E. (eds.) Lecture Notes in Physics, Vol. 651, pp. 1–32. Springer, Heildelberg (2004).
18. P. Descouvemont: *Theoretical Models in Nuclear Astrophysics* (Nova Science Publishers Inc., New York 2003)
19. M.S. Smith, K.E. Rehm: Annu. Rev. Nucl. Part. Sci. **51**, 91 (2001)
20. S. Kubono: Nucl. Phys. A **693**, 221 (2001)
21. E.E. Salpeter: Phys. Rev. **88**, 547 (1952)
22. R. Bonetti et al.: Phys. Rev. Lett. **82**, 5205 (1999)
23. P. Descouvemont, D. Baye, P.-H. Heenen: Z. Phys. A **306**, 79 (1982)
24. C. Arpesella et al.: Phys. Lett. B **389**, 452 (1996)
25. M. Junker et al.: Phys. Rev. C **57**, 2700 (1998)
26. P. Descouvemont: Phys. Rev. C **70**, 065802 (2004)
27. J.N. Bahcall, M.H. Pisonneault: Phys. Rev. Lett. **92**, 121301 (2004)
28. A. Coc et al: Astrophys. J. **600**, 544 (2004)
29. B.S. Nara Singh et al.: Phys. Rev. Lett. **93**, 262503 (2004) 262503.
30. D. Bemmerer et al.: Phys. Rev. Lett. **97**, 122502 (2006)
31. P. Descouvemont et al.: At. Data Nucl. Data Tables **88**, 203 (2004)
32. P.A. Bergbusch, B.A. Vanderberg: Astrophys. J. Suppl. **81**, 163 (1992)
33. B. Chaboyer et al.: Science **271**, 957 (1996)
34. B. Chaboyer et al.: Astrophys. J. **494**, 96 (1998)
35. R. C. Runkle et al.: Phys. Rev. Lett. **94**, 082503 (2005)
36. F. Herwig, S.M. Austin, J.C. Lattanzio: Phys. Rev. C **73**, 025802 (2006)
37. A. Formicola et al.: Phys. Lett. B **591**, 61 (2004)
38. C.W. Cook, W.A. Fowler, C.C. Lauritsen, T. Lauritsen: Phys. Rev. **107**, 508 (1957)

39. C.W. Cook, W.A. Fowler, C.C. Lauritsen, T. Lauritsen: Phys. Rev. **111**, 567 (1958)
40. H.O.U. Fynbo et al.: Nature **433**, 136 (2005)
41. L.R. Buchmann, C.A. Barnes: Nucl. Phys. A **777**, 254 (2006)
42. R. Gallino et al.: Astrophys. J. **497**, 388 (1998)
43. C.R. Brune, I. Licot, R.W. Kavanagh: Phys. Rev. C **48**, 3119 (1993)
44. H.W. Drotleff et al.: Astrophys. J. **414**, 735 (1993)
45. P. Descouvemont: Phys. Rev. C **36**, 2206 (1987)
46. J. José, A. Coc: 'Nucleosynthesis in Classical Nova Explosions: Modeling and Nuclear Uncertainties'. In: *Nuclear Physics News International*, Vol. 15, No. 4 (2005) p17.
47. H. Schatz et al.: Astrophys. J. **524**, 1014 (1999)
48. H. Schatz et al.: Nucl. Phys. A **688**, 150 (2001)
49. J. José: Nucl. Phys. A **752**, 540c (2005)
50. C. Fox et al.: Phys. Rev. Lett. **93**, 081102 (2004)
51. C. Fox et al.: Phys. Rev. C **71**, 055801 (2005)
52. A. Chafa et al.: Phys. Rev. Lett. **95**, 031101 (2005); Erratum: Phys. Rev. Lett. **96**, 0199902 (2006)
53. S. Bishop et al.: Phys. Rev. Lett. **90**, 162501 (2003); Erratum: Phys. Rev. Lett. **90**, 229902 (2003)
54. B. Davids et al.: Phys. Rev. C **68**, 055805 (2003)
55. J.M. D'Auria et al.: Phys. Rev. C **69**, 065803 (2004)
56. D.G. Jenkins et al.: Phys. Rev. Lett. **92**, 031101 (2004)
57. S.E. Hale et al.: Phys. Rev. C **70**, 045802 (2004)
58. D.G. Jenkins et al.: Phys. Rev. C **72**, 031303 (2005)
59. A. Coc, M. Hernánz, J. José J.-P. Thibaud: Astron. Astrophys. **357**, 561 (2000).
60. W. Galster et al.: Phys. Rev. C **44**, 2776 (1991)
61. K.E. Rehm et al.: Nucl. Instr. and Meth. in Phys. Res. **449**, 208 (1998)
62. F. Vanderbist et al.: Nucl. Instr. and Meth. in Phys. Res. B **197**, 165 (2002)
63. D.A. Hutcheon et al.: Nucl. Instr. and Meth. in Phys. Res. A **498**, 190 (2003)
64. G. Fiorentini, R.W. Kavanagh, C. Rolfs: Z. Phys. A **350**, 289 (1995)
65. T. Davinson et al.: Nucl. Instr. and Meth. in Phys. Res. A **454**, 350 (2000)
66. C. Angulo et al.: Nucl. Phys. A **716**, 213 (2003)
67. I.Y. Lee: Nucl. Phys. A **520**, 641c (1990)
68. C.N. Davids: Nucl. Instr. and Meth. in Phys. Res. B **204**, 124 (2003)
69. D. Schürman et al.: Nucl. Instr. and Meth. in Phys. Res. A **531**, 428 (2004).
70. R. Fitzgerald et al.: Nucl. Phys. A **748**, 351 (2005).
71. J. Görres: Proceedings of Science, (NIC-IX) 027.
72. D. Lunney, J.M. Pearson, C. Thibault: Rev. Mod. Phys. **75**, 1021 (2003)
73. Ultra-accurate mass spectrometry and related topics, Jürgen Kluge Special Issue, Intl. J Mass Spectrom. **251**, 85 (2006)
74. K. Blaum: Phys. Rep. **425**, 1 (2006)
75. R.A. Laubestein et al.: Phys. Rev. **84**, 12 (1951)
76. K.P. Artemov et al.: Yad. Fiz. **52**, 634 (1990) [Sov. J. Nucl. Phys. **52**, 408 (1990)]
77. Th. Delbar et al.: Nucl. Phys. A **542**, 263 (1992)
78. A.M. Lane, R.G. Thomas: Rev. Mod. Phys. **30**, 257 (1958)
79. G.V. Rogachev et al.: Phys. Rev. C **64**, 061601 (2001)
80. C. Angulo et al.: Phys. Rev. C **67**, 014303 (2003), and references therein

81. C. Angulo: Nucl. Phys. A **746** , 222c (2004)
82. C. Ruiz et al.: Phys. Rev. C **65**, 042801 (2002)
83. J. D'Auria et al.: Phys. Rev. C **69**, 065803 (2004)
84. B. Harss et al.: Phys. Rev. Lett. **82**, 3964 (1999)
85. J.C. Blackmon et al.: Nucl. Phys. A **688**, 142c (2001)
86. B. Harss et al.: Phys. Rev. C **65**, 035803 (2002)
87. J.C. Blackmon et al.: Nucl. Phys. A **718**, 127c (2003)
88. G.R. Satchler: *Direct Nuclear Reactions* (Clarendon Press, Oxford, 1983)
89. K.E. Rehm et al.: Phys. Rev. Lett. **80**, 676 (1998)
90. N. de Séréville et al.: Phys. Rev. C **67**, 052801 (2003)
91. N. de Séréville et al.: submitted to Phys. Rev. C
92. R.L. Kozub et al.: Phys. Rev. C **71**, 032801 (2005)
93. R.L. Kozub et al.: Phys. Rev. C **73**, 044307 (2006)
94. R.A. Paddock: Phys. Rev. C **5** , 485 (1972)
95. D.R. Osgood, J.R. Patterson, E.W. Titterton: Nucl. Phys. **60**, 503 (1964)
96. K.I. Hahn et al.: Phys. Rev. C **54**, 1999 (1996)
97. R. Mendelson, G.J. Wozniak, A.D. Bacher, J.M. Loiseaux, J. Cerny: Phys. Rev. Lett. **25**, 533 (1970)
98. J.A. Caggiano et al.: Phys. Rev. C **66**, 015804 (2002)
99. Y. Parpottas et al:, Phys. Rev. C **70**, 065805 (2004); Erratum: Phys. Rev. C **73**, 049907 (2006)
100. W. Benenson et al.: Phys. Rev. C **15**, 1187 (1977)
101. H. Schatz et al.: Phys. Rev. Lett. **79**, 3845 (1997) 3845
102. M. Wiescher et al.: Nucl. Phys. A **484**, 90 (1988)
103. J.A. Caggiano et al.: Phys. Rev. C **64** 025802 (2001)
104. J.A. Caggiano et al.: Phys. Rev. C **65** 055801 (2002)
105. U. Schröder et al.: Nucl. Phys. A **467**, 240 (1987)
106. C. Angulo, P. Descouvemont: Nucl. Phy. A **690**, 755 (2001)
107. P.F. Bertone et al.: Phys. Rev. Lett. **87**, 152501 (2001)
108. K. Yamada et al.: Phys. Lett. B **579**, 265 (2004)
109. C. Angulo, A.E. Champagne, H.P. Trautvetter: Nucl. Phys. A **758**, 391c (2005)
110. D.W. Bardayan et al.: Phys. Rev. C **62**, 042802 (2000)
111. D.W. Bardayan, et al.: Phys. Rev. C **63**, 065802 (2001)
112. D.W. Bardayan et al.: Phys. Rev. C **70**, 015804 (2004)
113. J.-S. Graulich et al.: Phys. Rev. C **63**, 011302 (2001)
114. R. Coszach et al.: Phys. Lett. B **353** 184 (1995)
115. K.E. Rehm et al: Phys. Rev. C **52**, 460 (1995)
116. J.S.Graulich et al.: Nucl. Phys. A **626**, 751 (1997)
117. D.W. Bardayan et al.: Phys. Rev. Lett. **89**, 262501 (2002)
118. F. de Oliveira et al.: Phys. Rev. C **55**, 3149 (1997)
119. S. Utku et al.: Phys. Rev. C **57**, 2731 (1998), Erratum: Phys. Rev. C **58**, 1354 (1998)
120. D.W. Visser et al.: Phys. Rev. C **69**, 048801 (2004)
121. S.G. Ryan et al.: Astrophys. J. **530**, L57 (2000)
122. D.N. Spergel et al.: Astrophys. J. Suppl. **148**, 175 (2003)
123. P.D. Parker: Astrophys. J. **175**, 261 (1972)
124. R.W. Kavanagh: Nucl. Phys. **18**, 492 (1960)
125. F. Ajzenberg-Selove: Nucl. Phys. A **490**, 1 (1988)
126. C. Angulo et al.: Astrophys. J. **630**, L105 (2005)
127. J.F. Ziegler, J.P. Biersack: SRIM program, v2003.26.

Index

Accelerator, 157, 205–207, 223, 224, 237, 264, 270, 273
Adiabatic cutoff, 42
AGATA, 50
Alder-Winther theory, 41
Allowed transitions, 106
Alpha particle, *see* α Particles
Angular momentum, 155, 158, 162, 163, 165, 167, 173, 259, 268, 270
Antisymmetrisation, 64
ANTOINE, 16, 72
Argonne, 157, 164, 169, 237, 266, 269, 270
Argonne Fragment Mass Analyser, 69

Back-bending, 75
Barrier penetration, 161, 162, 164, 165, 169, 172, 175, 176
^{11}Be, 44
Beam identification, 39
Beam velocity, 35
Berkeley, 205, 206, 213, 219
Beta decay, see β Decay
Big Bang, 253, 254, 259, 273, 275, 278
Binding energy, 172, 204, 232, 235, 236

α Capture, 255, 261
CDCC calculations, 45
Center-of-mass, 256
Charge exchange (CE) reaction, 112
Classical "ISOL" technique, 130
"Cluster Cube," 130
CNO cycle, 259–261, 271, 273
^{53}Co, 68

Cold fusion, 203, 205, 213, 215, 221–223, 225–227, 238, 239, 241, 246
Cold fusion reaction, 241, 242, 244
Compound nucleus, 172, 204, 215, 219, 221, 223, 231, 235, 238, 241, 242, 245, 267
Compton effect, 37
Compton suppression, 68
Configuration mixing, 8
Coulomb barrier, 154, 156, 161, 163–165, 187, 191, 192, 231, 241, 255, 256
Coulomb displacement energy, 64, 74, 79, 91
Coulomb energy differences, 66
Coulomb excitation, 28, 272
Coulomb nuclear interference, 43
Coulomb repulsion, 238, 240, 243, 244
Coulomb-nuclear interference, 45
^{49}Cr, 85
Cranked shell model, 77
Cross section, 170–172, 174, 203, 206, 210, 212, 213, 216, 219, 221–227, 229, 231, 232, 235–246, 254–262, 264, 265, 268–274, 276, 278, 279
Cross-conjugate nuclei, 85

Diagonalization of Hamiltonian, 11
α Decay, 171, 173, 205, 211–214, 216–219, 221, 224–232, 234–236
β Decay, 99–148, 154, 155, 162, 175, 176, 178, 184, 185, 190, 191, 203, 205,

206, 218, 223, 233, 234, 254, 255, 260–262, 266, 273

γ Decay, 154, 167, 169, 191, 220

Decay heat, 144

Deformation, 154, 166–171, 191, 192, 204, 232, 233, 237, 241, 244

Degrader, 176

β-Delayed proton, 180, 181, 191

Detector, 155, 157, 163, 164, 177–183, 185–187, 204, 205, 208–213, 216–218, 224, 225, 229, 230, 264–266, 268, 269, 273, 274, 276

Detector acceptance, 40

Detector response functions, 36

DGFRS, 224, 227, 229–231

Di-proton model, 174, 187

Doppler-shift, 35

Double-sided silicon strip detector (DSSSD), 157, 163, 164, 169, 177, 178, 186

Dubna, 203, 205, 206, 224–227, 229–231, 234, 236, 246

E1 Coulomb excitation, 43

E1 transition, 44, 59

E2 transition, 46, 59

Effective interaction, 14

Efficiency, 180, 212, 218, 222, 224, 227, 264–266, 273

Electromagnetic spin-orbit term, 65, 80

Electromagnetic transitions, 59

Electromagnetic transition strengths, 28

Electron, 203, 204, 206–210, 212, 218, 257

Electron capture (EC), 105, 210, 220, 221, 234, 256

α Emission, 171, 190, 203, 210, 219, 224, 228, 231

α Emitter, 210

Energy loss, 178–180, 206, 207

Energy resolution, 211, 217, 264, 268, 274

EUCLIDES array, 71

Euroball, 68

Euroball Neutron Wall, 71

Excitation energy, 167, 168, 216, 223, 226–229, 238, 239, 241, 242, 245, 265, 269, 271, 274

Excitation function, 206, 207, 213, 216, 219, 225, 226, 229, 231, 241, 242, 244, 246

Exotic nuclei, beta decay of, 99–148

^{50}Fe, 87

^{53}Fe, 68

Fermi decays, 108, 109

Fermi integral, 105

Fermi–Kurie plot, 126

Fermi transitions, 114

Fission, 203, 204, 206, 210, 211, 216, 218, 219, 222, 224, 226, 232, 235, 237–239, 241, 242, 245

Fission barrier, 232, 233, 235, 237–241, 245

FLNR, 203, 205, 224

Forbidden transitions, 106

Fragment Mass Analyser (FMA), 157, 164, 169, 170

Fragment Separator (FRS), 176–178

Fragmentation, 163, 169, 176, 183

Fusion, 156, 157, 163, 204, 206, 209, 210, 213, 223, 238, 239, 241, 243, 245, 246, 256, 260, 267

Fusion barrier, 241, 245

Fusion–evaporation, 68, 101, 156, 163, 169, 171, 172, 204, 228, 241

Gammasphere, 68

Gamow, 275

Gamow–Teller (GT) decays, 108

Gamow window, 254, 256, 258, 259, 261, 276

GANIL, 43, 176–181, 184–187, 247, 269

GASP array, 68

Germanium detector, 29, 50, 210, 212

Globular, 271

Globular cluster, 260, 271, 272

G-matrix, 18

GRETA, 50

Ground state, 154–158, 162, 163, 165, 167–174, 178, 182, 188, 191, 204, 218–220, 232, 233, 237, 239, 240, 266, 269, 272, 273, 276, 277

GSI, 43, 156, 176–180, 189, 192, 203, 205, 213, 219, 222, 229, 238, 247

GSI–TAS, 120

GT transitions, 114

Half life, 154–156, 158, 161, 162, 164, 165, 167, 172–176, 178–180, 182, 183, 187, 191, 204–206, 210, 213, 215, 216, 218, 220, 221, 223, 224, 226, 227, 230–236, 247, 261, 262, 266, 267, 273

Harmonic Oscillator, 4

Hartree-Fock, 46

Hot fusion, 203, 205, 213, 214, 223, 225, 226, 228, 231, 232 238–240, 246

Ikeda sum rule, *see* Model-independent rule

Impact parameter, 33, 43

Implantation, 155, 157, 163, 164, 169, 175, 177–183, 185, 186, 204, 205, 210–212, 216–218, 224, 225, 230

Inert core, 8

In-flight, 156, 163, 184

Ion source, 205–207, 209

ISAC, 29

ISIS array, 71

Island of Inversion, 45

Isobaric analogue states, 58

Isobaric multiplet, 60, 84

Isobaric multiplet mass equation, 60

Isobaric nuclei, 58

ISOL method, 123–130, 176

Isomer, 155, 157, 165, 167, 171, 173, 191, 219, 220, 224, 237, 238

Isospin, 190

Isospin quantum number, 58

Isospin symmetry, 57, 59, 61, 63, 65, 67, 69, 71, 73, 75, 77, 79, 81, 83, 85, 87, 89, 91, 93, 95, 97

Isospin symmetry breaking, 82

Isospin-symmetry-breaking, 58

JAERI tandem accelerator, 29

JYFLTRAP, 146

Kinematic focusing, 30

^{76}Kr, 29

Laboratory energy spectrum, 35

Ladder diagram, 14

LBNL, 205, 219, 221, 229, 238

Level scheme, 220, 267, 271, 276, 277

^{8}Li, 29

Liquid Drop model, 79, 204, 232

LISE3, 176–178, 183–185

Lorentz factor, 41

Macroscopic-microscopic model, 215, 232, 235, 238

Magic numbers, 6, 204, 215

Major shell, 8

Matrix element, 105

Mirror energy differences, 67, 74, 76, 78, 83

Mirror nuclei, 58

^{30,32,34}Mg, 45

^{48}Mn, 70

Model-independent rule, 112

Monte Carlo Shell Model, 16, 72

M-scheme, 10

MSHELL, 16

MSU, 270

Multi-step excitation, 34

α,n, 261

NaI detector, 273, 276

Neutrino, 256, 260, 271, 273

Neutron separation energy, 6

Neutron-deficient, 203, 210, 215, 267

Neutron-halo, 44

Neutron-rich, 203, 205, 214–216, 223, 236, 241, 246

Neutron-separation energy, 44

^{56}Ni, 91

NN interaction, 2

Nolen-Schiffer anomaly, 62, 65

NSCL, 36

Nuclear spectroscopy, 27

Nucleosynthesis, 253, 254, 257, 260–262, 266, 275, 279–281

Oak Ridge, 157, 168, 266

Oak Ridge National Laboratory, 29

One-proton(1p) radioactivity, 153–155, 157, 162, 163, 173, 190, 192

Optical model calculation, 33

Optical time projection chamber (OTPC), 183, 184

OXBASH, 16

p,α, 258, 274
p,γ, 257, 258
α,p, 269
Parity, 170, 219
α Particles, 153, 190, 210, 211, 216–218,
    221, 226, 231, 261, 264, 267, 274, 276
β Particles, 180, 181, 186, 187
Pauli principle, 60
Perturbation theory, 83
Positron, 178
p,α reaction, 262, 269, 270, 273, 274, 275
Projectile, 171, 172, 176, 203, 206–211,
    213, 216, 221, 224, 226, 236, 238, 239,
    241–245
Proton capture, 254, 255, 273
Proton–proton chain, 258, 260
Proton-rich, 265
Proton drip line, 153, 154, 162, 163,
    170–173, 176, 192, 262, 266
Proton emission, 155–158, 162, 165,
    167–169, 173, 175, 184, 189–191
Proton emitter, 156–158, 161–163,
    166–174, 184
Proton radioacrivity, 210

Quasi-particle, 14
Q value, 155, 226
Qp value, 161–163, 172
Q2p value, 174

Radioactive decay, 157, 180
γ Ray, 169, 170, 174, 179, 205, 212, 218,
    220, 245, 260, 262, 264, 273
RDT, 205, 245
α,p Reaction, 269
α,γ Reaction, 257, 258, 260, 261, 266,
    275
α,n Reaction, 261
d,p Reaction, 270, 274, 275, 277
p,α Reaction, 260–262
p,γ Reaction, 258, 259–262, 264, 266,
    269–273
Reactor decay heat, 144–147
Recoil decay tagging(RDT), 169
Recoil effects, 43
Recoil separation, 204, 208, 209,
    213, 247
Recoil separator, 169, 173, 207–210,
    224, 231, 265, 266, 274

Reduced transition probability, 42
Relativistic effects, 41
Resonance, 206
REX-Isolde, 29
RIKEN, 43, 203, 205, 213, 218, 219,
    221–223, 226, 227
R-matrix, 268, 272, 275
R-matrix model, 187–189
r-process, 254, 256, 279
rp-process, 267, 270, 271
Rutherford trajectory, 41

^{40}S, 32
Saddle point, 233, 240
Scattering angle, 40
Scintillation detectors, 50
S-factor, 256–259, 273–275, 277, 278
Shell closure, 204, 228, 235, 237
Shell correction energy, 215, 232, 233,
    236, 240, 241
Shell gap, 4
Shell model, 1, 46, 72, 82, 154, 164, 166,
    173, 187, 203, 204, 232
Shell model embedded in the continuum
    (SMEC), 188, 189
SHIP, 156, 209–213, 216–218, 221–223,
    226, 229–231, 242, 247
Si(Li) detector, 181
Silicon detector, 156, 176, 178–182, 186,
    209–212, 224
Single-particle energies, 3
SISSI, 177–179
Slater determinants, 9
Spectroscopic factor, 30, 165, 166, 187,
    191
Spin, 155, 158, 169–171, 174, 191,
    218–220, 237, 244
Spin-orbit coupling, 4
Spontaneous fission (SF), 205, 210, 213,
    214, 218–222, 224–235
S-process, 254, 261
Strip detector, 264, 269, 274, 276
Subshell closure, 235, 236
Super–allowed decays, 110
Superheavy element(SHE),
    203–205, 213, 223, 224, 229,
    231–233, 235–238, 242, 246,
    247, 250
Supernova, 254, 255, 261

Target, 156, 163, 169–173, 176,
    179, 184, 203–210, 213, 215, 219,
    222–229, 231, 232, 236, 238, 239,
    241–247, 257, 262–270, 273, 274,
    276, 278
Tensor force, 19
Time-of-flight, 36, 207, 209, 210, 212,
    213, 224, 225, 269, 270
Total absorption spectroscopy, 118–122
Transfer, 204, 241, 243, 244, 265, 267,
    269, 270, 273, 274, 279
Transfer reaction, 29
Transition matrix elements, 41
Triplet energy differences, 67, 83
Triple-α process, 254, 261
Two-body matrix element (TBME),
    12, 17
Two-proton (2p) radioactivity, 153–155,
    173, 174, 176–182, 184, 186–192
Two-proton (12p) radioactivity, 192

UNILAC, 206, 207, 209, 219, 247
"Unique" transitions, 110
USD interaction, 13

^{47}V, 86
^{48}V, 70
Valence shell, 8
VASSILISSA, 224, 227

Weizsäcker parabola, 64
Weizsäcker–Williams method, 42
Wigner-Eckart theorem, 61
WKB model, 172, 173
Woods–Saxon potential, 4

X-ray, 205
X-ray burst, 253, 255, 261

Yrast states, 93

Color Plate Section

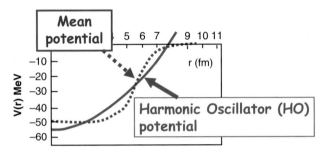

Plate 1. Comparison between a harmonic oscillator potential and a Woods–Saxon potential. HO is simpler, and can be treated analytically (See also Figure 5 on page 5)

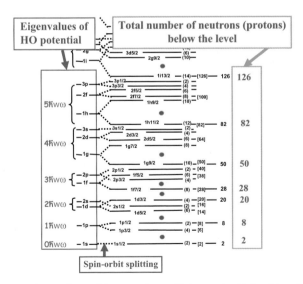

Plate 2. Energy eigenvalues of harmonic oscillator potential with spin–orbit force, and Mayer–Jensen's magic numbers (See also Figure 6 on page 5)

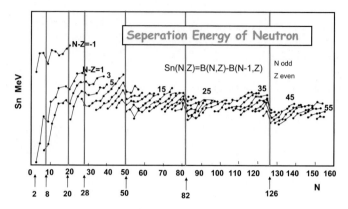

Plate 3. Observed neutron separation energy, partly taken from Ref. [3] (See also Figure 7 on page 7)

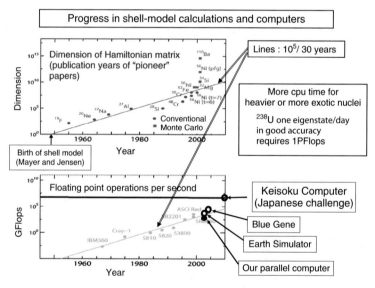

Plate 4. *Upper panel*: Maximum dimension of the Hamiltonian matrix feasible for the year of publication. *Blue* points are for conventional shell model, while *red* points are for Monte Carlo Shell Model. *Lower panel*: The computer capability (Flops) as a function of the year. The lines in the upper and lower panels indicate an increase of 10^5 times/30 years. (See also Figure 13 on page 16)

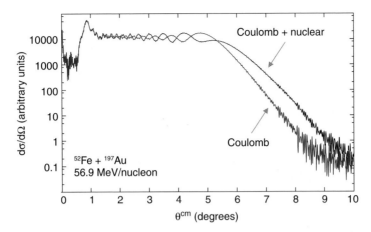

Plate 5. Calculated excitation cross sections versus center-of-mass scattering angle θ^{cm} for the reaction $^{52}\mathrm{Fe} + {}^{197}\mathrm{Au}$ at $56.9\,\mathrm{MeV/nucleon}$. Shown are the Coulomb excitation cross section and the Coulomb plus nuclear excitation cross sections. The Coulomb cross section dominates for small scattering angles. Optical model parameters from the $^{40}\mathrm{Ar} + {}^{208}\mathrm{Pb}$ reaction at $41\,\mathrm{MeV/nucleon}$ [30] were used to calculate the cross sections. Figure adapted from [31] (See also Figure 3 on page 34)

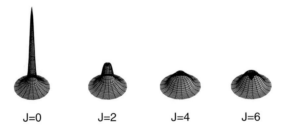

Plate 6. A calculation of the probability distribution for the relative distance of two like-particles in the $f_{\frac{7}{2}}$ shell as a function of their coupled angular momentum. The calculations were undertaken in [55]. The centre of each plot corresponds to zero separation (See also Figure 6 on page 75)

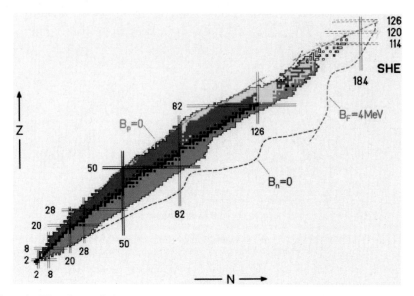

Plate 7. The chart of the nuclides showing the neutron and proton drip lines which are defined by demanding that the respective binding energies, B_n and B_p, be zero. The neutron and proton numbers for the closed shells are indicated by the *horizontal* and *vertical lines*. The figure also shows the line where the fission barrier B_F goes below 4 MeV and a prediction of the possible shell structure for the super–heavy elements (SHE) (See also Figure 1 on page 100)

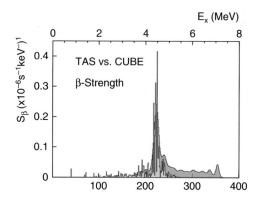

Plate 8. Beta–strength as a function of excitation energy in the daughter nucleus following the β–decay of the 2^- ground state in ^{150}Ho measured with the CLUSTER CUBE (*sharp lines*) and the GSI-TAS (continuous function). See the text for details (See also Figure 18 on page 139)

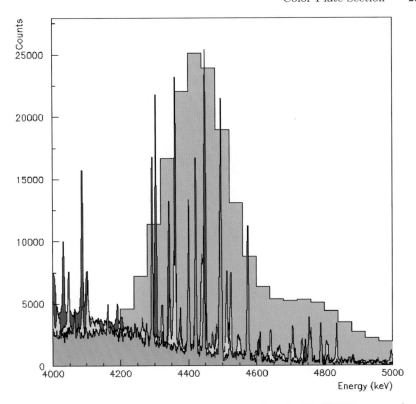

Plate 9. Comparison of part of the GSI–TAS and CLUSTER CUBE spectra (various coincidence gates) for the decay of the 2$^-$ ground state in ^{150}Ho (see text) (See also Figure 19 on page 140)

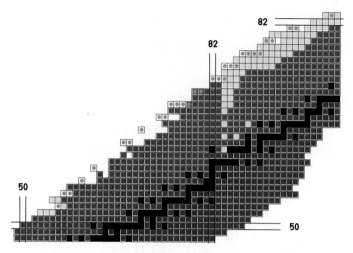

Plate 10. Chart of isotopes in the region of ground-state proton emitters. The isotopes for which proton emission from the ground state or from long-lived isomers was observed are indicated by *full circles*. From $Z=53$ up to $Z=83$, proton radioactivity was observed for all odd-Z elements but one (See also Figure 2 on page 157)

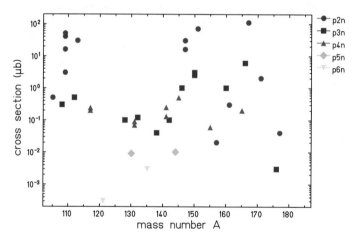

Plate 11. Production cross sections for fusion–evaporation reactions used to produce ground-state proton emitters. The cross sections vary from $10\,\mu b$ for the least exotic proton emitters produced via *p2n* reactions to $1\,nb$ for the most exotic proton emitters produced by means of *p6n* reactions. Each additional neutron to be emitted "costs" about an order of magnitude in cross section (See also Figure 9 on page 171)

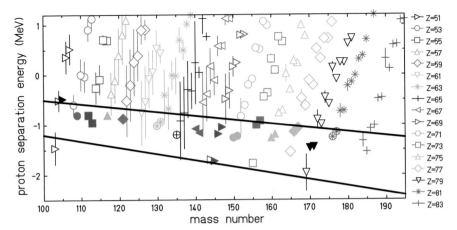

Plate 12. Ground-state proton-separation energies for proton-rich nuclei. The *open symbols* show values taken from the 2003 mass evaluation [2], whereas the *full symbols* show the known ground-state proton emitters. The *full lines* show calculations for barrier penetration half-lives of 100 ms ($\ell=0$) and 1 μs ($\ell=5$) with a spherical WKB model, a semiclassical approximation of quantum mechanics (See also Figure 10 on page 172)

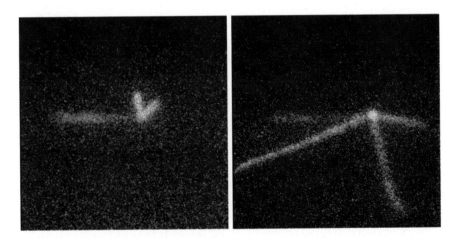

Plate 13. *Left-hand side*: Example of a CCD image of a two-proton decay event of ^{45}Fe taken in a 25 ms exposure. A track of a ^{45}Fe ion entering the chamber from the left is seen. The *two bright short tracks* are protons of about 0.6 MeV. They were emitted 0.62 ms after implantation of the ion (from [105]). *Right-hand side*: Example of a CCD image of a β-decay event of ^{45}Fe. A weak track of a ^{45}Fe ion entering from the left is seen. The *three long bright tracks* are consistent with high-energy protons escaping the active volume of the detector. The decay occurred 3.33 ms after the implantation. The event is interpreted as a β-delayed three-proton decay of ^{45}Fe (from [123]) (See also Figure 21 on page 185)

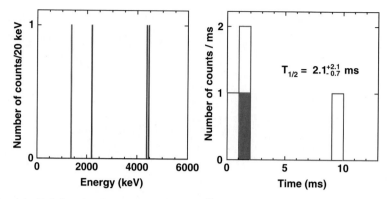

Plate 14. *Left-hand side*: Decay energy of ^{48}Ni obtained from a recent GANIL experiment [43]. The event with the lowest energy is most likely due to 2p radioactivity of ^{48}Ni. The other events were observed in coincidence with β particles in adjacent detectors. *Right-hand side*: The decay-time spectrum for the four decay events of ^{48}Ni is plotted. The event shown as a full histogram is the low-energy event. The half-life is $T_{1/2} = 2.1^{+2.1}_{-0.7}$ ms (See also Figure 24 on page 187)

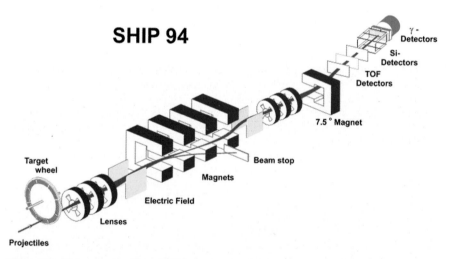

Plate 15. The velocity filter SHIP (separator for heavy ion reaction products) and its detection system [17–19], as it was used for the study of element 110 in 1994. The drawing is approximately to scale, however, the target wheel and the detectors are enlarged by a factor of 2. The length of SHIP from the target to the detector is 11 m. The target wheel has a radius up to the center of the targets of 155 mm. It rotates synchronously with beam macrostructure at 1,125 rpm [29]. The target thickness is usually $450\,\mu\text{g/cm}^2$. The detector system consists of three large area secondary-electron time-of-flight detectors [30] and a position-sensitive silicon-detector array (see text and Fig. 2). The flight time of the reaction products through SHIP is 1–$2\,\mu\text{s}$. The filter, consisting of two electric and four magnetic dipole fields plus two quadrupole triplets, was later extended by a fifth deflection magnet, allowing for positioning of the detectors away from the straight beam line and leading to further reduction of the background (See also Figure 1 on page 209)

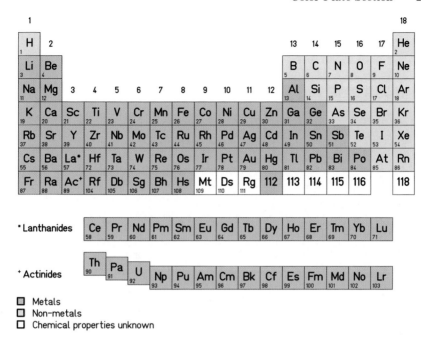

Plate 16. Periodic table of elements. The known transactinide elements 104–116 and 118 take the positions from below hafnium in group 4 to below radon in group 18. Elements 108, hassium (Hs), and element 112, the heaviest elements chemically investigated, are placed in groups 8 and 12, respectively. The arrangement of the actinides reflects the fact that the first actinide elements still resemble, to a decreasing extent, the chemistry of the other groups: Thorium (group 4 below hafnium), protactinium (group 5 below tantalum), and uranium (group 6 below tungsten) [43]. The name 'roentgenium', symbol 'Rg', was proposed for element 111 and recommended for acceptance by the Inorganic Chemistry Division of IUPAC in 2004 [44] (See also Figure 3 on page 214)

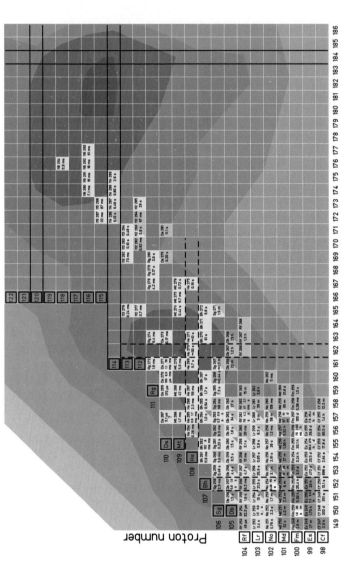

Plate 17. *Upper end* of the chart of nuclei showing the presently (2007) known nuclei. For each known isotope the element name, mass number, and half-life are given. The relatively neutron-deficient isotopes of the elements up to proton number 113 were created after evaporation of one neutron from the compound nucleus (cold-fusion reactions based on ^{208}Pb and ^{209}Bi targets). The more neutron-rich isotopes from element 112 to 118 were produced in reactions using a ^{48}Ca beam and targets of ^{238}U, ^{237}Np, ^{242}Pu, ^{244}Pu, ^{243}Am, ^{245}Cm, ^{248}Cm, and ^{249}Cf. The magic numbers for protons at element 114 and 120 are emphasized. The *bold dashed lines* mark proton number 108 and neutron numbers 152 and 162. Nuclei with that number of protons or neutrons have increased stability; however, they are deformed contrary to the spherical superheavy nuclei. At $Z=114$ and $N=162$ it is uncertain whether nuclei in that region are deformed or spherical. The background structure in gray shows the calculated shell correction energy according to the macroscopic–microscopic model. See Sect. 4 and Fig. 9 for details (See also Figure 4 on page 215)

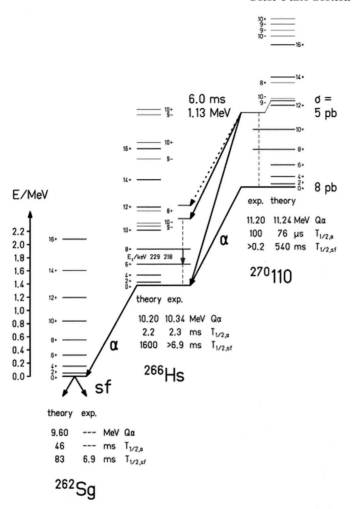

Plate 18. Assignment of measured α and γ decay and SF data observed in the reaction ^{64}Ni + ^{207}Pb → ^{271}Ds*. The data were assigned to the ground-state decays of the new isotopes ^{270}Ds, ^{266}Hs, and ^{262}Sg and to a high-spin K isomer in ^{270}Ds. *Arrows* in bold represent measured α-and γ-rays and SF. The data of the proposed partial level schemes are taken from theoretical studies of Muntian et al. [78] for the rotational levels, of Cwiok et al. [77] for the K isomers and of Smolanczuk [79] and Smolanczuk et al. [80] for the α energies and SF half-life of ^{262}Sg, respectively. For a detailed discussion see [76] (See also Figure 6 on page 220)

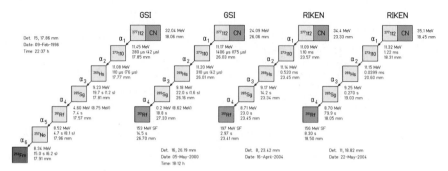

Plate 19. Decay chains measured in the cold-fusion reaction ^{70}Zn + ^{208}Pb → ^{278}112*. In the *left part*, the two chains are shown which were measured in 1996 and 2000 at SHIP [49, 59], in the *right part* those measured in 2004 at RIKEN [40]. The chains were assigned to the isotope ^{277}112 produced by evaporation of one neutron from the compound nucleus. The lifetimes given in brackets were calculated using the measured α energies. In the case of escaped α particles the α energies, given in brackets, were determined using the measured lifetimes (See also Figure 7 on page 222)

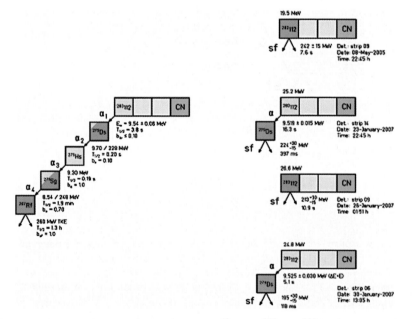

Plate 20. Events observed for the reaction ^{48}Ca + ^{238}U → ^{286}112* at SHIP (*right panel*) [116] and at the DGFRS (*left panel*) [101]. The latter data represent mean values obtained from a total of 22 decay chains measured at the DGFRS including those produced by α decay of heavier elements [88]. The two implantation–α–SF events from the SHIP work completely agree with the data assigned to the decay of ^{283}112 in [101]. A new result obtained at SHIP was the observation of two implantation–SF events which were assigned to a 50% SF branch of ^{283}112 (See also Figure 8 on page 230)

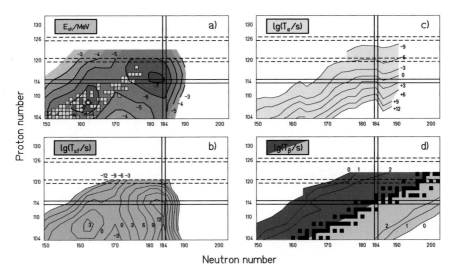

Plate 21. Shell-correction energy (**a**) and partial half-lives for SF, α and β decay (**b**)–(**d**). The calculated values in (**a**)−(**c**) were taken from [80, 124] and in (**d**) from [131]. The *squares* in (**a**) mark the nuclei presently known, the *filled squares* in (**d**) indicate the β stable nuclei (See also Figure 9 on page 233)

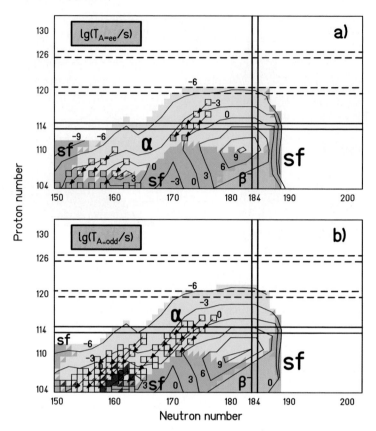

Plate 22. Dominating partial half-lives for α decay, β^+ decay/EC, β^- decay, and SF: (**a**) for even–even nuclei; (**b**) for odd-A nuclei. Nuclei and decay chains known at present are marked in (**a**) and, in the latter case (**b**) also the known odd–odd nuclei are included (See also Figure 10 on page 234)

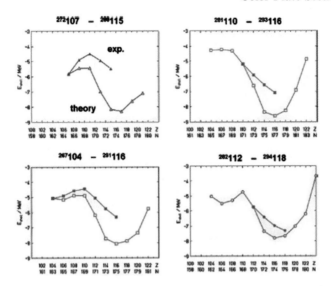

Plate 23. Comparison of shell correction energies as calculated in [132] to 'experimental' ones obtained as difference between the calculated liquid-drop binding energy taken from [132] and the 'experimental' total binding energy obtained by using the measured Q_α values of nuclei along a decay chain [88]. The unknown shell correction energy of the nucleus at the end of a chain is normalized to the theoretical value (See also Figure 11 on page 236)

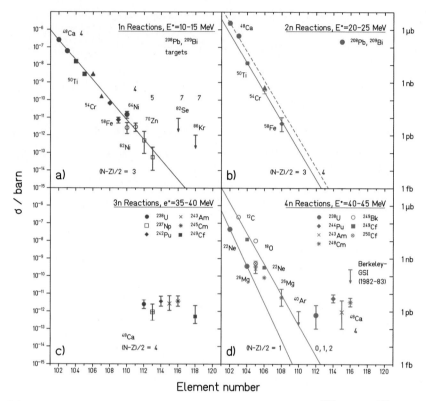

Plate 24. Measured cross-sections for fusion reactions with ^{208}Pb and ^{209}Bi targets and evaporation of one (**a**) and two (**b**) neutrons and for fusion reactions with actinide targets and evaporation of three (**c**) and four neutrons (**d**). Values $(N–Z)/2$ characterize the neutron excess of the projectile (See also Figure 12 on page 239)

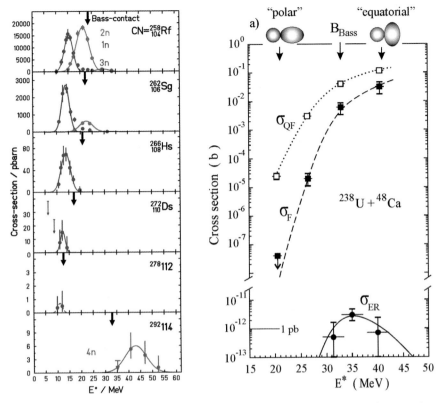

Plate 25. *Left side*: Measured excitation functions of even elements from ruther-fordium to element 112 produced in reactions with ^{208}Pb targets and beams from ^{50}Ti to ^{70}Zn. The data were measured in experiments at SHIP. For comparison the excitation function for synthesis of element 114 in the reaction ^{48}Ca + ^{244}Pu is plotted in the *bottom panel* [101]. The *arrows* mark the energy at reaching a contact configuration using the model by Bass [156]. *Right panel*: Comparison of the cross-sections as function of the excitation energy E^* for quasifission (σ_{QF}), compound-nucleus fission (σ_F), and evaporation residues (σ_{ER}) for the reaction ^{48}Ca + ^{238}U. The figure on the *right side* is taken from [101] (See also Figure 14 on page 242)

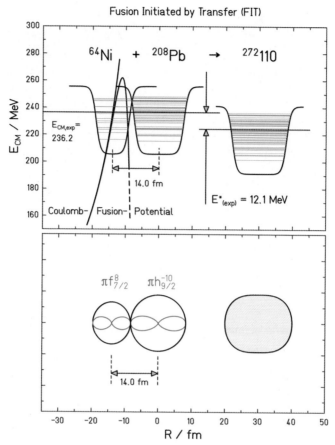

Plate 26. Energy-against-distance diagram for the reaction of an almost spherical ^{64}Ni projectile with a spherical ^{208}Pb target nucleus resulting in the deformed fusion product ^{271}110 after emission of one neutron. At the center-of-mass energy of 236.2 MeV, the maximum cross-section was measured. In the *top panel*, the reaction partners are represented by their nuclear potentials (Woods–Saxon) at the contact configuration where the initial kinetic energy is exhausted by the Coulomb potential. At this configuration projectile and target nuclei are 14 fm apart from each other. This distance is 2 fm larger than the Bass contact configuration [156], where the mean radii of projectile and target nucleus are in contact. In the *bottom panel*, the outermost proton orbitals are shown at the contact point. For the projectile ^{64}Ni, an occupied 1f$_{7/2}$ orbit is drawn, and for the target ^{208}Pb an empty 1h$_{9/2}$ orbit. The protons circulate in a plane perpendicular to the drawing. The Coulomb repulsion, and thus the probability for separation, is reduced by the transfer of protons. In this model, the fusion is initiated by transfer (see also [157, 158]). The figure is taken from [159] (See also Figure 15 on page 243)

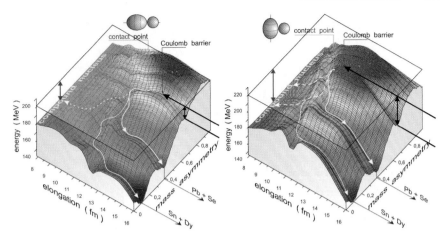

Plate 27. Three dimensional potential energy surface as function of elongation and mass asymmetry for the reaction ^{48}Ca + ^{248}Cm. Two scenarios are shown: On the *left side* for a reaction at low-projectile energy so that contact can occur only at polar orientation. On the *right side* the projectile energy is high enough so that contact occurs also at equatorial orientation. In both cases the projectile energies are chosen so that the kinetic energy in the center of mass system is zero at the contact configuration. The *rectangular plane* marks the total energy. It is drawn through the potential energy at contact. The difference between this plane and the curved potential energy surface is mainly kinetic energy in the entrance channel, after contact mainly intrinsic excitation energy of the system. The valleys in the potential energy surface are due to shell effects of the reaction partners. After contact the paths of the configuration in the plane of total kinetic energy is influenced by the structure of the potential energy surface. Trajectories above the valleys are favored due to enhanced conversion of potential energy into kinetic energy, which results in re-separation dominantly with the double magic ^{208}Pb as one of the reaction partners. Due to the longer distance from contact to the compound system (*boarder on the left*), the probability for re-separation is higher in the case of the elongated configuration. Therefore, the fusion cross-section is considerably smaller as in the case of the compact configuration, although the excitation energy of the compound nucleus, marked by the length of the *double arrow on the left*, is significantly smaller, about 20 MeV instead of 40 MeV. The main part of the figure has been provided by courtesy of V.I. Zagrebaev (2007) (See also Figure 16 on page 245)

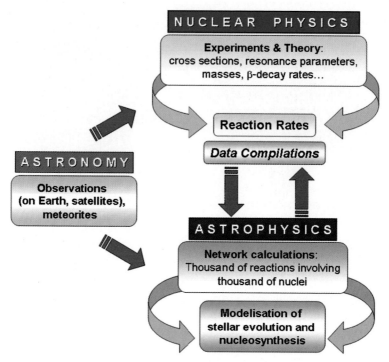

Plate 28. Link between nuclear physics, astrophysics and astronomy and their respective tasks within nuclear astrophysics [12] (See also Figure 1 on page 255)

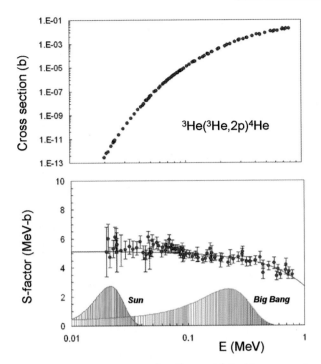

Plate 29. Cross section and S-factor of ^{3}He(^{3}He,2p)^{4}He (data are taken from [15]). The *solid curve* is a theoretical extrapolation. The Gamow window for $T_9 = 0.015$ (Sun) and $T_9 = 0.5$ (Big Bang) are indicated (See also Figure 2 on page 259)

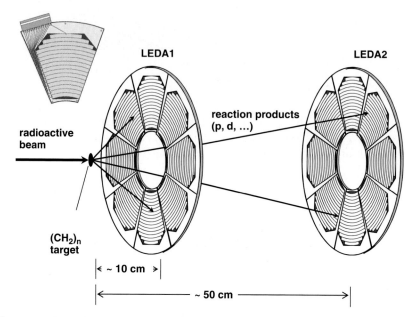

Plate 30. Typical experimental setup for the measurement of a transfer reaction by means of the LEDA detector [65]. The insert on the *top left* is a schematic drawing of one LEDA sector (See also Figure 4 on page 265)

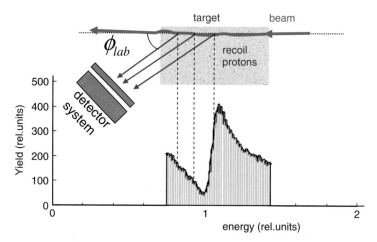

Plate 31. Principles of the elastic scattering method in inverse kinematics. The spectrum is a typical interference pattern for a $\ell = 0$ resonance (see text) (See also Figure 5 on page 268)

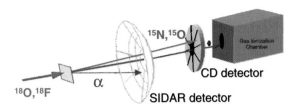

Plate 32. Experimental setup used at HRIBF to study the $^{18}\text{F}(d,p)^{19}\text{F}(\alpha)^{15}\text{N}$ reaction [92, 93]. The SIDAR and the CD detectors are composed of independent sixteen-strip silicon sectors (See also Figure 8 on page 274)

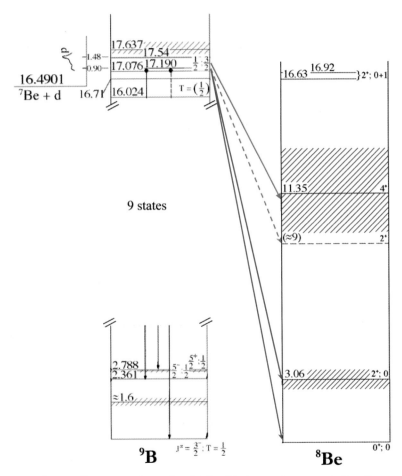

Plate 33. ^{9}B and ^{8}Be level scheme [125]. The (d,p) reactions populating the ground state and first excited state of ^{8}Be were investigated by Kavanagh [124]. Other ^{8}Be states of interest are also indicated (See also Figure 10 on page 277)

Printing: Krips bv, Meppel, The Netherlands
Binding: Stürtz, Würzburg, Germany